Traditional Functional-Discrete Methods for the
Problems of Mathematical Physics

Series Editor
Nikolaos Limnios

Traditional Functional-Discrete Methods for the Problems of Mathematical Physics

New Aspects

Volodymyr Makarov
Nataliya Mayko

WILEY

First published 2023 in Great Britain and the United States by ISTE Ltd and John Wiley & Sons, Inc.

Apart from any fair dealing for the purposes of research or private study, or criticism or review, as permitted under the Copyright, Designs and Patents Act 1988, this publication may only be reproduced, stored or transmitted, in any form or by any means, with the prior permission in writing of the publishers, or in the case of reprographic reproduction in accordance with the terms and licenses issued by the CLA. Enquiries concerning reproduction outside these terms should be sent to the publishers at the undermentioned address:

ISTE Ltd
27-37 St George's Road
London SW19 4EU
UK

www.iste.co.uk

John Wiley & Sons, Inc.
111 River Street
Hoboken, NJ 07030
USA

www.wiley.com

© ISTE Ltd 2023

The rights of Volodymyr Makarov and Nataliya Mayko to be identified as the authors of this work have been asserted by them in accordance with the Copyright, Designs and Patents Act 1988.

Any opinions, findings, and conclusions or recommendations expressed in this material are those of the author(s), contributor(s) or editor(s) and do not necessarily reflect the views of ISTE Group.

Library of Congress Control Number: 2023943904

British Library Cataloguing-in-Publication Data
A CIP record for this book is available from the British Library
ISBN 978-1-78630-933-4

Contents

Preface ix

Introduction xi

Chapter 1. Elliptic Equations in Canonical Domains with the Dirichlet Condition on the Boundary or its Part 1

1.1. A standard finite-difference scheme for Poisson's equation with mixed boundary conditions 1
    1.1.1. Discretization of the BVP 1
    1.1.2. Properties of the finite-difference operators 3
    1.1.3. Discrete Green's function 8
    1.1.4. Accuracy with the boundary effect 10
    1.1.5. Conclusion. 17
1.2. A nine-point finite-difference scheme for Poisson's equation with the Dirichlet boundary condition 18
    1.2.1. Discretization of the BVP 19
    1.2.2. Properties of the finite-difference operators 20
    1.2.3. Discrete Green's function 24
    1.2.4. Accuracy with the boundary effect 27
    1.2.5. Conclusion. 31
1.3. A finite-difference scheme of the higher order of approximation for Poisson's equation with the Dirichlet boundary condition 31
    1.3.1. Auxiliary results. 32
    1.3.2. Accuracy with the boundary effect 43
    1.3.3. Conclusion. 46
1.4. A finite-difference scheme for the equation with mixed derivatives 46
    1.4.1. Discretization of the BVP 47
    1.4.2. Properties of the finite-difference operators 49
    1.4.3. Discrete Green's function 54

1.4.4. Accuracy with the boundary effect 57
1.4.5. Conclusion. 67

Chapter 2. Parabolic Equations in Canonical Domains with the Dirichlet Condition on the Boundary or its Part 69

2.1. A standard finite-difference scheme for the one-dimensional heat equation with mixed boundary conditions 69
    2.1.1. Discretization of the problem 69
    2.1.2. Accuracy with the boundary effect 71
    2.1.3. Accuracy with the initial effect 77
    2.1.4. Conclusion. 81
2.2. A standard finite-difference scheme for the two-dimensional heat equation with mixed boundary conditions 82
    2.2.1. Discretization of the differential problem and properties of the finite-difference operators 82
    2.2.2. Discrete Green's function 85
    2.2.3. Accuracy with the boundary effect 86
    2.2.4. Conclusion. 101
2.3. A standard finite-difference scheme for the two-dimensional heat equation with the Dirichlet boundary condition 102
    2.3.1. Discretization of the differential problem. 102
    2.3.2. Accuracy with the boundary effect 103
    2.3.3. Accuracy with the initial effect 111
    2.3.4. Conclusion. 112

Chapter 3. Differential Equations with Fractional Derivatives. 115

3.1. BVP for a differential equation with constant coefficients and a fractional derivative of order $½$ 115
    3.1.1. A weighted estimate for the exact solution 115
    3.1.2. Weighted estimates for approximate solutions 118
    3.1.3. Conclusion. 123
3.2. BVP for a differential equation with constant coefficients and a fractional derivative of order $\alpha \in (0,1)$ 124
    3.2.1. A scale of weighted estimates for the exact solution. 124
    3.2.2. The scale of weighted estimates for approximate solutions 136
    3.2.3. A numerical example and conclusion 143
3.3. BVP for a differential equation with variable coefficients and a fractional derivative of order $\alpha \in (0,1)$ 145
    3.3.1. Differential properties of the exact solution 145
    3.3.2. The accuracy of the mesh scheme 162
    3.3.3. Conclusion. 166
3.4. Two-dimensional differential equation with a fractional derivative. 166

3.4.1. A weighted estimate for the exact solution 166
3.4.2. A mesh scheme of the first order of accuracy 172
3.4.3. A mesh scheme of the second order of accuracy. 177
3.4.4. Conclusion. 181
3.5. The Goursat problem with fractional derivatives 181
3.5.1. Properties of the exact solution 181
3.5.2. The accuracy of the mesh scheme 198
3.5.3. Conclusion. 212

Chapter 4. The Abstract Cauchy Problem 213

4.1. The approximation of the operator exponential
function in a Hilbert space 213
4.2. Inverse theorems for the operator sine and cosine functions. 230
4.3. The approximation of the operator exponential function in a
Banach space 236
4.4. Conclusion 247

Chapter 5. The Cayley Transform Method for
Abstract Differential Equations 249

5.1. Exact and approximate solutions of the BVP in a Hilbert space 249
5.1.1. Auxiliary results 250
5.1.2. The exact solution of the BVP 257
5.1.3. The approximate method without saturation of accuracy 267
5.1.4. The approximate method with a super-exponential rate of convergence. . 274
5.1.5. Conclusion. 281
5.2. Exact and approximate solutions of the BVP in a Banach space 282
5.2.1. The BVP for the homogeneous equation 282
5.2.2. The BVP for the inhomogeneous equation 292
5.2.3. Conclusion. 305

References 307

Index 315

Preface

New Aspects of the Traditional Functional-Discrete Methods for the Problems of Mathematical Physics

This book is based on the authors' latest research focusing on obtaining weighted accuracy estimates of numerical methods for solving boundary value and initial value problems. The idea of such estimates is based on Volodymyr Makarov's observation that due to the Dirichlet boundary condition for a differential equation in a canonical domain (e.g. on an interval or in a rectangle), the accuracy of the approximate solution in the mesh nodes near the boundary of the domain is higher compared to the accuracy in the mesh nodes away from the boundary. The study commenced about 30 years ago with the finite-difference scheme for the two-dimensional elliptic equation with the generalized solution from Sobolev spaces and later continued for other types of problems: quasilinear stationary and non-stationary equations with boundary conditions, boundary value problems for equations with fractional derivatives, the Cauchy problem and boundary value problems for abstract differential equations in Hilbert and Banach spaces, etc. For brevity, to name the influence that boundary and initial conditions have on the accuracy of the approximate solution, we choose to use the wording *boundary effect* or *initial effect*. Thus, we obtain a priori accuracy weighted estimates, taking into account the boundary and initial effects. These effects are quantitatively described by means of a suitable weight function, which characterizes the distance of a point to the boundary of the domain.

To our best knowledge, there are very few publications addressing these issues. It is our hope that the present book will meet this need and thus help to inspire new generations of students, researchers and practitioners. We also sincerely hope that our approach, methods and techniques developed in the book will contribute not only to the theory of the numerical analysis but also to its applications, since awareness of the boundary and initial effects makes it possible to use a greater mesh step near the boundary of the domain. Since the finite-difference approximations and the mesh schemes proposed and studied in this book are traditional and not exotic, they can be used for solving a wide range of problems in physics, engineering, chemistry, biology, finance, etc.

The target audience of our book is graduate and postgraduate students, specialists in numerical analysis, computational and applied mathematics, and engineers. As in books like ours, the analytical and numerical components are closely intertwined, we expect the potential reader to have fluency in both univariate and multivariate analysis, familiarity with ordinary and partial differential equations, basic knowledge of functional analysis, advanced knowledge of numerical analysis, and be at ease with modern scientific computing. These mathematical prerequisites will make the text much easier to understand.

We are deeply grateful to Professor Nikolaos Limnios and Professor Dmytro Koroliuk for their suggestion to submit the manuscript, to Professor Ivan Gavrilyuk for many fruitful discussions, and to Professor Vyacheslav Ryabichev for his valuable software advice and constant professional assistance. We also express our gratitude to the team at ISTE Group for useful recommendations and careful preparation of our book for publication. We are immensely thankful to our families for everyday understanding, support and encouragement.

Volodymyr MAKAROV,
Institute of Mathematics of the National Academy of Sciences of Ukraine,
Nataliya MAYKO,
Taras Shevchenko National University of Kyiv,
Kyiv, Ukraine
July 2023

Introduction

It is well known that the vast majority of boundary value and initial value problems cannot be solved exactly and require the use of appropriate approximate methods. An important characteristic of any approximate method is its accuracy. To estimate the accuracy, we traditionally use a certain discretization parameter: a mesh step, the number of terms of the partial sum of the series, etc.

However, for both theoretical and practical reasons, it is also important to take into account the influence of other factors, for example, the so-called boundary and initial effects. Precisely, the boundary effect means that due to the Dirichlet boundary condition for a differential equation in the canonical domain, the accuracy of the approximate solution near the boundary of the domain is higher compared to the accuracy further from the boundary. A similar situation is observed for non-stationary equations near those mesh nodes where the initial condition is set.

For the quantitative characteristics of the boundary or initial effect, we can take an a priori error estimate (in a certain mesh norm) with a certain weight function, which characterizes the distance of a point inside the domain to the boundary of the domain. The idea of such estimates was first announced by Volodymyr Makarov in Makarov (1987) for an elliptic equation in case of generalized solutions from Sobolev spaces and developed further in publications for quasilinear stationary and non-stationary equations. Since the concept was quite new, there were (and still are) very few publications on this subject. In some respects, the same issue is studied in the works of Galba (1985) and Molchanov and Galba (1990). However, they assume only the classical smoothness of solutions and do not consider time-dependent problems.

In this book, we develop our previous studies and present some new results on the impact of initial and boundary conditions on the accuracy of the following methods: the finite-difference method for elliptic and parabolic equations, the discrete method for solving equations with fractional derivatives, and the Cayley transform method for abstract differential equations in Hilbert and Banach spaces. Regardless of the type of problem or method, our main focus is always on obtaining weighted estimates with a proper weight function.

For a better understanding of the reasoning and easier navigation through the computation, some information is assumed to be known to the reader from classical mathematics courses, while the rest is provided directly in the text. Some of the formulas may seem a bit long and cumbersome, but this is partly because we are trying to be as detailed as possible and help the reader follow the calculations with ease.

This book consists of five chapters. Chapters 1 and 2 are devoted to the study of the accuracy of finite-difference schemes for stationary and non-stationary equations respectively, taking into account the influence of boundary and initial conditions (in the sense of Makarov as mentioned above).

The finite-difference method is historically one of the first and most recognized numerical methods for solving problems of mathematical physics, mainly due to its universality and convenience in practical implementation. In recent decades, it has gained considerable popularity due to growing interest in the study of nonlinear processes in various fields of physics, chemistry, seismology, ecology, etc. Mathematical models of such phenomena involve nonlinear partial differential equations. For example, in aerodynamics and hydrodynamics, the one-dimensional quasilinear Burgers parabolic equation arises as an adequate mathematical model of turbulence. A special case of the Burgers equation is the quasilinear transport equation (the Hopf equation), which is the simplest equation describing discontinuous flows or flows with shock waves. In biology, ecology, physiology, combustion theory, crystallization theory, plasma physics, etc., the Fisher–Kolmogorov–Petrovsky–Piskunov equation (the Fisher–KPP equation) plays an important role as the simplest semi-linear parabolic equation. The propagation of shallow water waves that weakly and nonlinearly interact, ion acoustic waves in plasma, acoustic waves on crystal lattices, etc. are often modeled by the Korteweg–de Vries equation (the KdV equation). Many publications are devoted to finite-difference schemes for solving problems for elliptic and parabolic equations with dynamic conjugation conditions at the contact boundary (which is associated with the presence of concentrated heat capacities in a heat-conducting medium) and/or dynamic boundary conditions (which model heat conduction in a solid body in contact with fluid, as well as processes in semiconductor devices). In the mathematical modeling of some processes in ecology, physics and technology, when it is impossible to set the exact values of the desired solution at the boundary of a domain, problems with non-local boundary conditions usually arise.

These and many other examples demonstrate that the finite-difference method is actively developing and is widely used to solve current scientific and technical problems. At the same time, there are very few publications dedicated to the study of the initial and boundary effects in the above sense, and our book is a certain step towards filling this gap. One of the first such works is the announcement (Makarov 1989) that deals with the problem

$$Lu \equiv -\sum_{i,j=1}^{2} \frac{\partial}{\partial x_i}\left(a_{ij}(x)\frac{\partial u}{\partial x_j}\right) + q(x)u(x) = f(x), \quad x \in \Omega,$$

$$u(x) = 0, \quad x \in \Gamma = \partial\Omega,$$

where

$$\nu \sum_{i=1}^{2} \xi_i^2 \le \sum_{i,j=1}^{2} a_{ij}(x)\xi_i\xi_j \le \mu \sum_{i=1}^{2} \xi_i^2 \quad \forall x = (x_1,x_2) \in \Omega, \ \forall \xi = (\xi_1,\xi_2) \in \mathbb{R}^2,$$

$$\nu > 0, \ a_{ij}(x) = a_{ji}(x), \ q(x) \in C(\overline{\Omega}), \ q(x) \ge 0,$$

and $\Omega = \{(x_1,x_2) : 0 < x_\alpha < 1, \alpha = 1,2\}$ is a unit square. The problem is discretized by the finite-difference scheme

$$Ay \equiv -\frac{1}{2}\sum_{i,j=1}^{2}\left[\left(a_{ij}^{-0.5i}y_{\bar{x}_j}\right)_{x_i} + \left(a_{ij}^{+0.5i}y_{x_j}\right)_{\bar{x}_i}\right] + qy = \varphi(x), \quad x \in \omega, \qquad [\text{I.1}]$$

$$y(x) = 0, \quad x \in \gamma,$$

where $\omega = \omega_1 \times \omega_2$, $\omega_\alpha = \{x_\alpha = i_\alpha h_\alpha : i_\alpha = 1,2,\ldots,N_\alpha - 1, h_\alpha = 1/N_\alpha\}$, $\alpha = 1,2$; γ is a boundary of the mesh ω. Some traditional notations for finite-difference schemes from Samarskii (2001) are used here, for example: $a_{11}^{\pm 0.5_1} = a_{11}(x_1 \pm 0.5h_1, x_2)$,

$$y_{x_1} = \frac{y(x_1+h_1,x_2)-y(x_1,x_2)}{h_1}, \ y_{\bar{x}_1} = \frac{y(x_1,x_2)-y(x_1-h_1,x_2)}{h_1}, \text{ etc.}$$

The main result was presented in the following statement.

THEOREM.– *Let* $\varphi(x), f(x), a_{ij}(x) \in W_2^3(\Omega)$ *and* $u(x) \in W_2^4(\Omega)$. *Then, there exists* $h_0 > 0$ *such that for all* $h \in (0, h_0]$ *the accuracy of the finite-difference scheme [I.1] is characterized by the weighted estimate*

$$\left\| \rho^{-1/2}(x)[y(x) - u(x)] \right\|_{C(\omega)} \leq M h^2 \| u \|_{W_2^4(\Omega)},$$

with the weight function $\rho(x) = \min\{x_1 x_2, x_1(1-x_2), (1-x_1)x_2, (1-x_1)(1-x_2)\}$.

This idea is further developed in the present book for other types of boundary conditions for elliptic and parabolic equations. It is worth mentioning that the important stages in obtaining such weighted estimates are the evaluation of discrete Green's functions and the analysis of approximation errors. Each time, when it is necessary to estimate discrete Green's functions, we apply the following proposition, which is formulated and proved in Samarskii et al. (1987, p. 54).

MAIN LEMMA.– *Let the following assumptions be fulfilled: 1)* $A : H \to H$ *is a self-adjoint operator acting in a Hilbert space H; 2)* $B : H^* \to H$ *is a linear operator; 3) the inverse operator* A^{-1} *exists; 4)* $\| B^* v \|_* \leq \gamma \| Av \|$ *for all* $v \in H$, *where* $B^* : H \to H^*$ *is the adjoint operator of* B, $(y, v)_*$ *and* $\| v \|_* = \sqrt{(v, v)_*}$ *are an inner product and an associate norm in* H^* *respectively. Then,* $\| A^{-1} B v \| \leq \gamma \| v \|_*$ *for all* $v \in H^*$.

Similarly, when it comes to estimating an approximation error for a generalized solution from Sobolev spaces, we refer to the Bramble–Hilbert lemma (e.g. Samarskii et al. (1987, p. 29)). We recall it here for convenience.

LEMMA (BRAMBLE–HILBERT).– *Let* $\Omega \subset \mathbb{R}^n$ *be an open convex bounded set of the diameter* $d > 0$, *let* $l(u)$ *be a bounded linear functional in the space* $W_2^m(\Omega)$ *with* $0 < m = \bar{m} + \lambda$, *where* $\bar{m}$ *is a positive non-negative number and* $0 < \lambda \leq 1$, *namely:*

$$| l(u) | \leq M \left\{ \sum_{j=0}^{\bar{m}} d^{2j} | u |_{W_2^j(\Omega)}^2 + d^{2m} | u |_{W_2^m(\Omega)}^2 \right\}^{1/2},$$

and let $l(u)$ turn into zero on polynomials of degree $\bar{m}$ of variables $x_1, x_2, \ldots, x_n$. Then, there exists a positive constant $\bar{M}$, which is dependent on Ω and independent of $u(x)$, such that the following inequality holds true:

$$|l(u)| \leq M\bar{M}d^m \, |u|_{W_2^m(\Omega)} \quad \forall u \in W_2^m(\Omega).$$

The study of the boundary and initial effects is also of great interest for new classes of problems, for example, related to the application of fractional integro-differentiation. In Chapter 3, we address the accuracy of the mesh methods for solving boundary value problems for differential equations with fractional derivatives.

For almost 300 years (from 1695 until recently) this branch of classical analysis was no more than an abstract mathematical theory. However, over the past several decades, fractional analysis has found wide applications in the construction of adequate mathematical models of many natural and social phenomena, as evidenced by a considerable number of publications (e.g. Kilbas et al. (2006); Sabatier et al. (2007); Nakagawa et al. (2010), to mention a few). Due to the ability to model hereditary phenomena with long memory, fractional analysis is widely used in viscoelasticity problems, models of anomalous diffusion (in particular, subdiffusion), control theory, electrodynamics and nonlinear hydroacoustics, for multidimensional signal processing in radiophysics, etc. However, exact solutions of such problems can be found only in a few (mostly linear) cases. The integral nature of the fractional derivative (in contrast to the classical derivative, which is local in nature) complicates the construction, analysis and implementation of approximate methods. For example, one of such problems is a considerable increase in costs associated with large data storage due to systems of linear equations with large, densely filled matrices. This requires adaptation of known and development of new approaches in the field of fractional numerical analysis, which is actively developing and constantly updated (Li and Zeng 2012; Jin et al. 2017; Jovanović et al. 2019).

Throughout this chapter, we use exclusively the *left Riemann–Liouville derivative of order* $\alpha > 0$ for a function $f(x)$:

$$^{RL}_{a}D_x^\alpha f(x) = \frac{1}{\Gamma(n-\alpha)} \frac{d^n}{dx^n} \int_a^x (x-t)^{n-\alpha-1} f(t)dt, \quad n = \lfloor \alpha \rfloor + 1.$$

It is quite natural that discretization of the fractional derivative is an important step in the construction of an effective approximate method for solving any fractional differential equation. The most widely used approximations of fractional derivatives can be roughly divided into two groups. The first group includes convolution-type quadrature formulas, while the second group includes the so-called L1 and L2 schemes. Each group has its advantages and disadvantages, which is discussed in detail by Jin et al. (2019). For instance, quadrature formulas are flexible, convenient for analysis, have good stability properties, but are applicable mainly to uniform meshes. The strengths of the L1 and L2 methods are flexibility, easy implementation and the possibility of generalization for non-uniform meshes, while the disadvantages are sophisticated analysis and the first order of accuracy (in the case of direct application without proper correction).

However, in our book, we take another approach. Whether it is a one-dimensional equation with constant or variable coefficients, the Dirichlet boundary value problem for Poisson's equation with the fractional derivative for one of the two variables in a unit square, or the two-dimensional Goursat problem, we reduce each of them to an integral equation of the second kind. Then, we study the kernel and apply the fixed-point iteration to show that a solution of the problem belongs to a particular Sobolev class. After that we propose a mesh scheme and study its convergence in some discrete norm with a weight function, taking into account the boundary condition.

Chapters 4 and 5 are devoted to the Cayley transform method first proposed in Arov and Gavrilyuk (1993) and Arov et al. (1995). This method is designed for the constructive representation of exact and approximate solutions of abstract differential equations in Hilbert and Banach spaces. One of the advantages of this method is the automatic dependence of its accuracy on the smoothness of the input data. This means that the Cayley transform method belongs to the methods *without saturation of accuracy* according to Babenko (2002), and is therefore optimal in a certain sense. The construction of such methods is a topical issue of numerical analysis.

The importance of the Cayley transform method is also explained by the observation that mathematical models of many processes studied in science and technology can be written in the form of differential equations in Banach and Hilbert spaces, namely in the form of the Cauchy problem for the first-order differential equation:

$$u'(t) + Au(t) = f(t), \quad x \in (0, T],$$
$$u(0) = u_0,$$

[I.2]

the Cauchy problem for the second-order differential equation:

$$u''(t) + Au(t) = f(t), \quad x \in (0, T],$$
$$u(0) = u_0, \quad u'(0) = u_1,$$

[I.3]

and the boundary value problem for the second-order differential equation:

$$u''(x) - Au(x) = -f(x), \quad x \in (0,1),$$
$$u(0) = u_0, \quad u(1) = u_1. \qquad [\text{I.4}]$$

Here, A is a closed linear operator with the dense domain $D(A)$ in a Banach space E (or a self-adjoint positive definite operator with the dense domain $D(A)$ in a Hilbert space H), u_0 and u_1 are given vectors from E (or from H), $f(\cdot)$ and $u(\cdot)$ are respectively a given function and an unknown solution with values in E (or in H).

For example, in the case of a Hilbert space $H = L_2(0,1)$ and the operator $Au(x) = -u''(x)$, $D(A) = H^2(0,1) \cap \overset{\circ}{H}{}^1(0,1)$, the Cauchy problems [I.2] and [I.3] turn into the initial–boundary value problems for a parabolic and hyperbolic equations respectively:

$$u_t(x,t) = u_{xx}(x,t) + f(x,t), \quad x \in (0,1), \ t \in (0,T],$$
$$u(0,t) = 0, \ u(1,t) = 0, \ t \in [0,T],$$
$$u(x,0) = u_0(x), \ x \in [0,1],$$

$$u_{tt}(x,t) = u_{xx}(x,t) + f(x,t), \quad x \in (0,1), \ t \in (0,T],$$
$$u(0,t) = 0, \ u(1,t) = 0, \ t \in [0,T],$$
$$u(x,0) = u_0(x), \ u_t(x,0) = u_1(x), \ x \in [0,1].$$

Similarly, in the case of a Hilbert space $H = L_2(0,1)$ and the operator

$$Au(y) = -u''(y), \quad D(A) = H^2(0,1) \cap \overset{\circ}{H}{}^1(0,1),$$

the boundary value problem [I.4] takes the form of the Dirichlet boundary value problem for Poisson's equation:

$$u_{xx} + u_{yy} = -f(x,y), \quad (x,y) \in \Omega = (0,1)^2,$$
$$u(x,y) = 0, \quad (x,y) \in \partial\Omega.$$

For convenience, we briefly recall the results of the pioneering publication (Arov and Gavrilyuk 1993). In a Hilbert space H and for a bounded operator A, it studies the Cauchy problem

$$x'(t) + Ax(t) = 0, \quad t > 0,$$
$$x(0) = x_0, \qquad \text{[I.5]}$$

and proves that the solution $x(t)$ can be represented by the series

$$x(t) = e^{-\gamma t} \sum_{p=0}^{\infty} (-1)^p L_p^{(0)}(2\gamma t) \left[y_{\gamma,p} + y_{\gamma,p+1} \right], \qquad \text{[I.6]}$$

where $\gamma > 0$ is an arbitrary number, $L_p^{(\alpha)}(t) = \sum_{k=0}^{p} \binom{p+\alpha}{p-k} \frac{(-t)^k}{k!}$ are the Laguerre polynomials and the sequence $(y_{\gamma,p})$ satisfies the recurrence relation

$$y_{\gamma,p+1} = T_\gamma y_{\gamma,p} = T_\gamma^{p+1} y_{\gamma,0}, \quad p = 0, 1, \ldots, \quad y_{\gamma,0} = x_0,$$

and therefore $y_{\gamma,p}$ can be effectively found from the recurrent sequence of the operator equations (with the same operator and different right-hand sides):

$$(\gamma I + A) y_{\gamma,p+1} = (\gamma I - A) y_{\gamma,p}, \quad p = 0, 1, \ldots, \quad y_{\gamma,0} = x_0.$$

The partial sum of series [I.6] is then taken as an approximate solution of problem [I.5]:

$$x_N(t) = e^{-\gamma t} \sum_{p=0}^{N} (-1)^p L_p^{(0)}(2\gamma t) \left[y_{\gamma,p} + y_{\gamma,p+1} \right].$$

The accuracy of this approximation is characterized by the estimate

$$\sup_{t \geq 0} \| x(t) - x_N(t) \| \leq \frac{q_\gamma^{N+1}}{1 - q_\gamma} \| x_0 \| \quad (0 < q_\gamma < 1),$$

which indicates that the proposed Cayley transform method is exponentially convergent.

Other approaches to the construction of approximate solutions of operator differential equations are used in Gorodnii (1998) and Kashpirovskii and Mytnik (1998).

The results obtained in Arov and Gavrilyuk (1993) and Arov et al. (1995) were then extended to other abstract problems in Hilbert and Banach spaces and subsequently summarized in a monograph (Gavrilyuk and Makarov 2004). Our book continues this tradition and develops the Cayley transform method even further – now taking into account the influence of boundary and initial conditions. Therefore, the proposed technique of obtaining weighted estimates with a proper weight function meets both challenges – taking into account the boundary and initial effects and also the smoothness of input data (e.g. coefficients and the right-hand side of the equation, initial vectors, etc.).

With this brief introduction, we sincerely hope that the reader will share our interest in the issues discussed above and embark on a journey of new research and discovery.

1

Elliptic Equations in Canonical Domains with the Dirichlet Condition on the Boundary or its Part

1.1. A standard finite-difference scheme for Poisson's equation with mixed boundary conditions

We consider here the following boundary value problem:

$$-\Delta u = f(x), \quad x \in D,$$
$$-\frac{\partial u}{\partial x_1} + \sigma u(x) = 0, \quad x \in \Gamma_{-1}, \qquad [1.1]$$
$$u(x) = 0, \quad x \in \Gamma \setminus \Gamma_{-1},$$

where $x = (x_1, x_2)$, $\Delta = \dfrac{\partial^2}{\partial x_1^2} + \dfrac{\partial^2}{\partial x_2^2}$ is the Laplace operator in a Cartesian coordinate system, $D = \{x = (x_1, x_2): 0 < x_\alpha < l_\alpha, \alpha = 1, 2\}$ is a rectangle, $\Gamma = \partial D$ is a boundary of D, $\Gamma_{-1} = \{x = (0, x_2): 0 < x_2 < l_2\}$ is the left side of D, $f(x)$ is a given function, $\sigma = \text{const} \geq 0$.

1.1.1. *Discretization of the BVP*

To construct and study the discrete analogue of problem [1.1], we use the traditional notation of the theory of finite-difference schemes (e.g. Samarskii (2001)). We introduce the following sets of nodes:

$\omega_\alpha = \{i_\alpha h_\alpha, \ i_\alpha = 1, \ldots, N_\alpha - 1, \ h_\alpha = l_\alpha/N_\alpha\}, \ N_\alpha \geq 2$ is an integer number,

$\omega_\alpha^- = \omega_\alpha \cup \{0\}, \quad \omega_\alpha^+ = \omega_\alpha \cup \{1\}, \quad \bar{\omega}_\alpha = \omega_\alpha \cup \{0\} \cup \{1\},$

$\omega = \omega_1 \times \omega_2, \quad \bar{\omega} = \bar{\omega}_1 \times \bar{\omega}_2, \quad \gamma = \bar{\omega} \setminus \omega;$ [1.2]

$\gamma_{-\alpha} = \{x_\alpha = 0, \ x_{3-\alpha} \in \omega_{3-\alpha}\}, \quad \gamma_{+\alpha} = \{x_\alpha = l_\alpha, \ x_{3-\alpha} \in \omega_{3-\alpha}\}, \quad \alpha = 1, 2.$

We also use the operators of the exact finite-difference schemes:

$$T_2 v(x_1, x_2) = \frac{1}{h_2^2} \int_{x_2 - h_2}^{x_2 + h_2} (h_2 - |x_2 - \xi_2|) v(x_1, \xi_2) d\xi_2, \quad x \in \omega \cup \gamma_{-1},$$

$$T_1 v(x_1, x_2) = \begin{cases} \dfrac{1}{h_1^2} \displaystyle\int_{x_1 - h_1}^{x_1 + h_1} (h_1 - |x_1 - \xi_1|) v(\xi_1, x_2) d\xi_1, & x \in \omega, \\[2pt] \dfrac{2}{h_1^2} \displaystyle\int_0^{h_1} (h_1 - \xi_1) v(\xi_1, x_2) d\xi_1, & x \in \gamma_{-1}, \end{cases}$$

Using the relations

$$T_\alpha \frac{\partial^2 u}{\partial x_\alpha^2} = u_{\bar{x}_\alpha x_\alpha}, \quad x \in \omega,$$

$$T_1 1 = 1, \ T_1 x_1 = \frac{h_1}{3}, \ T_1 x_2 = x_2, \ T_1 \frac{\partial^2 u}{\partial x_1^2} = \frac{2}{h_1}\left(u_{x_1} - \frac{\partial u}{\partial x_1}\right), \quad x \in \gamma_{-1},$$

we approximate problem [1.1] by the finite-difference scheme

$$\begin{aligned} -\Lambda y(y) &= \varphi(x), \quad x \in \omega \cup \gamma_{-1}, \\ y(x) &= 0, \quad x \in \gamma \setminus \gamma_{-1}, \end{aligned}$$ [1.3]

where $\varphi(x) = T_1 T_2 f(x), \ \Lambda = \Lambda_1 + \Lambda_2,$

$$\Lambda_1 y(x) = \begin{cases} y_{\bar{x}_1 x_1}, & x \in \omega, \\ \dfrac{2}{h_1}(y_{x_1} - \sigma y), & x \in \gamma_{-1}, \end{cases} \quad \Lambda_2 y(x) = \begin{cases} y_{\bar{x}_2 x_2}, & x \in \omega, \\ \left(1 + \dfrac{h_1 \sigma}{3}\right) y_{\bar{x}_2 x_2}, & x \in \gamma_{-1}. \end{cases}$$

For the error $z(x) = y(x) - u(x)$, we have the problem

$$-\Lambda z(x) = \psi(x), \quad x \in \omega \cup \gamma_{-1},$$
$$z(x) = 0, \quad x \in \gamma \setminus \gamma_{-1}, \qquad [1.4]$$

where $\psi(x)$ is the approximation error:

$$\psi(x) = T_1 T_2 f(x) + \Lambda u(x) = -\Lambda_1 \eta_1(x) - \eta_{2\bar{x}_2 x_2}(x), \qquad [1.5]$$

$$\eta_1(x) = (T_2 u)(x) - u(x), \ x \in \omega \cup \gamma_{-1}, \quad \eta_2(x) = \begin{cases} (T_1 u)(x) - u(x), & x \in \omega, \\ (T_1 u)(x) - u(x) - \dfrac{h_1}{3} \sigma u(x), & x \in \gamma_{-1}. \end{cases}$$

1.1.2. *Properties of the finite-difference operators*

We denote by H a set of mesh functions defined on $\bar{\omega}$ and equal to zero on $\gamma \setminus \gamma_{-1}$. The inner product and the associate norm in H are defined by the formulas

$$(y, v) = \sum_{x \in \omega} h_1 h_2 y(x) v(x) + \frac{h_1}{2} \sum_{x \in \gamma_{-1}} h_2 y(x) v(x),$$

$$\|v\| = \|v\|_{L_2(\omega)} = \sqrt{(v,v)} = \left(\sum_{x \in \omega} h_1 h_2 v^2(x) + \frac{h_1}{2} \sum_{x \in \gamma_{-1}} h_2 v^2(x) \right)^{1/2}.$$

We introduce the difference operators

$$A_\alpha, A : H \to H, \quad A_\alpha = -\Lambda_\alpha, \quad \alpha = 1, 2, \quad A = A_1 + A_2 = -\Lambda.$$

(If necessary, a function defined on ω is set equal to zero for $x \in \gamma \setminus \gamma_{-1}$ and equal to arbitrary values for $x \in \gamma_{-1}$.)

Then, the finite-difference scheme [1.3] can be written as the operator equation

$$Ay = \varphi, \quad y, \varphi \in H, \qquad [1.6]$$

and similarly problem [1.4] can be written as the operator equation

$$Az = \psi, \quad z, \psi \in H.$$

LEMMA 1.1.– *The difference operator A is symmetric and positive definite.*

PROOF.– The difference operators A_1 and A_2 are symmetric and positive definite in H. Indeed, for A_1, we have

$$(A_1 y, v) = \sum_{x \in \omega} h_1 h_2 (-y_{\bar{x}_1 x_1}) v + \frac{h_1}{2} \sum_{x \in \gamma_{-1}} h_2 \frac{-2}{h_1}(y_{x_1} - \sigma y) v =$$

$$= \sum_{x \in \omega_1^+ \times \omega_2} h_1 h_2 y_{\bar{x}_1} v_{\bar{x}_1} + \sigma \sum_{x \in \gamma_{-1}} h_2 yv$$

and therefore

$$(A_1 y, y) = \sum_{x \in \omega_1^+ \times \omega_2} h_1 h_2 y_{\bar{x}_1}^2 + \sigma \sum_{x \in \gamma_{-1}} h_2 y^2 \geq \sum_{x_2 \in \omega_2} h_2 \sum_{x_1 \in \omega_1^+} h_1 y_{\bar{x}_1}^2 \geq$$

$$\geq \sum_{x_2 \in \omega_2} h_2 \frac{2}{l_1^2} \left(\sum_{x_1 \in \omega_1} h_1 y^2(x) + \frac{h_1}{2} y(0, x_2) \right) = \frac{2}{l_1^2} \| y \|^2 .$$

And similarly for A_2, we obtain

$$(A_2 y, v) = \sum_{x \in \omega} h_1 h_2 (-y_{\bar{x}_2 x_2}) v + \frac{h_1}{2} \sum_{x \in \gamma_{-1}} h_2 \left(1 + \frac{h_1 \sigma}{3}\right)(-y_{\bar{x}_2 x_2}) v =$$

$$= \sum_{x \in \omega_1 \times \omega_2^+} h_1 h_2 y_{\bar{x}_2} v_{\bar{x}_2} + \left(1 + \frac{h_1 \sigma}{3}\right) \frac{h_1}{2} \sum_{\substack{x_2 \in \omega_2^+ \\ (x_1 = 0)}} h_2 y_{\bar{x}_2} v_{\bar{x}_2}$$

which gives

$$(A_2 y, y) = \sum_{x \in \omega_1 \times \omega_2^+} h_1 h_2 y_{\bar{x}_2}^2 + \left(1 + \frac{h_1 \sigma}{3}\right) \frac{h_1}{2} \sum_{\substack{x_2 \in \omega_2^+ \\ (x_1 = 0)}} h_2 y_{\bar{x}_2}^2 \geq$$

$$\geq \sum_{x_1 \in \omega_1} h_1 \sum_{x_2 \in \omega_2^+} h_2 y_{\bar{x}_2}^2 + \frac{h_1}{2} \sum_{\substack{x_2 \in \omega_2^+ \\ (x_1 = 0)}} h_2 y_{\bar{x}_2}^2 \geq$$

$$\geq \sum_{x_1 \in \omega_1} h_1 \frac{8}{l_2^2} \sum_{x_2 \in \omega_2} h_2 y^2(x) + \frac{h_1}{2} \frac{8}{l_2^2} \sum_{x_2 \in \omega_2} h_2 y^2(0, x_2) = \frac{8}{l_2^2} \| y \|^2.$$

Then, the difference operator $A = A_1 + A_2$ is also symmetric and positive definite. The lemma is proved.

It follows from Lemma 1.1 that there exist the inverse operator A^{-1} and thus the discrete problem [1.6] is uniquely solvable. We can now prove the following proposition.

LEMMA 1.2.– *It holds*

$$\| A^{-1} B_k v \| \leq \sqrt{\frac{6}{11}} \| v \|_k \ \text{for all } v \in H_k \quad (k = 1, 2), \qquad [1.7]$$

where the operator B_k, the space H_k and the norm $\|v\|_k$ are defined further in the text.

PROOF.– Applying summation by parts and the ε-inequality for $\varepsilon = 1/(4\sigma)$, we have

$$\| Ay \|^2 = (Ay, Ay) = (A_1 y + A_2 y, A_1 y + A_2 y) = \| A_1 y \|^2 + \| A_2 y \|^2 + 2(A_1 y, A_2 y) \geq$$

$$\geq 2(A_1 y, A_2 y) = 2 \sum_{x \in \omega} h_1 h_2 y_{\bar{x}_1 x_1} y_{\bar{x}_2 x_2} + 2 \frac{h_1}{2} \sum_{x \in \gamma_{-1}} h_2 \frac{2}{h_1} (y_{x_1} - \sigma y) \left(1 + \frac{h_1 \sigma}{3} \right) y_{\bar{x}_2 x_2} =$$

$$= 2 \sum_{x \in \omega_1^- \times \omega_2^-} h_1 h_2 y_{x_1 x_2}^2 - 2 \frac{h_1 \sigma}{3} \sum_{\substack{x_2 \in \omega_2^- \\ (x_1 = 0)}} h_2 y_{x_1 x_2} y_{x_2} + 2\sigma \left(1 + \frac{h_1 \sigma}{3} \right) \sum_{\substack{x_2 \in \omega_2^- \\ (x_1 = 0)}} h_2 y_{x_2}^2 \geq$$

$$\geq 2 \sum_{x \in \omega_1^- \times \omega_2^-} h_1 h_2 y_{x_1 x_2}^2 - 2 \frac{h_1 \sigma}{3} \left(\frac{1}{4\sigma} \sum_{\substack{x_2 \in \omega_2^- \\ (x_1 = 0)}} h_2 y_{x_1 x_2}^2 + \sigma \sum_{\substack{x_2 \in \omega_2^- \\ (x_1 = 0)}} h_2 y_{x_2}^2 \right) +$$

$$+ 2\sigma \left(1 + \frac{h_1 \sigma}{3} \right) \sum_{\substack{x_2 \in \omega_2^- \\ (x_1 = 0)}} h_2 y_{x_2}^2 =$$

$$= 2 \sum_{x \in \omega_1^- \times \omega_2^-} h_1 h_2 y_{x_1 x_2}^2 - 2\frac{h_1 \sigma}{3}\frac{1}{4\sigma} \sum_{\substack{x_2 \in \omega_2^- \\ (\tilde{x}_1 = 0)}} h_2 y_{x_1 x_2}^2 + 2\sigma \sum_{\substack{x_2 \in \omega_2^- \\ (\tilde{x}_1 = 0)}} h_2 y_{x_2}^2 \geq$$

$$\geq 2 \sum_{x \in \omega_1^- \times \omega_2^-} h_1 h_2 y_{x_1 x_2}^2 - \frac{h_1}{6} \sum_{\substack{x_2 \in \omega_2^- \\ (\tilde{x}_1 = 0)}} h_2 y_{x_1 x_2}^2 \geq \left(2 - \frac{1}{6}\right) \sum_{x \in \omega_1^- \times \omega_2^-} h_1 h_2 y_{x_1 x_2}^2 = \frac{11}{6} \| B_1^* y \|_1^2 ,$$

where

$$B_1^* : H \to H_1, \quad B_1^* y = -y_{x_1 x_2}, \quad x \in \omega_1^- \times \omega_2^- ,$$

is a difference operator acting from H into the space H_1 of mesh functions defined on $\omega_1^- \times \omega_2^-$ with the inner product and the associate norm

$$(y, v)_1 = \sum_{x \in \omega_1^- \times \omega_2^-} h_1 h_2 y(x) v(x), \quad \| y \|_1 = \sqrt{(y, y)_1} = \left(\sum_{x \in \omega_1^- \times \omega_2^-} h_1 h_2 y^2(x) \right)^{1/2}.$$

Applying summation by parts, we have

$$(B_1^* y, w)_1 = - \sum_{x \in \omega_1^- \times \omega_2^-} h_1 h_2 y_{x_1 x_2} w = -\sum_{x \in \omega} h_1 h_2 y w_{\bar{x}_1 \bar{x}_2} - \frac{h_1}{2} \sum_{x \in \gamma_{-1}} h_2 y \frac{2}{h_1} w_{\bar{x}_2} = (y, B_1 w),$$

where $B_1 : H_1 \to H$ is the adjoint operator of $B_1^* : H \to H_1$,

$$B_1 w(x) = - \begin{cases} w_{\bar{x}_1 \bar{x}_2}, & x \in \omega, \\ \dfrac{2}{h_1} w_{\bar{x}_2}, & x \in \gamma_{-1}. \end{cases}$$

Similarly, we have

$$\| Ay \|^2 = \| A_1 y \|^2 + \| A_2 y \|^2 + 2(A_1 y, A_2 y) \geq 2(A_1 y, A_2 y) =$$

$$= 2 \sum_{x \in \omega} h_1 h_2 y_{\bar{x}_1 x_1} y_{\bar{x}_2 x_2} + 2\frac{h_1}{2} \sum_{x \in \gamma_{-1}} h_2 \frac{2}{h_1} (y_{x_1} - \sigma y) \left(1 + \frac{h_1 \sigma}{3}\right) y_{\bar{x}_2 x_2} =$$

$$= 2 \sum_{x \in \omega_1^- \times \omega_2^+} h_1 h_2 y_{x_1 \bar{x}_2}^2 - 2 \frac{h_1 \sigma}{3} \sum_{\substack{x_2 \in \omega_2^+ \\ (\bar{x}_1 = 0)}} h_2 y_{x_1 \bar{x}_2} y_{\bar{x}_2} + 2\sigma \left(1 + \frac{h_1 \sigma}{3}\right) \sum_{\substack{x_2 \in \omega_2^+ \\ (\bar{x}_1 = 0)}} h_2 y_{\bar{x}_2}^2 \geq$$

$$\geq 2 \sum_{x \in \omega_1^- \times \omega_2^+} h_1 h_2 y_{x_1 \bar{x}_2}^2 - 2 \frac{h_1 \sigma}{3} \left(\frac{1}{4\sigma} \sum_{\substack{x_2 \subset \omega_2^+ \\ (\bar{x}_1 = 0)}} h_2 y_{x_1 \bar{x}_2}^2 + \sigma \sum_{\substack{x_2 \in \omega_2^+ \\ (\bar{x}_1 = 0)}} h_2 y_{\bar{x}_2}^2 \right) +$$

$$+ 2\sigma \left(1 + \frac{h_1 \sigma}{3}\right) \sum_{\substack{x_2 \in \omega_2^+ \\ (\bar{x}_1 = 0)}} h_2 y_{\bar{x}_2}^2 =$$

$$= 2 \sum_{x \in \omega_1^- \times \omega_2^+} h_1 h_2 y_{x_1 \bar{x}_2}^2 - 2 \frac{h_1 \sigma}{3} \frac{1}{4\sigma} \sum_{\substack{x_2 \in \omega_2^+ \\ (\bar{x}_1 = 0)}} h_2 y_{x_1 \bar{x}_2}^2 + 2\sigma \sum_{\substack{x_2 \in \omega_2^+ \\ (\bar{x}_1 = 0)}} h_2 y_{\bar{x}_2}^2 \geq$$

$$\geq 2 \sum_{x \in \omega_1^- \times \omega_2^+} h_1 h_2 y_{x_1 \bar{x}_2}^2 - \frac{h_1}{6} \sum_{\substack{x_2 \in \omega_2^+ \\ (\bar{x}_1 = 0)}} h_2 y_{x_1 \bar{x}_2}^2 \geq \left(2 - \frac{1}{6}\right) \sum_{x \in \omega_1^- \times \omega_2^+} h_1 h_2 y_{x_1 \bar{x}_2}^2 = \frac{11}{6} \| B_2^* y \|_2^2,$$

where

$$B_2^* : H \to H_2, \quad B_2^* y = -y_{x_1 \bar{x}_2}, \quad x \in \omega_1^- \times \omega_2^+,$$

is a difference operator acting from H into the space H_2 of mesh functions defined on $\omega_1^- \times \omega_2^+$ with the inner product and the associate norm

$$(y, v)_2 = \sum_{x \in \omega_1^- \times \omega_2^+} h_1 h_2 y(x) v(x), \quad \| y \|_2 = \sqrt{(y, y)_2} = \left(\sum_{x \in \omega_1^- \times \omega_2^+} h_1 h_2 y^2(x) \right)^{1/2}.$$

We find

$$(B_2^* y, w)_2 = - \sum_{x \in \omega_1^- \times \omega_2^+} h_1 h_2 y_{x_1 \bar{x}_2} w = - \sum_{x \in \omega} h_1 h_2 y w_{\bar{x}_1 \bar{x}_2} - \frac{h_1}{2} \sum_{x \in \gamma_{-1}} h_2 y \frac{2}{h_1} w_{x_2} = (y, B_2 w),$$

where $B_2 : H_2 \to H$ is the adjoint operator of $B_2^* : H \to H_2$,

$$B_2 w(x) = -\begin{cases} w_{\bar{x}_1 x_2}, & x \in \omega, \\ \dfrac{2}{h_1} w_{x_2}, & x \in \gamma_{-1}. \end{cases}$$

Applying the main lemma from Samarskii et al. (1987, p. 54) to the operators A, B_1, B_2, we obtain estimate [1.7] and thus complete the proof.

1.1.3. Discrete Green's function

We denote by $G(x,\xi)$ Green's function of the finite-difference problem [1.4]:

$$-G_{\bar{\xi}_1 \xi_1}(x,\xi) - G_{\bar{\xi}_2 \xi_2}(x,\xi) = \frac{\delta(x_1,\xi_1)\delta(x_2,\xi_2)}{h_1 h_2}, \quad \xi \in \omega, \qquad [1.8]$$

$$-\frac{2}{h_1}\left(G_{\xi_1}(x,\xi) - \sigma G(x,\xi)\right) - \left(1 + \frac{h_1 \sigma}{3}\right) G_{\bar{\xi}_2 \xi_2}(x,\xi) = \frac{2}{h_1} \frac{\delta(x_1,\xi_1)\delta(x_2,\xi_2)}{h_2}, \quad \xi \in \gamma_{-1},$$

$$G(x,\xi) = 0, \quad \xi \in \gamma \setminus \gamma_{-1},$$

where $\delta(m,n)$ is the Kronecker delta symbol and $\xi = (\xi_1, \xi_2)$.

LEMMA 1.3.– *For the error $z(x)$, the following estimate holds true:*

$$|z(x)| \leq \sqrt{\frac{6}{11}} \rho(x) \|\psi\|, \quad x \in \omega \cup \gamma_{-1},$$

where $\rho(x) = \min\left\{\sqrt{(l_1 - x_1)(l_2 - x_2)}, \sqrt{(l_1 - x_1)x_2}\right\}$.

PROOF.– Using the Heaviside step function

$$H(s) = \begin{cases} 1, & s \geq 0, \\ 0, & s < 0, \end{cases}$$

we rewrite problem [1.8] as follows:

$$-G_{\bar{\xi}_1\xi_1}(x,\xi) - G_{\bar{\xi}_2\xi_2}(x,\xi) = \left(H(\xi_1 - x_1)H(\xi_2 - x_2)\right)_{\bar{\xi}_1\bar{\xi}_2}, \quad \xi \in \omega,$$

$$-\frac{2}{h_1}\left(G_{\xi_1}(x,\xi) - \sigma G(x,\xi)\right) - \left(1 + \frac{h_1\sigma}{3}\right)G_{\bar{\xi}_2\xi_2}(x,\xi) =$$

$$= \frac{2}{h_1}\left(H(\xi_1 - x_1)H(\xi_2 - x_2)\right)_{\bar{\xi}_2}, \quad \xi \in \gamma_{-1},$$

$$G(x,\xi) = 0, \quad \xi \in \gamma \setminus \gamma_{-1},$$

which takes the operator form

$$A_\xi G(x,\xi) = -B_{1\xi}\left(H(\xi_1 - x_1)H(\xi_2 - x_2)\right).$$

Taking [1.7] into account, we then have

$$\|G(x,\cdot)\| = \|-A_\xi^{-1} B_{1\xi}\left(H(\cdot - x_1)H(\cdot - x_2)\right)\| \le \sqrt{\frac{6}{11}} \|H(\cdot - x_1)H(\cdot - x_2)\|_{1*} =$$

$$= \sqrt{\frac{6}{11}}\left(\sum_{\xi \in \omega_1^- \times \omega_2^-} h_1 h_2 H^2(\xi_1 - x_1) H^2(\xi_2 - x_2)\right)^{1/2} =$$

$$= \sqrt{\frac{6}{11}}\left(\sum_{\xi_1=0}^{l_1-h_1} h_1 H^2(\xi_1 - x_1)\right)^{1/2} \left(\sum_{\xi_2=0}^{l_2-h_2} h_2 H^2(\xi_2 - x_2)\right)^{1/2} = \quad [1.9]$$

$$= \sqrt{\frac{6}{11}}\left(\sum_{\xi_1=x_1}^{l_1-h_1} h_1\right)^{1/2} \left(\sum_{\xi_2=x_2}^{l_2-h_2} h_2\right)^{1/2} = \sqrt{\frac{6}{11}}\sqrt{(l_1 - x_1)(l_2 - x_2)}.$$

Problem [1.8] can also be written in a different way:

$$-G_{\bar{\xi}_1\xi_1}(x,\xi) - G_{\bar{\xi}_2\xi_2}(x,\xi) = -\left(H(\xi_1 - x_1)H(x_2 - \xi_2)\right)_{\bar{\xi}_1\bar{\xi}_2}, \quad \xi \in \omega,$$

$$-\frac{2}{h_1}\left(G_{\xi_1}(x,\xi) - \sigma G(x,\xi)\right) - \left(1 + \frac{h_1\sigma}{3}\right)G_{\bar{\xi}_2\xi_2}(x,\xi) =$$

$$= -\frac{2}{h_1}\left(H(\xi_1 - x_1)H(x_2 - \xi_2)\right)_{\bar{\xi}_2}, \quad \xi \in \gamma_{-1},$$

$$G(x,\xi) = 0, \quad \xi \in \gamma \setminus \gamma_{-1},$$

which takes the operator form

$$A_\xi G(x,\xi) = B_{2\xi}\left(H(\xi_1 - x_1)H(x_2 - \xi_2)\right).$$

This leads to the relation

$$\|G(x,\cdot)\| = \|A_\xi^{-1} B_{2\xi}\left(H(\cdot - x_1)H(x_2 - \cdot)\right)\| \leq \sqrt{\frac{6}{11}}\, \|H(\cdot - x_1)H(x_2 - \cdot)\|_{2*} =$$

$$= \sqrt{\frac{6}{11}}\left(\sum_{\xi \in \omega_1^- \times \omega_2^+} h_1 h_2 H^2(\xi_1 - x_1)H^2(x_2 - \xi_2)\right)^{1/2} =$$

$$= \sqrt{\frac{6}{11}}\left(\sum_{\xi_1=0}^{l_1-h_1} h_1 H^2(\xi_1 - x_1)\right)^{1/2}\left(\sum_{\xi_2=h_2}^{l_2} h_2 H^2(x_2 - \xi_2)\right)^{1/2} = \quad [1.10]$$

$$= \sqrt{\frac{6}{11}}\left(\sum_{\xi_1=x_1}^{l_1-h_1} h_1\right)^{1/2}\left(\sum_{\xi_2=h_2}^{x_2} h_2\right)^{1/2} = \sqrt{\frac{6}{11}}\sqrt{(l_1 - x_1)x_2}\,.$$

Inequalities [1.9] and [1.10] yield the estimate

$$\|G(x,\cdot)\| \leq \sqrt{\frac{6}{11}}\, \rho(x),$$

with $\rho(x) = \min\left\{\sqrt{(l_1 - x_1)(l_2 - x_2)},\, \sqrt{(l_1 - x_1)x_2}\right\}$. Then, we have

$$|z(x)| = \left|\left(G(x,\cdot), \psi(\cdot)\right)\right| \leq \|G(x,\cdot)\| \cdot \|\psi\|$$

and prove the lemma.

1.1.4. Accuracy with the boundary effect

First, we prove the following auxiliary result.

LEMMA 1.4.– *Let the solution $u(x)$ of problem [1.1] satisfy the condition $u \in W_2^4(D)$. Then, for the approximation error $\psi(x)$, the following estimate holds true:*

$$\|\psi\| \leq 16\,(h_1^2 + h_2^2)\left(\iint\limits_D \left|\frac{\partial^4 u(x_1,x_2)}{\partial x_1^2 \partial x_2^2}\right|^2 dx_1 dx_2\right)^{1/2}.$$

PROOF.– We obviously have

$$\|\psi\| \leq \|\Lambda_1\eta_1\| + \|\eta_{2\bar{x}_2 x_2}\|.$$

Next, we separately consider each summand $\|\Lambda_1\eta_1\|$ and $\|\eta_{2\bar{x}_2 x_2}\|$.

For the functional $\eta_1(x)$, $x \in \omega \cup \gamma_{-1}$, we find

$$\eta_1(x) = (T_2 u)(x) - u(x) = \frac{1}{h_2^2}\int\limits_{x_2-h_2}^{x_2+h_2}(h_2 - |x_2-\xi|)u(x_1,\xi)d\xi - u(x_1,x_2) =$$

$$= \frac{1}{h_2^2}\int\limits_{x_2-h_2}^{x_2+h_2}(h_2-|x_2-\xi|)\bigl(u(x_1,\xi)-u(x_1,x_2)\bigr)d\xi = \qquad [1.11]$$

$$= \frac{1}{h_2^2}\int\limits_{x_2-h_2}^{x_2+h_2}(h_2-|x_2-\xi|)d\xi\int\limits_{x_2}^{\xi}\frac{\partial u(x_1,\xi_1)}{\partial \xi_1}d\xi_1 =$$

$$= \frac{1}{h_2^2}\int\limits_{x_2-h_2}^{x_2+h_2}(h_2-|x_2-\xi|)d\xi\int\limits_{x_2}^{\xi}\left(\frac{\partial u(x_1,\xi_1)}{\partial \xi_1} - \frac{1}{h_2}\int\limits_{x_2-\frac{h_2}{2}}^{x_2+\frac{h_2}{2}}\frac{\partial u(x_1,\xi_2)}{\partial \xi_2}d\xi_2\right)d\xi_1 =$$

$$= \frac{1}{h_2^3}\int\limits_{x_2-h_2}^{x_2+h_2}(h_2-|x_2-\xi|)d\xi\int\limits_{x_2}^{\xi}d\xi_1\int\limits_{x_2-\frac{h_2}{2}}^{x_2+\frac{h_2}{2}}\left(\frac{\partial u(x_1,\xi_1)}{\partial \xi_1} - \frac{\partial u(x_1,\xi_2)}{\partial \xi_2}\right)d\xi_2 =$$

$$= \frac{1}{h_2^3}\int\limits_{x_2-h_2}^{x_2+h_2}(h_2-|x_2-\xi|)d\xi\int\limits_{x_2}^{\xi}d\xi_1\int\limits_{x_2-\frac{h_2}{2}}^{x_2+\frac{h_2}{2}}d\xi_2\int\limits_{\xi_2}^{\xi_1}\frac{\partial^2 u(x_1,\xi_3)}{\partial \xi_3^2}d\xi_3.$$

Taking into account the relation $\left(T_1 \dfrac{\partial^2 u}{\partial x_1^2}\right)(x) = u_{\bar{x}_1 x_1}(x)$, $x \in \omega$, we then have

$$\eta_{1\bar{x}_1 x_1}(x) = \frac{1}{h_1^2 h_2^3} \int_{x_2-h_2}^{x_2+h_2} (h_2 - |x_2 - \xi|) d\xi \int_{x_2}^{\xi} d\xi_1 \int_{x_2-\frac{h_2}{2}}^{x_2+\frac{h_2}{2}} d\xi_2 \times$$

$$\times \int_{\xi_2}^{\xi_1} d\xi_3 \int_{x_1-h_1}^{x_1+h_1} (h_1 - |x_1 - \xi_4|) \frac{\partial^4 u(\xi_4, \xi_3)}{\partial \xi_4^2 \partial \xi_3^2} d\xi_4, \quad x \in \omega,$$

which yields the inequality

$$|\eta_{1\bar{x}_1 x_1}(x)| \le \frac{h_2 h_1}{h_1^2 h_2^3} \int_{x_2-h_2}^{x_2+h_2} d\xi \int_{x_2-h_2}^{x_2+h_2} d\xi_1 \int_{x_2-\frac{h_2}{2}}^{x_2+\frac{h_2}{2}} d\xi_2 \int_{x_2-h_2}^{x_2+h_2} d\xi_3 \int_{x_1-h_1}^{x_1+h_1} \left|\frac{\partial^4 u(\xi_4, \xi_3)}{\partial \xi_4^2 \partial \xi_3^2}\right| d\xi_4 \le$$

$$\le \frac{h_2 h_1}{h_1^2 h_2^3} 2h_2 \cdot 2h_2 \cdot h_2 \int_{x_2-h_2}^{x_2+h_2} d\xi_3 \int_{x_1-h_1}^{x_1+h_1} \left|\frac{\partial^4 u(\xi_4, \xi_3)}{\partial \xi_4^2 \partial \xi_3^2}\right| d\xi_4 \le \qquad [1.12]$$

$$\le \frac{h_2 h_1}{h_1^2 h_2^3} 2h_2 \cdot 2h_2 \cdot h_2 \sqrt{2h_2 \cdot 2h_1} \left(\int_{x_2-h_2}^{x_2+h_2} d\xi_3 \int_{x_1-h_1}^{x_1+h_1} \left|\frac{\partial^4 u(\xi_4, \xi_3)}{\partial \xi_4^2 \partial \xi_3^2}\right|^2 d\xi_4\right)^{1/2} =$$

$$= 8\sqrt{\frac{h_2^3}{h_1}} \left(\int_{x_2-h_2}^{x_2+h_2} d\xi_3 \int_{x_1-h_1}^{x_1+h_1} \left|\frac{\partial^4 u(\xi_4, \xi_3)}{\partial \xi_4^2 \partial \xi_3^2}\right|^2 d\xi_4\right)^{1/2}, \quad x \in \omega.$$

Formula [1.11] and the relation $T_1 \dfrac{\partial^2 u}{\partial x_1^2} = \dfrac{2}{h_1}\left(u_{x_1} - \dfrac{\partial u}{\partial x_1}\right)$, $x \in \gamma_{-1}$, produce the representation

$$\frac{2}{h_1}\left(\eta_{1x_1}(x)-\sigma\eta_1(x)\right)=\frac{2}{h_1^2 h_2^3}\int_{x_2-h_2}^{x_2+h_2}(h_2-|x_2-\xi|)d\xi\int_{x_2}^{\xi}d\xi_1\int_{x_2-\frac{h_2}{2}}^{x_2+\frac{h_2}{2}}d\xi_2\times$$

$$\times\int_{\xi_2}^{\xi_1}d\xi_3\int_0^{h_1}(h_1-\xi_4)\frac{\partial^4 u(\xi_4,\xi_3)}{\partial\xi_4^2\partial\xi_3^2}d\xi_4,\quad x\in\gamma_{-1},$$

which leads to the estimate

$$\left|\frac{2}{h_1}\left(\eta_{1x_1}(x)-\sigma\eta_1(x)\right)\right|\leq$$

$$\leq 8\sqrt{\frac{2h_2^3}{h_1}}\left(\int_{x_2-h_2}^{x_2+h_2}d\xi_3\int_0^{h_1}\left|\frac{\partial^4 u(\xi_4,\xi_3)}{\partial\xi_4^2\partial\xi_3^2}\right|^2 d\xi_4\right)^{1/2},\quad x\in\gamma_{-1}. \qquad [1.13]$$

Now, from inequalities [1.12] and [1.13], we obtain

$$\|\Lambda_1\eta_1\|^2=\sum_{x\in\omega}h_1 h_2\eta_{1\bar{x}_1 x_1}^2(x)+\frac{h_1}{2}\sum_{x\in\gamma_{-1}}h_2\left(\frac{2}{h_1}\eta_{1x_1}(x)-\sigma\eta_1(x)\right)^2\leq$$

$$\leq 64h_2^4\sum_{x\in\omega}\int_{x_2-h_2}^{x_2+h_2}d\xi_3\int_{x_1-h_1}^{x_1+h_1}\left|\frac{\partial^4 u(\xi_4,\xi_3)}{\partial\xi_4^2\partial\xi_3^2}\right|^2 d\xi_4+$$

$$+64h_2^4\sum_{x_2\in\omega_2}\int_{x_2-h_2}^{x_2+h_2}d\xi_3\int_0^{h_1}\left|\frac{\partial^4 u(\xi_4,\xi_3)}{\partial\xi_4^2\partial\xi_3^2}\right|^2 d\xi_4\leq$$

$$\leq 4\cdot 64h_2^4\iint_D\left|\frac{\partial^4 u(\xi_4,\xi_3)}{\partial\xi_4^2\partial\xi_3^2}\right|^2 d\xi_3 d\xi_4=256h_2^4\iint_D\left|\frac{\partial^4 u(x_1,x_2)}{\partial x_1^2\partial x_2^2}\right|^2 dx_1 dx_2,$$

which eventually means the estimate

$$\|\Lambda_1\eta_1\| \le 16h_2^2 \left(\iint_D \left| \frac{\partial^4 u(x_1,x_2)}{\partial x_1^2 \partial x_2^2} \right|^2 dx_1 dx_2 \right)^{1/2}.$$ [1.14]

Next, we consider the summand $\|\eta_{2\bar{x}_2 x_2}\|$. For the nodes $x \in \omega$, we have

$$\eta_2(x) = (T_1 u)(x) - u(x) = \frac{1}{h_1^2} \int_{x_1-h_1}^{x_1+h_1} (h_1 - |x_1 - \xi|)u(\xi, x_2) d\xi - u(x_1, x_2) =$$

$$= \frac{1}{h_1^3} \int_{x_1-h_1}^{x_1+h_1} (h_1 - |x_1-\xi|)d\xi \int_{x_1}^{\xi} d\xi_1 \int_{x_1-\frac{h_1}{2}}^{x_1+\frac{h_1}{2}} d\xi_2 \int_{\xi_2}^{\xi_1} \frac{\partial^2 u(\xi_3, x_2)}{\partial \xi_3^2} d\xi_3,$$

which produces the representation

$$\eta_{2\bar{x}_2 x_2}(x) = \frac{1}{h_2^2 h_1^3} \int_{x_1-h_1}^{x_1+h_1} (h_1 - |x_1-\xi|)d\xi \int_{x_1}^{\xi} d\xi_1 \int_{x_1-\frac{h_1}{2}}^{x_1+\frac{h_1}{2}} d\xi_2 \times$$

$$\times \int_{\xi_2}^{\xi_1} d\xi_3 \int_{x_2-h_2}^{x_2+h_2} (h_2 - |x_2-\xi_4|) \frac{\partial^4 u(\xi_3, \xi_4)}{\partial \xi_3^2 \partial \xi_4^2} d\xi_4, \quad x \in \omega,$$

And similarly to [1.12], we obtain

$$\left|\eta_{2\bar{x}_2 x_2}(x)\right| \le 8 \sqrt{\frac{h_1^3}{h_2}} \left(\int_{x_1-h_1}^{x_1+h_1} d\xi_3 \int_{x_2-h_2}^{x_2+h_2} \left| \frac{\partial^4 u(\xi_3,\xi_4)}{\partial \xi_3^2 \partial \xi_4^2} \right|^2 d\xi_4 \right)^{1/2}, \quad x \in \omega.$$ [1.15]

For the nodes $x \in \gamma_{-1}$, we perform the transformations

$$\eta_2(x) = (T_1 u)(x) - u(x) - \frac{h_1}{3}\sigma u(x) =$$

$$= \frac{2}{h_1^2} \int_0^{h_1} (h_1 - \xi)(u(\xi, x_2) - u(0, x_2)) d\xi - \frac{h_1}{3} \frac{\partial u(0, x_2)}{\partial x_1} =$$

$$= \frac{2}{h_1^2} \int_0^{h_1} (h_1 - \xi) d\xi \int_0^{\xi} \frac{\partial u(\xi_1, x_2)}{\partial \xi_1} d\xi_1 - \frac{h_1}{3} \frac{\partial u(0, x_2)}{\partial x_1} =$$

$$= \frac{2}{h_1^2} \int_0^{h_1} (h_1 - \xi) d\xi \int_0^{\xi} \left(\frac{\partial u(\xi_1, x_2)}{\partial \xi_1} - \frac{\partial u(0, x_2)}{\partial x_1} \right) d\xi_1 =$$

$$= \frac{2}{h_1^2} \int_0^{h_1} (h_1 - \xi) d\xi \int_0^{\xi} d\xi_1 \int_0^{\xi_1} \frac{\partial^2 u(\xi_2, x_2)}{\partial \xi_2^2} d\xi_2 \ .$$

Due to the relation $\left(T_2 \frac{\partial^2 u}{\partial x_2^2} \right)(x) = u_{\bar{x}_2 x_2}(x)$, $x \in \omega \cup \gamma_{-1}$, we then have

$$\eta_{2\bar{x}_2 x_2}(x) =$$

$$= \frac{2}{h_1^2 h_2^2} \int_0^{h_1} (h_1 - \xi) d\xi \int_0^{\xi} d\xi_1 \int_0^{\xi_1} d\xi_2 \int_{x_2 - h_2}^{x_2 + h_2} (h_2 - |x_2 - \xi_3|) \frac{\partial^4 u(\xi_2, \xi_3)}{\partial \xi_2^2 \partial \xi_3^2} d\xi_3, \ x \in \gamma_{-1}.$$

This representation provides the estimate

$$|\eta_{2\bar{x}_2 x_2}(x)| \le \frac{2 h_1 h_2}{h_1^2 h_2^2} \int_0^{h_1} d\xi \int_0^{h_1} d\xi_1 \int_0^{h_1} d\xi_2 \int_{x_2 - h_2}^{x_2 + h_2} \left| \frac{\partial^4 u(\xi_2, \xi_3)}{\partial \xi_2^2 \partial \xi_3^2} \right| d\xi_3 \le$$

$$\le \frac{2 h_1 h_2}{h_1^2 h_2^2} h_1^2 \sqrt{h_1 \cdot 2 h_2} \left(\int_0^{h_1} d\xi_2 \int_{x_2 - h_2}^{x_2 + h_2} \left| \frac{\partial^4 u(\xi_2, \xi_3)}{\partial \xi_2^2 \partial \xi_3^2} \right|^2 d\xi_3 \right)^{1/2} = \quad [1.16]$$

$$= 2 \sqrt{\frac{2 h_1^3}{h_2}} \left(\int_0^{h_1} d\xi_2 \int_{x_2 - h_2}^{x_2 + h_2} \left| \frac{\partial^4 u(\xi_2, \xi_3)}{\partial \xi_2^2 \partial \xi_3^2} \right|^2 d\xi_3 \right)^{1/2}, \ x \in \gamma_{-1}.$$

Combining formulas [1.15] and [1.16], we obtain

$$\|\eta_{2\bar{x}_2 x_2}\|^2 = \sum_{x\in\omega} h_1 h_2 \eta^2_{2\bar{x}_2 x_2}(x) + \frac{h_1}{2}\sum_{x\in\gamma_{-1}} h_2 \eta^2_{2\bar{x}_2 x_2}(x) \leq$$

$$\leq 64 h_1^4 \sum_{x\in\omega} \int_{x_1-h_1}^{x_1+h_1} d\xi_3 \int_{x_2-h_2}^{x_2+h_2} \left|\frac{\partial^4 u(\xi_3,\xi_4)}{\partial \xi_3^2 \partial \xi_4^2}\right|^2 d\xi_4 +$$

$$+ 4 h_1^4 \sum_{x_2\in\omega_2} \int_0^{h_1} d\xi_2 \int_{x_2-h_2}^{x_2+h_2} \left|\frac{\partial^4 u(\xi_2,\xi_3)}{\partial \xi_2^2 \partial \xi_3^2}\right|^2 d\xi_3 \leq$$

$$\leq 4\cdot 64 h_1^4 \iint_D \left|\frac{\partial^4 u(\xi_3,\xi_4)}{\partial \xi_3^2 \partial \xi_4^2}\right|^2 d\xi_3 d\xi_4 = 256 h_1^4 \iint_D \left|\frac{\partial^4 u(x_1,x_2)}{\partial x_1^2 \partial x_2^2}\right|^2 dx_1 dx_2,$$

i.e.

$$\|\eta_{2\bar{x}_2 x_2}\| \leq 16 h_1^2 \left(\iint_D \left|\frac{\partial^4 u(x_1,x_2)}{\partial x_1^2 \partial x_2^2}\right|^2 dx_1 dx_2\right)^{1/2}. \qquad [1.17]$$

Taking into account inequalities [1.14] and [1.17], we finally come to the estimate

$$\|\psi\| \leq \|\Lambda_1 \eta_1\| + \|\eta_{2\bar{x}_2 x_2}\| \leq 16(h_1^2 + h_2^2)\left(\iint_D \left|\frac{\partial^4 u(x_1,x_2)}{\partial x_1^2 \partial x_2^2}\right|^2 dx_1 dx_2\right)^{1/2},$$

which completes the proof of the lemma.

Lemmas 1.3 and 1.4 help to substantiate the following proposition.

THEOREM 1.1.– *Let the solution $u(x)$ of problem [1.1] satisfy the condition $u \in W_2^4(D)$. Then, the accuracy of the finite-difference scheme [1.3] is characterized by the weighted estimate*

$$\|\rho^{-1} z\| \leq 16\sqrt{\frac{6}{11}} |h|^2 |u|_{W_2^4(D)},$$

where $\rho(x) = \min\left\{\sqrt{(l_1-x_1)(l_2-x_2)}, \sqrt{(l_1-x_1)x_2}\right\}$ and $|h|^2 = h_1^2 + h_2^2$.

PROOF.— Applying Lemmas 1.3 and 1.4, we easily obtain the estimate

$$|z(x)| \leq \sqrt{\frac{6}{11}} \rho(x) \|\psi\| \leq \sqrt{\frac{6}{11}} \rho(x) 16(h_1^2 + h_2^2) \left(\iint_D \left| \frac{\partial^4 u(x_1, x_2)}{\partial x_1^2 \partial x_2^2} \right|^2 dx_1 dx_2 \right)^{1/2} \leq$$

$$\leq 16 \sqrt{\frac{6}{11}} \rho(x) |h|^2 |u|_{W_2^4(D)}, \quad x \in \omega \cup \gamma_{-1}.$$

The theorem is proved.

1.1.5. Conclusion

We will summarize and comment on the above results.

The weighted a priori estimate obtained in Theorem 1.1 shows that, due to the Dirichlet boundary condition, the error of the difference scheme [1.3] is $O(\sqrt{h_1} |h|^2)$ near the side Γ_{+1} and $O(\sqrt{h_2} |h|^2)$ near the sides $\Gamma_{\pm 2}$ of the rectangle D, while it is $O(|h|^2)$ further away from them.

REMARK 1.1.— *To evaluate the functionals η_1 and η_2 in [1.14] and [1.17], we can also use the Bramble–Hilbert lemma. However, such estimates contain undefined constants independent of $|h|$ and $u(x)$.*

REMARK 1.2.— *The accuracy of the traditional discretization of problem [1.1] by the finite-difference scheme*

$$-y_{\bar{x}_1 x_1} - y_{\bar{x}_2 x_2} = (T_1 T_2 f)(x), \quad x \in \omega,$$

$$-\frac{2}{h_1}(y_{x_1} - \sigma y) - y_{\bar{x}_2 x_2} = (T_1 T_2 f)(x), \quad x \in \gamma_{-1},$$

$$y = 0, \quad x \in \gamma \setminus \gamma_{-1},$$

is $O(|h|^{3/2})$, which means the loss of $|h|^{1/2}$ due to the approximation of the boundary condition containing a partial derivative.

REMARK 1.3.– *The analogue of Theorem 1.1 can be proved for the BVP with the inhomogeneous boundary condition of the third kind, namely:*

$$-\Delta u = f(x), \quad x \in D,$$
$$-\frac{\partial u}{\partial x_1} + \sigma u(x) = g(x), \quad x \in \Gamma_{-1},$$
$$u(x) = 0, \quad x \in \Gamma \setminus \Gamma_{-1}.$$

To discretize this problem, we can take scheme [1.3] with the right-hand side

$$\varphi(x) = (T_1 T_2 f)(x) + \frac{2}{h_1} T_2 g - \frac{h_1}{3} (S_2 g)_{\bar{x}_2 x_2}, \quad x \in \gamma_{-1}.$$

Then, for the error $z(x) = y(x) - u(x)$, we have problem [1.4] with

$$\eta_2(x) = (T_1 u)(x) - \left(1 + \frac{h_1 \sigma}{3}\right) u(x) + \frac{h_1}{3}(S_2 g)(x), \quad x \in \gamma_{-1}.$$

Applying here the Bramble–Hilbert lemma and following Samarskii et al. (1987, p. 165), we can obtain the weighted estimate

$$\|\rho^{-1} z\| \le M |h|^2 \|u\|_{W_2^4(D)}$$

with the positive constant M independent of $|h|$ and $u(x)$.

1.2. A nine-point finite-difference scheme for Poisson's equation with the Dirichlet boundary condition

We consider here the boundary value problem

$$\Delta u(x) = -f(x), \quad x \in D,$$
$$u(x) = 0, \quad x \in \Gamma, \qquad [1.18]$$

where $x = (x_1, x_2)$, $D = \{x = (x_1, x_2) : 0 < x_\alpha < l_\alpha, \alpha = 1, 2\}$ is a rectangle, $\Gamma = \partial D$ is a boundary of D, $\Delta = \frac{\partial^2}{\partial x_1^2} + \frac{\partial^2}{\partial x_2^2}$ is the Laplace operator in a Cartesian coordinate system, and $f(x)$ is a known function.

1.2.1. *Discretization of the BVP*

We introduce here sets of nodes [1.2] and denote

$$\gamma_\alpha = \gamma_{-\alpha} \cup \gamma_{+\alpha}, \quad \alpha = 1, 2.$$

Using the operators of the exact finite-difference schemes (Samarskii et al. 1987)

$$T_1 v(x) = \frac{1}{h_1^2} \int_{x_1 - h_1}^{x_1 + h_1} (h_1 - |x_1 - \xi|) v(\xi, x_2) d\xi, \quad x \in \omega,$$

$$T_2 v(x) = \frac{1}{h_2^2} \int_{x_2 - h_2}^{x_2 + h_2} (h_2 - |x_2 - \xi|) v(x_1, \xi) d\xi, \quad x \in \omega,$$

[1.19]

we approximate problem [1.18] by the difference scheme

$$\Lambda y(x) + \frac{h_1^2 + h_2^2}{12} \Lambda_1 \Lambda_2 y(x) = -T_1 T_2 f(x), \quad x \in \omega,$$

$$y(x) = 0, \quad x \in \gamma,$$

[1.20]

where $\Lambda = \Lambda_1 + \Lambda_2$, $\Lambda_\alpha y(x) = y_{\bar{x}_\alpha x_\alpha}(x)$, $x \in \omega$, $\alpha = 1, 2$.

For the error $z(x) = y(x) - u(x)$, we have the problem

$$\Lambda z(x) + \frac{h_1^2 + h_2^2}{12} \Lambda_1 \Lambda_2 z(x) = -\psi(x), \quad x \in \omega,$$

$$z(x) = 0, \quad x \in \gamma,$$

[1.21]

where $\psi(x)$ is the approximation error:

$$\psi(x) = T_1 T_2 f(x) + \Lambda u(x) + \frac{h_1^2 + h_2^2}{12} \Lambda_1 \Lambda_2 u(x) =$$

$$= \Lambda_1 u + \Lambda_2 u + \frac{h_1^2 + h_2^2}{12} \Lambda_1 \Lambda_2 u - \Lambda_1 (T_2 u) - \Lambda_2 (T_1 u) =$$

$$= \Lambda_1 \left(u - T_2 u + \frac{h_2^2}{12} \Lambda_2 u \right) + \Lambda_2 \left(u - T_1 u + \frac{h_1^2}{12} \Lambda_1 u \right) = \Lambda_1 \eta_1 + \Lambda_2 \eta_2,$$

$$\eta_\alpha(x) = u(x) - T_{3-\alpha}u(x) + \frac{h_{3-\alpha}^2}{12}\Lambda_{3-\alpha}u(x), \quad x \in \omega \quad (\alpha = 1,2).$$

Note that we use here the relation $\Lambda_1(T_2 u) + \Lambda_2(T_1 u) = -T_1 T_2 f$, which is obtained by applying the operator $T = T_1 T_2$ to equation [1.18] at the nodes $x \in \omega$ and taking into account the formula $T_\alpha \dfrac{\partial^2 u}{\partial x_\alpha^2}(x) = u_{\bar{x}_\alpha x_\alpha}$, $x \in \omega$ $(\alpha = 1,2)$.

1.2.2. Properties of the finite-difference operators

Next, we introduce the space $\overset{0}{H}$ of mesh functions defined on $\bar{\omega}$ and equal to zero on γ with the inner product and the associate norm

$$(y,v) = (y,v)_{L_2(\omega)} = \sum_{x \in \omega} h_1 h_2 y(x) v(x),$$

$$\|v\| = \|v\|_{L_2(\omega)} = \sqrt{(v,v)} = \left(\sum_{x \in \omega} h_1 h_2 v^2(x)\right)^{1/2}. \qquad [1.22]$$

Note that the mesh function $\eta_\alpha(x)$ is defined at the nodes $x \in \omega \cup \gamma_\alpha$ and equal to zero at the nodes $x \in \gamma_\alpha$. We will consider it (and, if necessary, other mesh functions as well) equal to zero at the remaining nodes of the set γ. Then, $\eta_\alpha \in \overset{0}{H}$, and we can rewrite problem [1.21] as the operator equation

$$A'z = A_1 \eta_1 + A_2 \eta_2, \quad z, \eta_\alpha \in \overset{0}{H}, \qquad [1.23]$$

where

$$A_\alpha y = -\Lambda_\alpha y, \alpha = 1,2, \quad A = A_1 + A_2, \quad A' = A_1 + A_2 - \frac{h_1^2 + h_2^2}{12} A_1 A_2,$$

$$A_\alpha, A, A' : \overset{0}{H} \to \overset{0}{H}, \quad \alpha = 1,2.$$

Next, we study the properties of these finite-difference operators.

LEMMA 1.5.– *The operator* A' *is symmetric and positive definite in* $\overset{0}{H}$. *The following estimates hold true:*

$$\frac{2}{3}A \leq A' \leq A,\qquad\qquad [1.24]$$

$$\frac{2}{3}\|Ay\| \leq \|A'y\| \leq \|Ay\| \quad \forall y \in \overset{0}{H}.\qquad\qquad [1.25]$$

PROOF.– The operators A_1 and A_2 are symmetric since

$$(A_\alpha y, v) = (-y_{x_\alpha \bar{x}_\alpha}, v) = (y, -v_{x_\alpha \bar{x}_\alpha}) = (y, A_\alpha v) \quad \forall y, v \in \overset{0}{H} \quad (\alpha = 1, 2).$$

They commute: $A_1 A_2 = A_2 A_1$, since

$$A_1 A_2 y = (y_{\bar{x}_2 x_2})_{\bar{x}_1 x_1} = (y_{\bar{x}_1 x_1})_{\bar{x}_2 x_2} = A_2 A_1 y \quad \forall y \in \overset{0}{H}.$$

In addition, the operators A_1 and A_2 satisfy the inequalities

$$(A_\alpha y, y) \leq \frac{4}{h_\alpha^2}\cos^2\frac{\pi h_\alpha}{2l_\alpha}\|y\|^2 \leq \frac{4}{h_\alpha^2}\|y\|^2 \quad \forall y \in \overset{0}{H},$$

$$(A_\alpha y, y) \geq \frac{4}{h_\alpha^2}\sin^2\frac{\pi h_\alpha}{2l_\alpha}\|y\|^2 \geq \frac{8}{l_\alpha^2}\|y\|^2 \quad \forall y \in \overset{0}{H},$$

which can be combined in the operator inequality

$$\frac{8}{l_\alpha^2}I \leq A_\alpha \leq \frac{4}{h_\alpha^2}I,\ \alpha=1,2,$$

where I is the identity operator. This means that they are positive definite.

Then, the operator $A_1 A_2$ is self-adjoint, i.e. $(A_1 A_2)^* = A_1 A_2$, since

$$(A_1 A_2)^* = A_2^* A_1^* = A_2 A_1 = A_1 A_2,$$

and positive definite, namely, $A_1 A_2 \geq \dfrac{64}{l_1^2 l_2^2} I$, since

$$(A_1 A_2 y, y) = (A_1 A_2^{1/2} A_2^{1/2} y, y) = (A_2^{1/2} A_1 A_2^{1/2} y, y) = (A_1 A_2^{1/2} y, A_2^{1/2} y) \geq$$

$$\geq \dfrac{8}{l_1^2}(A_2^{1/2} y, A_2^{1/2} y) = \dfrac{8}{l_1^2}(A_2 y, y) \geq \dfrac{8}{l_1^2} \cdot \dfrac{8}{l_2^2} \| y \|^2 \quad \forall y \in \overset{0}{H}.$$

Therefore, the operators A_1 and A_2 are self-adjoint, the operator A is positive definite, and the operator inequality $A' \leq A$ in [1.24] holds true.

Next, we obtain the inequality $\dfrac{2}{3} A \leq A'$ в [1.24]:

$$A' = A_1 + A_2 - \dfrac{h_1^2 + h_2^2}{12} A_1 A_2 = A_1 \left(I - \dfrac{h_2^2}{12} A_2 \right) + A_2 \left(I - \dfrac{h_1^2}{12} A_1 \right) \geq$$

$$\geq A_1 \left(I - \dfrac{h_2^2}{12} \cdot \dfrac{4}{h_2^2} I \right) + A_2 \left(I - \dfrac{h_1^2}{12} \cdot \dfrac{4}{h_1^2} I \right) = \dfrac{2}{3}(A_1 + A_2) = \dfrac{2}{3} A,$$

which means that the operator A' is positive definite.

Estimates [1.24] lead to estimates [1.25]:

$$\| Ay \|^2 = (Ay, Ay) = (A A^{1/2} y, A^{1/2} y) \leq \dfrac{3}{2}(A' A^{1/2} y, A^{1/2} y) =$$

$$= \dfrac{3}{2}(A'y, Ay) \leq \dfrac{3}{2} \| A'y \| \cdot \| Ay \|,$$

which yield the inequality $\dfrac{2}{3} \| Ay \| \leq \| A'y \|$.

Similarly, we obtain $\| A'y \| \leq \| Ay \|$. The lemma is proved.

It follows from Lemma 1.5 that there exists the inverse operator A'^{-1} and therefore problem [1.23] is uniquely solvable.

LEMMA 1.6.– *The following inequality holds true:*

$$\|(A')^{-1} B_k y\| \le \frac{3}{2\sqrt{2}} \|y\|_k \quad \forall y \in H_k \quad (k=1,2,3,4), \qquad [1.26]$$

where the operator B_k, the space H_k and the norm $\|y\|_k$ are defined further in the text.

PROOF.– First, we find

$$\|A'y\|^2 \ge \frac{4}{9}\|Ay\|^2 = \frac{4}{9}(A_1 y + A_2 y, A_1 y + A_2 y) =$$

$$= \frac{4}{9}\left(\|A_1 y\|^2 + \|A_2 y\|^2 + (A_1 y, A_2 y)\right) \ge \frac{8}{9}(A_1 y, A_2 y) =$$

$$= \frac{8}{9}\sum_{x \in \omega} h_1 h_2 y_{\bar{x}_1 x_1} y_{\bar{x}_2 x_2} = \frac{8}{9}\|B_k^* y\|_k^2 \quad \forall y \in \overset{0}{H},$$

where we use the following notations:

1) $B_1^* : \overset{0}{H} \to H_1$, $B_1^* y = -y_{\bar{x}_1 \bar{x}_2}$, H_1 is a space of mesh functions defined on $\tilde{\omega} = \omega_1^+ \times \omega_2^+$;

2) $B_2^* : \overset{0}{H} \to H_2$, $B_2^* y = -y_{x_1 x_2}$, H_2 is a space of mesh functions defined on $\tilde{\omega} = \omega_1^- \times \omega_2^-$;

3) $B_3^* : \overset{0}{H} \to H_3$, $B_3^* y = -y_{\bar{x}_1 x_2}$, H_3 is a set of mesh functions defined on $\tilde{\omega} = \omega_1^+ \times \omega_2^-$;

4) $B_4^* : \overset{0}{H} \to H_4$, $B_4^* y = -y_{x_1 \bar{x}_2}$, H_4 is a space of mesh functions defined on $\tilde{\omega} = \omega_1^- \times \omega_2^+$;

$$(y,v)_k = \sum_{x \in \tilde{\omega}} h_1 h_2 y(x) v(x), \quad \|y\|_k = \sqrt{(y,y)_k} = \left(\sum_{x \in \tilde{\omega}} h_1 h_2 y(x) v(x)\right)^{1/2}$$

are the inner product and the associate norm in the space H_k, $k = \overline{1,4}$.

For the operator $B_k : H_k \to \overset{0}{H}$, which is the adjoint of $B_k^* : \overset{0}{H} \to H_k$, we have

$$(B_k y, v) = (y, B_k^* v)_k \quad \forall y \in H_k \quad \forall v \in \overset{0}{H},$$

where $B_1 y = -y_{x_1 x_2}$, $B_2 y = -y_{\bar{x}_1 \bar{x}_2}$, $B_3 y = -y_{x_1 \bar{x}_2}$, $B_4 y = -y_{\bar{x}_1 x_2}$.

Applying the main lemma from Samarskii et al. (1987) to the operators

$$A': \overset{0}{H} \to \overset{0}{H}, \ B_k : H_k \to \overset{0}{H}, \ B_k^* : \overset{0}{H} \to H_k \quad (k = \overline{1,4}),$$

due to the inequality

$$\| B_k^* y \|_k \leq \frac{3}{2\sqrt{2}} \| A' y \| \quad \forall y \in \overset{0}{H}$$

we obtain estimate [1.26]. The lemma is proved.

1.2.3. *Discrete Green's function*

We denote by $G(x,\xi)$ Green's function of the finite-difference problem [1.21]:

$$\Lambda_\xi G(x,\xi) + \frac{h_1^2 + h_2^2}{12} \Lambda_{1\xi} \Lambda_{2\xi} G(x,\xi) = -\frac{\delta(x_1,\xi_1)\delta(x_2,\xi_2)}{h_1 h_2}, \quad \xi \in \omega, \quad [1.27]$$
$$G(x,\xi) = 0, \quad \xi \in \gamma,$$

where $\delta(m,n)$ is the Kronecker delta symbol, and the subscript ξ means the finite-difference derivatives in the variables ξ_1 and ξ_2.

LEMMA 1.7.– *For Green's function $G(x,\xi)$, the following estimate holds true:*

$$\| G(x,\cdot) \| \leq \frac{3}{2\sqrt{2}} \rho(x), \quad x \in \omega,$$

where $\rho(x) = \min \left\{ \sqrt{x_1 x_2}, \ \sqrt{x_1 (l_2 - x_2)}, \ \sqrt{(l_1 - x_1) x_2}, \ \sqrt{(l_1 - x_1)(l_2 - x_2)} \right\}$.

PROOF.– Let $H(s)$ be the Heaviside step function

$$H(s) = \begin{cases} 1, & s \geq 0, \\ 0, & s < 0. \end{cases}$$

First, we can rewrite problem [1.27] in the form

$$\Lambda_\xi G(x,\xi) + \frac{h_1^2 + h_2^2}{12}\Lambda_{1\xi}\Lambda_{2\xi}G(x,\xi) = -\left(H(x_1-\xi_1)H(x_2-\xi_2)\right)_{\xi_1\xi_2}, \quad \xi \in \omega,$$

$$G(x,\xi) = 0, \quad \xi \in \gamma,$$

or as the operator equation

$$A'_\xi G(x,\xi) = -B_{1\xi}\left(H(x_1-\xi_1)H(x_2-\xi_2)\right).$$

Due to estimate [1.26], this gives the relation

$$\|G(x,\cdot)\| = \|-A'^{-1}_\xi B_{1\xi}\left(H(x_1-\cdot)H(x_2-\cdot)\right)\| \le \frac{3}{2\sqrt{2}}\|H(x_1-\cdot)H(x_2-\cdot)\|_1 =$$

$$= \frac{3}{2\sqrt{2}}\left(\sum_{\xi \in \omega_1^+ \times \omega_2^+} h_1 h_2 H^2(x_1-\xi_1)H^2(x_2-\xi_2)\right)^{1/2} =$$

$$= \frac{3}{2\sqrt{2}}\left(\sum_{\xi_1 \in \omega_1^+} h_1 H^2(x_1-\xi_1)\right)^{1/2}\left(\sum_{\xi_2 \in \omega_2^+} h_2 H^2(x_2-\xi_2)\right)^{1/2} = \qquad [1.28]$$

$$= \frac{3}{2\sqrt{2}}\left(\sum_{\xi_1=h_1}^{x_1} h_1\right)^{1/2}\left(\sum_{\xi_2=h_2}^{x_2} h_2\right)^{1/2} = \frac{3}{2\sqrt{2}}\sqrt{x_1 x_2}.$$

We can also rewrite problem [1.27] as follows:

$$\Lambda_\xi G(x,\xi) + \frac{h_1^2 + h_2^2}{12}\Lambda_{1\xi}\Lambda_{2\xi}G(x,\xi) = -\left(H(\xi_1-x_1)H(\xi_2-x_2)\right)_{\bar\xi_1\bar\xi_2}, \quad \xi \in \omega,$$

$$G(x,\xi) = 0, \quad \xi \in \gamma,$$

that is in the operator form

$$A'_\xi G(x,\xi) = -B_{2\xi}\left(H(\xi_1-x_1)H(\xi_2-x_2)\right).$$

Applying here estimate [1.26], we obtain the relation

$$\|G(x,\cdot)\| = \|-A_\xi'^{-1} B_{2\xi}\left(H(\cdot - x_1)H(\cdot - x_2)\right)\| \le \frac{3}{2\sqrt{2}} \|H(\cdot - x_1)H(\cdot - x_2)\|_2 =$$

$$= \frac{3}{2\sqrt{2}} \left(\sum_{\xi \in \omega_1^- \times \omega_2^-} h_1 h_2 H^2(\xi_1 - x_1) H^2(\xi_2 - x_2) \right)^{1/2} =$$

$$= \frac{3}{2\sqrt{2}} \left(\sum_{\xi_1 \in \omega_1^-} h_1 H^2(\xi_1 - x_1) \right)^{1/2} \left(\sum_{\xi_2 \in \omega_2^-} h_2 H^2(\xi_2 - x_2) \right)^{1/2} = \qquad [1.29]$$

$$= \frac{3}{2\sqrt{2}} \left(\sum_{\xi_1 = x_1}^{l_1 - h_1} h_1 \right)^{1/2} \left(\sum_{\xi_2 = x_2}^{l_2 - h_2} h_2 \right)^{1/2} = \frac{3}{2\sqrt{2}} \sqrt{(l_1 - x_1)(l_2 - x_2)}.$$

Next, we can represent problem [1.27] once again in a different way:

$$\Lambda_\xi G(x,\xi) + \frac{h_1^2 + h_2^2}{12} \Lambda_{1\xi} \Lambda_{2\xi} G(x,\xi) = \left(H(x_1 - \xi_1)H(\xi_2 - x_2)\right)_{\xi_1 \xi_2}, \quad \xi \in \omega,$$

$$G(x,\xi) = 0, \quad \xi \in \gamma,$$

which takes the form of the operator equation

$$A_\xi' G(x,\xi) = B_{3\xi}\left(H(x_1 - \xi_1)H(\xi_2 - x_2)\right).$$

Then, due to estimate [1.26], we obtain the relation

$$\|G(x,\cdot)\| = \| A_\xi'^{-1} B_{3\xi}\left(H(x_1 - \cdot)H(\cdot - x_2)\right)\| \le \frac{3}{2\sqrt{2}} \|H(x_1 - \cdot)H(\cdot - x_2)\|_3 =$$

$$= \frac{3}{2\sqrt{2}} \left(\sum_{\xi \in \omega_1^+ \times \omega_2^-} h_1 h_2 H^2(x_1 - \xi_1) H^2(\xi_2 - x_2) \right)^{1/2} =$$

$$= \frac{3}{2\sqrt{2}} \left(\sum_{\xi_1 \in \omega_1^+} h_1 H^2(x_1 - \xi_1) \right)^{1/2} \left(\sum_{\xi_2 \in \omega_2^-} h_2 H^2(\xi_2 - x_2) \right)^{1/2} = \qquad [1.30]$$

$$= \frac{3}{2\sqrt{2}} \left(\sum_{\xi_1 = h_1}^{x_1} h_1 \right)^{1/2} \left(\sum_{\xi_2 = x_2}^{l_2 - h_2} h_2 \right)^{1/2} = \frac{3}{2\sqrt{2}} \sqrt{x_1 (l_2 - x_2)}.$$

Finally, we can rewrite problem [1.27] as follows:

$$\Lambda_\xi G(x,\xi) + \frac{h_1^2 + h_2^2}{12}\Lambda_{1\xi}\Lambda_{2\xi}G(x,\xi) = \left(H(\xi_1 - x_1)H(x_2 - \xi_2)\right)_{\overline{\xi_1\xi_2}}, \quad \xi \in \omega,$$

$$G(x,\xi) = 0, \quad \xi \in \gamma,$$

that is as the operator equation

$$A'_\xi G(x,\xi) = B_{4\xi}\left(H(\xi_1 - x_1)H(x_2 - \xi_2)\right).$$

Applying estimate [1.26], we come to the relation

$$\|G(x,\cdot)\| = \|A'^{-1}_\xi B_{4\xi}\left(H(\cdot - x_1)H(x_2 - \cdot)\right)\| \le \frac{3}{2\sqrt{2}}\|H(\cdot - x_1)H(x_2 - \cdot)\|_4 =$$

$$= \frac{3}{2\sqrt{2}}\left(\sum_{\xi \in \omega_1^- \times \omega_2^+} h_1 h_2 H^2(\xi_1 - x_1)H^2(x_2 - \xi_2)\right)^{1/2} =$$

$$= \frac{3}{2\sqrt{2}}\left(\sum_{\xi_1 \in \omega_1^-} h_1 H^2(\xi_1 - x_1)\right)^{1/2}\left(\sum_{\xi_2 \in \omega_2^+} h_2 H^2(x_2 - \xi_2)\right)^{1/2} = \qquad [1.31]$$

$$= \frac{3}{2\sqrt{2}}\left(\sum_{\xi_1 = x_1}^{l_1 - h_1} h_1\right)^{1/2}\left(\sum_{\xi_2 = h_2}^{x_2} h_2\right)^{1/2} = \frac{3}{2\sqrt{2}}\sqrt{(l_1 - x_1)x_2}.$$

Combining four estimates [1.28]–[1.31], we prove the lemma.

1.2.4. Accuracy with the boundary effect

We begin with the important auxiliary result.

LEMMA 1.8.– *Let the solution $u(x)$ of problem [1.18] satisfy the condition $u \in W_2^6(D)$. Then, for the approximation error, the following estimate holds true:*

$$\|\psi\| \le \widetilde{M}\,|h|^4\,|u|_{W_2^6(D)},$$

where $\widetilde{M}$ is a positive constant independent of $u(x), h_1, h_2$; $|u|_{W_2^6(D)}$ is a seminorm in $W_2^6(D)$:

$$|u|_{W_2^6(D)} = \left\{ \sum_{\substack{\alpha_1+\alpha_2=6 \\ (\alpha_1\geq 0, \alpha_2\geq 0)}} \iint_D \left(\frac{\partial^{\alpha_1+\alpha_2} u(x_1,x_2)}{\partial x_1^{\alpha_1} \partial x_2^{\alpha_2}}\right)^2 dx_1 dx_2 \right\}^{1/2}.$$

PROOF.– We will study each of the two terms $\eta_{1\bar{x}_1 x_1}(x)$ and $\eta_{2\bar{x}_2 x_2}(x)$ of the approximation error $\psi(x)$ in formula [1.21]:

$$\psi(x) = \eta_{1\bar{x}_1 x_1}(x) + \eta_{2\bar{x}_2 x_2}(x),$$

where $\eta_1(x) = u(x) - T_2 u(x) + \dfrac{h_2^2}{12} u_{\bar{x}_2 x_2}(x)$, $\eta_2(x) = u(x) - T_1 u(x) + \dfrac{h_1^2}{12} u_{\bar{x}_1 x_1}(x)$.

First, we consider the functional $\eta_{1\bar{x}_1 x_1}(x)$. By linear substitution

$$\frac{\xi_\alpha - x_\alpha}{h_\alpha} = s_\alpha, \quad \alpha = 1, 2,$$

we map the square cell $e(x) = \{\xi = (\xi_1, \xi_2) : |\xi_\alpha - x_\alpha| < h_\alpha, \alpha = 1, 2\}$ onto the square

$$E = \{s = (s_1, s_2) : |s_\alpha| < 1, \alpha = 1, 2\}$$

and denote $u(\xi_1, \xi_2) = u(x_1 + h_1 s_1, x_2 + h_2 s_2) = U(s_1, s_2)$. Then, the functional $\eta_{1\bar{x}_1 x_1}(x)$ takes the form

$$\eta_{1\bar{x}_1 x_1}(x) = -\frac{1}{h_1^2}\left[\int_{-1}^{1}(1-|s_2|)\big(U(1,s_2) - 2U(0,s_2) + U(-1,s_2)\big)ds_2 -\right.$$

$$- \big(U(1,0) - 2U(0,0) + U(-1,0)\big) -$$

$$-\frac{1}{12}\big(U(1,1) - 2U(1,0) + U(1,-1)\big) + \frac{1}{6}\big(U(0,1) - 2U(0,0) + U(0,-1)\big) -$$

$$\left. -\frac{1}{12}\big(U(-1,1) - 2U(-1,0) + U(-1,-1)\big)\right].$$

This functional is bounded in $W_2^6(E)$ due to the embedding $W_2^6(E) \subset C(\overline{E})$ and becomes zero on the polynomials of the fifth degree. For example, for the polynomial x_2^2, we have

$$\eta_{1\overline{x}_1 x_1}(x) = -\left(\frac{1}{h_2^2}\int_{x_2-h_2}^{x_2+h_2}(h_2-|\xi-x_2|)\xi^2 d\xi - x_2^2 - \frac{h_2^2}{12}(x_2^2)_{\overline{x}_2 x_2}\right)_{\overline{x}_1 x_1} =$$

$$= -\left(\frac{h_2^2}{6} + x_2^2 - x_2^2 - \frac{h_2^2}{12}\cdot 2\right)_{\overline{x}_1 x_1} = 0.$$

Then, due to the Bramble–Hilbert lemma, we have the estimate

$$|\eta_{1\overline{x}_1 x_1}(x)| \le \frac{M}{h_1^2}|U|_{W_2^6(E)},$$

where M is a positive constant independent of $U(s_1,s_2), h_1, h_2$; $|U|_{W_2^6(E)}$ is a seminorm in $W_2^6(E)$:

$$|U|^2_{W_2^6(E)} = \sum_{\substack{\alpha_1+\alpha_2=6 \\ (\alpha_1\ge 0,\,\alpha_2\ge 0)}}\iint_E \left(\frac{\partial^{\alpha_1+\alpha_2}U(s_1,s_2)}{\partial s_1^{\alpha_1}\partial s_2^{\alpha_2}}\right)^2 ds_1 ds_2.$$

Returning to the variable ξ_1, ξ_2 and taking into account the relations

$$ds_1 ds_2 = \frac{d\xi_1 d\xi_2}{h_1 h_2}, \quad \frac{\partial U}{\partial s_\alpha}(s_1,s_2) = h_\alpha \frac{\partial u}{\partial \xi_\alpha}(\xi_1,\xi_2), \quad \alpha=1,2,$$

we come to the inequality

$$|\eta_{1\overline{x}_1 x_1}(x)| \le \frac{M}{h_1^2}\frac{|h|^6}{\sqrt{h_1 h_2}}|u|_{W_2^6(e(x))}.$$

Due to the condition $0 < C_1 \le h_1/h_2 \le C_2$, this leads to the estimate

$$|\eta_{1\overline{x}_1 x_1}(x)| \le M_1 \frac{|h|^4}{\sqrt{h_1 h_2}}|u|_{W_2^6(e(x))}, \quad x\in\omega. \qquad [1.32]$$

Similarly, for the functional $\eta_{2\bar{x}_2 x_2}(x)$, we obtain

$$|\eta_{1\bar{x}_2 x_2}(x)| \le M_2 \frac{|h|^4}{\sqrt{h_1 h_2}} |u|_{W_2^6(e(x))}, \quad x \in \omega. \qquad [1.33]$$

The positive constants M_1 and M_2 in [1.32] and [1.33] are independent of $h_1, h_2, u(x)$; $|u|_{W_2^6(e(x))}$ is a seminorm in $W_2^6(e(x))$.

Inequalities [1.32] and [1.33] finally lead to the estimate

$$\|\psi\| = \|\eta_{1\bar{x}_1 x_1} + \eta_{2\bar{x}_2 x_2}\| \le \|\eta_{1\bar{x}_1 x_1}\| + \|\eta_{2\bar{x}_2 x_2}\| \le$$

$$\le M_1 |h|^4 \left(\sum_{x \in \omega} |u|^2_{W_2^6(e(x))} \right)^{1/2} + M_2 |h|^4 \left(\sum_{x \in \omega} |u|^2_{W_2^6(e(x))} \right)^{1/2} = \widetilde{M} |h|^4 |u|_{W_2^6(D)},$$

which completes the proof of the lemma.

Lemmas 1.7 and 1.8 provide the main result of this section.

THEOREM 1.2.– *Let the solution $u(x)$ of problem [1.18] satisfy the condition $u \in W_2^6(D)$. Then, the accuracy of the finite-difference scheme [1.20] is characterized by the weighted estimate*

$$\|\rho^{-1} z\| \le M |h|^4 |u|_{W_2^6(D)} \qquad [1.34]$$

with the weight function

$$\rho(x) = \min\left\{ \sqrt{x_1 x_2}, \sqrt{x_1(l_2 - x_2)}, \sqrt{(l_1 - x_1) x_2}, \sqrt{(l_1 - x_1)(l_2 - x_2)} \right\},$$

where M is a positive constant independent of $u(x), h_1, h_2$; $|h| = \sqrt{h_1^2 + h_2^2}$, $|u|_{W_2^6(D)}$ is a seminorm in $W_2^6(D)$:

$$|u|_{W_2^6(D)} = \left(\sum_{\substack{\alpha_1 + \alpha_2 = 6 \\ (\alpha_1 \ge 0, \alpha_2 \ge 0)}} \iint_D \left(\frac{\partial^{\alpha_1 + \alpha_2} u(x_1, x_2)}{\partial x_1^{\alpha_1} \partial x_2^{\alpha_2}} \right)^2 dx_1 dx_2 \right)^{1/2}.$$

PROOF.— The solution $u(x)$ of problem [1.21] can be written in the form

$$z(x) = \big(G(x,\cdot), \psi(\cdot)\big) = \sum_{\xi \in \omega} h_1 h_2 G(x,\xi)\psi(\xi), \quad x \in \omega.$$

Applying Lemmas 1.7 and 1.8, we obtain the inequality

$$|z(x)| = \big|\big(G(x,\cdot), \psi(\cdot)\big)\big| \leq \|G(x,\cdot)\| \cdot \|\psi\| \leq \frac{3}{2\sqrt{2}} \rho(x)\widetilde{M} \,|h|^4 |u|_{W_2^6(D)} =$$

$$= M\rho(x) \,|h|^4 |u|_{W_2^6(D)}, \quad x \in \omega,$$

which finally gives the estimate [1.34]. The theorem is proved.

1.2.5. Conclusion

The weighted a priori estimate obtained in Theorem 1.2 shows that, due to the Dirichlet boundary condition, the error of the difference scheme [1.20] is $O\!\left(\sqrt{h_\alpha}\,|h|^4\right)$ near the sides $\Gamma_{\pm\alpha}$ ($\alpha = 1,2$) and $O\!\left(\sqrt{h_1 h_2}\,|h|^4\right)$ near the vertices of the rectangle D, while it is $O(|h|^4)$ further away from them.

Following the reasoning of Theorem 1.2, we can prove the scale of estimates

$$\|\rho^{-1} z\| \leq M\,|h|^{p-2}|u|_{W_2^p(D)}, \quad 4 \leq p \leq 6.$$

1.3. A finite-difference scheme of the higher order of approximation for Poisson's equation with the Dirichlet boundary condition

Here again we consider BVP [1.18]:

$$\begin{aligned}\Delta u(x) &= -f(x), \quad x \in D, \\ u(x) &= 0, \quad x \in \Gamma,\end{aligned} \qquad [1.35]$$

where $x = (x_1, x_2)$, $D = \{x = (x_1, x_2): 0 < x_\alpha < l_\alpha, \alpha = 1,2\}$ is a rectangle, $\Gamma = \partial D$ is a boundary of D, $\Delta = \dfrac{\partial^2}{\partial x_1^2} + \dfrac{\partial^2}{\partial x_2^2}$ is the Laplace operator in a Cartesian coordinate system, and $f(x)$ is a known function.

We will use sets of nodes [1.2] and the space $\overset{0}{H}$ of mesh functions [1.22]. To approximate problem [1.35], we again use the finite-difference scheme [1.20]:

$$\Lambda y(x) + \frac{h_1^2 + h_2^2}{12} \Lambda_1 \Lambda_2 y(x) = -T_1 T_2 f(x), \quad x \in \omega,$$

$$y(x) = 0, \quad x \in \gamma,$$

[1.36]

where $\Lambda = \Lambda_1 + \Lambda_2$, $\Lambda_\alpha y = y_{\bar{x}_\alpha x_\alpha}$, $x \in \omega$, $\alpha = 1, 2$, $T = T_1 T_2$ is an averaging operator, T_1, T_2 are the operators defined in [1.19].

For the error $z(x) = y(x) - u(x)$, we have problem [1.21]:

$$\Lambda z(x) + \frac{h_1^2 + h_2^2}{12} \Lambda_1 \Lambda_2 z(x) = -\psi(x), \quad x \in \omega,$$

$$z(x) = 0, \quad x \in \gamma,$$

[1.37]

with the approximation error $\psi(x)$:

$$\psi(x) = \Lambda_1 \eta_1(x) + \Lambda_2 \eta_2(x),$$

$$\eta_\alpha(x) = u(x) - T_{3-\alpha} u(x) + \frac{h_{3-\alpha}^2}{12} \Lambda_{3-\alpha} u(x), \quad x \in \omega, \quad \alpha = 1, 2.$$

We will also need the mesh norm $\|v\|_{C(\omega)} = \max\limits_{x \in \omega} |v(x)|$ and the difference operators $A_1, A_2, A, A' : \overset{0}{H} \to \overset{0}{H}$, where

$$A_\alpha y = -\Lambda_\alpha y, \quad \alpha = 1, 2, \quad A = A_1 + A_2, \quad A' = A_1 + A_2 - \frac{h_1^2 + h_2^2}{12} A_1 A_2.$$

Although problem [1.35] and its discrete analogue [1.36] are studied above in section 1.2, we will take here a different approach and some other techniques.

1.3.1. *Auxiliary results*

It is well known (e.g. Samarskii (2001)) that the one-dimensional spectral problem

$$\Lambda_\alpha w^{(\alpha)} + \lambda^{(\alpha)} w^{(\alpha)} = 0, \quad x_\alpha \in \omega_\alpha,$$
$$w^{(\alpha)}(0) = w^{(\alpha)}(l_\alpha) = 0,$$
[1.38]

has the eigenvalues

$$\lambda_{k_\alpha}^{(\alpha)} = \frac{4}{h_\alpha^2} \sin^2 \frac{k_\alpha \pi h_\alpha}{2 l_\alpha}, \quad k_\alpha = 1, \ldots, N_\alpha - 1,$$

and the corresponding eigenfunctions

$$w_{k_\alpha}^{(\alpha)} = w_{k_\alpha}^{(\alpha)}(x_\alpha) = \sqrt{\frac{2}{l_\alpha}} \sin \frac{k_\alpha \pi x_\alpha}{l_\alpha}, \quad \sum_{x_\alpha \in \omega_\alpha} h_\alpha \left(w_{k_\alpha}^{(\alpha)}(x_\alpha) \right)^2 = 1,$$
$$k_\alpha = 1, \ldots, N_\alpha - 1, \quad \alpha = 1, 2.$$

The eigenvalues $\lambda_{k_\alpha}^{(\alpha)}$ have the following properties:

$$\lambda_1^{(\alpha)} < \lambda_2^{(\alpha)} < \ldots < \lambda_{N_\alpha - 1}^{(\alpha)},$$

$$\lambda_1^{(\alpha)} = \frac{4}{h_\alpha^2} \sin^2 \frac{\pi h_\alpha}{2 l_\alpha} \geq \frac{8}{l_\alpha^2}, \quad \lambda_{N_\alpha - 1}^{(\alpha)} = \frac{4}{h_\alpha^2} \sin^2 \frac{(N_\alpha - 1)\pi h_\alpha}{2 l_\alpha} = \frac{4}{h_\alpha^2} \cos^2 \frac{\pi h_\alpha}{2 l_\alpha} \leq \frac{4}{h_\alpha^2},$$

$$\frac{4 k_\alpha^2}{l_\alpha^2} \leq \lambda_{k_\alpha}^{(\alpha)} \leq \frac{4}{h_\alpha^2}, \quad k_\alpha = 1, \ldots, N_\alpha - 1, \quad \alpha = 1, 2.$$

The operator A' satisfies the assumptions of Lemma 1.5, and therefore problem [1.36] is uniquely solvable.

We will also recall that the two-dimensional problem

$$\Lambda w + \lambda w = 0, \quad x \in \omega,$$
$$w = 0, \quad x \in \gamma,$$
[1.39]

has the eigenvalues

$$\lambda_{k_1 k_2} = \lambda_{k_1}^{(1)} + \lambda_{k_2}^{(2)} = \frac{4}{h_1^2} \sin^2 \frac{k_1 \pi h_1}{2 l_1} + \frac{4}{h_2^2} \sin^2 \frac{k_2 \pi h_2}{2 l_2}$$

and the corresponding eigenfunctions

$$w_{k_1k_2}(x) = w_{k_1}^{(1)}(x_1)w_{k_2}^{(2)}(x_2) = \frac{2}{\sqrt{l_1 l_2}} \sin\frac{k_1\pi x_1}{l_1} \sin\frac{k_2\pi x_2}{l_2}, \quad \|w_{k_1k_2}\| = 1,$$

$$k_\alpha = 1,\ldots, N_\alpha - 1, \quad \alpha = 1,2.$$

The eigenfunctions $w_{k_1k_2}$ satisfy the relations

$$\Lambda_\alpha w_{k_1k_2} = \Lambda_\alpha \left(w_{k_\alpha}^{(\alpha)} w_{k_{3-\alpha}}^{(3-\alpha)}\right) = w_{k_{3-\alpha}}^{(3-\alpha)} \Lambda_\alpha \left(w_{k_\alpha}^{(\alpha)}\right) =$$

$$= w_{k_{3-\alpha}}^{(3-\alpha)} \left(-\lambda_{k_\alpha}^{(\alpha)} w_{k_\alpha}^{(\alpha)}\right) = -\lambda_{k_\alpha}^{(\alpha)} w_{k_1k_2}, \quad \alpha = 1,2,$$

$$\Lambda_1\Lambda_2 w_{k_1k_2} = \Lambda_1\Lambda_2 \left(w_{k_1}^{(1)} w_{k_2}^{(2)}\right) =$$

$$= \Lambda_1\left(-w_{k_1}^{(1)} \lambda_{k_2}^{(2)} w_{k_2}^{(2)}\right) = \lambda_{k_1}^{(1)} w_{k_1}^{(1)} \lambda_{k_2}^{(2)} w_{k_2}^{(2)} = \lambda_{k_1}^{(1)} \lambda_{k_2}^{(2)} w_{k_1k_2}.$$

Now we consider the auxiliary problem

$$\Lambda v(x) + \frac{h_1^2 + h_2^2}{12}\Lambda_1\Lambda_2 v(x) = -1, \quad x \in \omega, \qquad [1.40]$$

$$v(x) = 0, \quad x \in \gamma.$$

We look for its solution $v(x)$ in the form of the sum

$$v(x) = \sum_{k_1=1}^{N_1-1}\sum_{k_2=1}^{N_2-1} v_{k_1k_2} w_{k_1k_2}(x). \qquad [1.41]$$

It is obvious that $v(x) = 0$ for $x \in \gamma$. Due to the maximum principle (Samarskii 2001), it holds $v(x) > 0$ for all $x \in \bar{\omega}$. To find the coefficients, we will use the representation

$$1 = \sum_{k_1=1}^{N_1-1}\sum_{k_2=1}^{N_2-1} c_{k_1k_2} w_{k_1k_2}(x), \qquad [1.42]$$

where

$$c_{k_1k_2} = (1, w_{k_1k_2}) = \sum_{x\in\omega} h_1 h_2 w_{k_1k_2}(x) = \sum_{x\in\omega} h_1 h_2 \frac{2}{\sqrt{l_1 l_2}} \sin\frac{k_1\pi x_1}{l_1} \sin\frac{k_2\pi x_2}{l_2} =$$

$$= \frac{2h_1 h_2}{\sqrt{l_1 l_2}} \sum_{i=1}^{N_1-1} \sin\frac{k_1\pi i h_1}{l_1} \cdot \sum_{j=1}^{N_2-1} \sin\frac{k_2\pi j h_2}{l_2} =$$

$$= \frac{2h_1 h_2}{\sqrt{l_1 l_2}} \frac{\sin\dfrac{(N_1-1)k_1\pi h_1}{2l_1}\sin\dfrac{N_1 k_1\pi h_1}{2l_1}}{\sin\dfrac{k_1\pi h_1}{2l_1}} \frac{\sin\dfrac{(N_2-1)k_2\pi h_2}{2l_2}\sin\dfrac{N_2 k_2\pi h_2}{2l_2}}{\sin\dfrac{k_2\pi h_2}{2l_2}} =$$

$$= \frac{2h_1 h_2}{\sqrt{l_1 l_2}} \frac{\cos\dfrac{k_1\pi h_1}{2l_1} - \cos\left(\dfrac{N_1 k_1\pi h_1}{l_1} - \dfrac{k_1\pi h_1}{2l_1}\right)}{2\sin\dfrac{k_1\pi h_1}{2l_1}} \frac{\cos\dfrac{k_2\pi h_2}{2l_2} - \cos\left(\dfrac{N_2 k_2\pi h_2}{l_2} - \dfrac{k_2\pi h_2}{2l_2}\right)}{2\sin\dfrac{k_2\pi h_2}{2l_2}} =$$

$$= \frac{2h_1 h_2}{\sqrt{l_1 l_2}} \frac{\cos\dfrac{k_1\pi h_1}{2l_1} - (-1)^{k_1}\cos\dfrac{k_1\pi h_1}{2l_1}}{h_1\sqrt{\lambda^{(1)}_{k_1}}} \cdot \frac{\cos\dfrac{k_2\pi h_2}{2l_2} - (-1)^{k_2}\cos\dfrac{k_2\pi h_2}{2l_2}}{h_2\sqrt{\lambda^{(2)}_{k_2}}} =$$

$$= \frac{2}{\sqrt{l_1 l_2}} \frac{\left(1-(-1)^{k_1}\right)\left(1-(-1)^{k_2}\right)\cos\dfrac{k_1\pi h_1}{2l_1}\cos\dfrac{k_2\pi h_2}{2l_2}}{\sqrt{\lambda^{(1)}_{k_1}\lambda^{(2)}_{k_2}}};$$

here we used the formula

$$\sum_{k=1}^{n}\sin k\theta = \frac{\sin\dfrac{n\theta}{2}\sin\dfrac{(n+1)\theta}{2}}{\sin\dfrac{\theta}{2}}, \quad \theta \notin \{2\pi m, m\in\mathbb{Z}\}, \quad n\in\mathbb{N}.$$

Substituting sums [1.41] and [1.42] into equation [1.40], we have

$$\sum_{k_1=1}^{N_1-1}\sum_{k_2=1}^{N_2-1}\left[\left(-\lambda_{k_1 k_2} + \frac{h_1^2+h_2^2}{12}\lambda^{(1)}_{k_1}\lambda^{(2)}_{k_2}\right)v_{k_1 k_2} + c_{k_1 k_2}\right]w_{k_1 k_2}(x) = 0, \quad x\in\omega.$$

This yields the formula

$$v_{k_1 k_2} = \frac{c_{k_1 k_2}}{\lambda_{k_1 k_2} - \frac{h_1^2 + h_2^2}{12} \lambda_{k_1}^{(1)} \lambda_{k_2}^{(2)}}, \quad k_\alpha = 1, \ldots, N_\alpha - 1, \quad \alpha = 1, 2,$$

since

$$\lambda_{k_1 k_2} - \frac{h_1^2 + h_2^2}{12} \lambda_{k_1}^{(1)} \lambda_{k_2}^{(2)} = \lambda_{k_1}^{(1)} \left(1 - \frac{h_2^2}{12} \lambda_{k_2}^{(2)}\right) + \lambda_{k_2}^{(2)} \left(1 - \frac{h_1^2}{12} \lambda_{k_1}^{(1)}\right) \geq$$

$$\geq \lambda_{k_1}^{(1)} \left(1 - \frac{h_2^2}{12} \frac{4}{h_2^2}\right) + \lambda_{k_2}^{(2)} \left(1 - \frac{h_1^2}{12} \frac{4}{h_2^2}\right) = \frac{2}{3} (\lambda_{k_1}^{(1)} + \lambda_{k_2}^{(2)}) > 0.$$

Therefore,

$$v(x) = \sum_{k_1=1}^{N_1-1} \sum_{k_2=1}^{N_2-1} \frac{c_{k_1 k_2}}{\lambda_{k_1 k_2} - \frac{h_1^2 + h_2^2}{12} \lambda_{k_1}^{(1)} \lambda_{k_2}^{(2)}} w_{k_1 k_2}(x), \quad x \in \overline{\omega}, \qquad [1.43]$$

with $\lambda_{k_1}^{(1)}$, $\lambda_{k_2}^{(2)}$, $w_{k_1 k_2}(x)$, $\lambda_{k_1 k_2}$, $c_{k_1 k_2}$ defined in [1.38], [1.39] and [1.42].

Next we study the properties of the function $v(x)$.

LEMMA 1.9.– *The function $v(x)$ in [1.43] can be estimated as follows:*

$$\begin{aligned} v(x) &\leq \frac{l_1 l_2 \pi^4}{48} = \text{const}, \quad x \in \omega, \\ v(x_1, x_2) &\leq M_1 |h|, \quad x_\alpha \in \{h_\alpha, l_\alpha - h_\alpha\}, \quad x_{3-\alpha} \in \omega_{3-\alpha} \ (\alpha = 1, 2), \\ v(x_1, x_2) &\leq M_2 |h|^2 \ln \frac{M_3}{|h|}, \quad x_\alpha \in \{h_\alpha, l_\alpha - h_\alpha\} \ (\alpha = 1, 2), \end{aligned} \qquad [1.44]$$

with the positive constants M_1, M_2, M_3 independent of $|h| = \sqrt{h_1^2 + h_2^2}$. Estimate [1.44] is unimprovable in the order of $|h|$.

PROOF.– Taking into account the inequality $\lambda_{k_1}^{(1)} + \lambda_{k_2}^{(2)} \geq 2\sqrt{\lambda_{k_1}^{(1)} \lambda_{k_2}^{(2)}}$ and the estimates

$$\lambda_{k_1 k_2} - \frac{h_1^2 + h_2^2}{12} \lambda_{k_1}^{(1)} \lambda_{k_2}^{(2)} \geq \frac{2}{3}(\lambda_{k_1}^{(1)} + \lambda_{k_2}^{(2)}), \quad \lambda_{k_\alpha}^{(\alpha)} \geq \frac{4 k_\alpha^2}{l_\alpha^2},$$

$$k_\alpha = 1, \ldots, N_\alpha - 1, \quad \alpha = 1, 2,$$

we have for all $x \in \omega$

$$v(x) = \sum_{k_1=1}^{N_1-1} \sum_{k_2=1}^{N_2-1} \frac{\frac{2}{\sqrt{l_1 l_2}} \left(1 - (-1)^{k_1}\right)\left(1 - (-1)^{k_2}\right) \cos \frac{k_1 \pi h_1}{2 l_1} \cos \frac{k_2 \pi h_2}{2 l_2}}{\sqrt{\lambda_{k_1}^{(1)} \lambda_{k_2}^{(2)}} \left(\lambda_{k_1 k_2} - \frac{h_1^2 + h_2^2}{12} \lambda_{k_1}^{(1)} \lambda_{k_2}^{(2)} \right)} \times \quad [1.45]$$

$$\times \frac{2}{\sqrt{l_1 l_2}} \sin \frac{k_1 \pi x_1}{l_1} \sin \frac{k_2 \pi x_2}{l_2} \leq$$

$$\leq \frac{24}{l_1 l_2} \sum_{k_1=1}^{N_1-1} \sum_{k_2=1}^{N_2-1} \frac{1}{\sqrt{\lambda_{k_1}^{(1)} \lambda_{k_2}^{(2)}} \left(\lambda_{k_1}^{(1)} + \lambda_{k_2}^{(2)}\right)} \leq \frac{12}{l_1 l_2} \sum_{k_1=1}^{N_1-1} \sum_{k_2=1}^{N_2-1} \frac{1}{\lambda_{k_1}^{(1)} \lambda_{k_2}^{(2)}} \leq$$

$$\leq \frac{3 l_1 l_2}{4} \sum_{k_1=1}^{N_1-1} \frac{1}{k_1^2} \sum_{k_2=1}^{N_2-1} \frac{1}{k_2^2} \leq \frac{3 l_1 l_2}{4} \left(\frac{\pi^2}{6}\right)^2 = \frac{l_1 l_2 \pi^4}{48} = \text{const}.$$

Next, we consider $v(x)$ at the nodes

$$(h_1, x_2), \ (l_1 - h_1, x_2), \ x_2 \in \omega_2, \ (x_1, h_2), \ (x_1, l_2 - h_2), \ x_1 \in \omega_1,$$

which are near the sides of the rectangle D. For example, at the node (h_1, x_2), $x_2 \in \omega_2$, we find

$$v(h_1, x_2) = \sum_{k_1=1}^{N_1-1} \sum_{k_2=1}^{N_2-1} \frac{\frac{2}{\sqrt{l_1 l_2}} \left(1 - (-1)^{k_1}\right)\left(1 - (-1)^{k_2}\right) \cos \frac{k_1 \pi h_1}{2 l_1} \cos \frac{k_2 \pi h_2}{2 l_2}}{\sqrt{\lambda_{k_1}^{(1)} \lambda_{k_2}^{(2)}} \left(\lambda_{k_1 k_2} - \frac{h_1^2 + h_2^2}{12} \lambda_{k_1}^{(1)} \lambda_{k_2}^{(2)} \right)} \times$$

$$\times \frac{2}{\sqrt{l_1 l_2}} \sin \frac{k_1 \pi h_1}{l_1} \sin \frac{k_2 \pi x_2}{l_2} \leq$$

$$\leq \frac{16}{l_1 l_2} \sum_{k_1=1}^{N_1-1} \sum_{k_2=1}^{N_2-1} \frac{1}{\sqrt{\lambda_{k_1}^{(1)} \lambda_{k_2}^{(2)}} \frac{2}{3}\left(\lambda_{k_1}^{(1)} + \lambda_{k_2}^{(2)}\right)} \sin \frac{k_1 \pi h_1}{l_1} = \quad [1.46]$$

$$= \frac{16}{l_1 l_2} \sum_{k_1=1}^{N_1-1} \sum_{k_2=1}^{N_2-1} \frac{2 \sin \frac{k_1 \pi h_1}{2l_1} \cos \frac{k_1 \pi h_1}{2l_1}}{\frac{2}{h_1} \sin \frac{k_1 \pi h_1}{2l_1} \frac{2}{h_2} \sin \frac{k_2 \pi h_2}{2l_2} \frac{2}{3}\left(\frac{4}{h_1^2} \sin^2 \frac{k_1 \pi h_1}{2l_1} + \frac{4}{h_2^2} \sin^2 \frac{k_2 \pi h_2}{2l_2}\right)} \leq$$

$$\leq h_1 \frac{24}{l_1 l_2} \sum_{k_1=1}^{N_1-1} \sum_{k_2=1}^{N_2-1} \frac{1}{\frac{2}{h_2} \sin \frac{k_2 \pi h_2}{2l_2}\left(\frac{4}{h_1^2} \sin^2 \frac{k_1 \pi h_1}{2l_1} + \frac{4}{h_2^2} \sin^2 \frac{k_2 \pi h_2}{2l_2}\right)} \leq$$

$$\leq h_1 \frac{24}{l_1 l_2} \sum_{k_1=1}^{N_1-1} \sum_{k_2=1}^{N_2-1} \frac{1}{\frac{2k_2}{l_2}\left(\frac{4k_1^2}{l_1^2} + \frac{4k_2^2}{l_2^2}\right)} < h_1 3 l_1 \sum_{k_2=1}^{N_2-1} \frac{1}{k_2} \sum_{k_1=1}^{\infty} \frac{1}{k_1^2 + \left(\frac{k_2 l_1}{l_2}\right)^2} =$$

$$= h_1 3 l_1 \sum_{k_2=1}^{N_2-1} \frac{1}{k_2} \frac{\frac{\pi k_2 l_1}{l_2} \operatorname{cth} \frac{\pi k_2 l_1}{l_2} - 1}{2\left(\frac{k_2 l_1}{l_2}\right)^2} = h_1 \frac{3}{2} \frac{l_2^2}{l_1} \sum_{k_2=1}^{N_2-1}\left(\frac{\pi l_1}{l_2} \frac{\operatorname{cth} \frac{\pi k_2 l_1}{l_2}}{k_2^2} - \frac{1}{k_2^3}\right) \leq$$

$$\leq h_1 \frac{3}{2} \frac{l_2^2}{l_1}\left(\frac{\pi l_1}{l_2} \operatorname{cth} \frac{\pi l_1}{l_2} \sum_{k_2=1}^{\infty} \frac{1}{k_2^2} - \sum_{k_2=1}^{\infty} \frac{1}{k_2^3}\right) = M_1 h_1 < M_1 |h|$$

with $M_1 = \frac{3}{2} \frac{l_2^2}{l_1}\left(\frac{l_1}{l_2} \operatorname{cth} \frac{\pi l_1}{l_2} \frac{\pi^3}{6} - \zeta(3)\right)$, $\zeta(\cdot)$ is the Riemann zeta function.

Finally, we consider $v(x)$ near the vertices of the rectangle D. For example, at the node (h_1, h_2), we have

$$v(h_1, h_2) = \sum_{k_1=1}^{N_1-1} \sum_{k_2=1}^{N_2-1} \frac{\frac{2}{\sqrt{l_1 l_2}}\left(1-(-1)^{k_1}\right)\left(1-(-1)^{k_2}\right) \cos \frac{k_1 \pi h_1}{2l_1} \cos \frac{k_2 \pi h_2}{2l_2}}{\sqrt{\lambda_{k_1}^{(1)} \lambda_{k_2}^{(2)}}\left(\lambda_{k_1 k_2} - \frac{h_1^2 + h_2^2}{12} \lambda_{k_1}^{(1)} \lambda_{k_2}^{(2)}\right)} \times$$

$$\times \frac{2}{\sqrt{l_1 l_2}} \sin \frac{k_1 \pi h_1}{l_1} \sin \frac{k_2 \pi h_2}{l_2} \leq$$

$$\leq \frac{16}{l_1 l_2} \sum_{k_1=1}^{N_1-1} \sum_{k_2=1}^{N_2-1} \frac{1}{\sqrt{\lambda_{k_1}^{(1)} \lambda_{k_2}^{(2)}} \frac{2}{3}\left(\lambda_{k_1}^{(1)} + \lambda_{k_2}^{(2)}\right)} \sin\frac{k_1 \pi h_1}{l_1} \sin\frac{k_2 \pi h_2}{l_2} =$$

$$= \frac{16}{l_1 l_2} \sum_{k_1=1}^{N_1-1} \sum_{k_2=1}^{N_2-1} \frac{2\sin\frac{k_1\pi h_1}{2l_1}\cos\frac{k_1\pi h_1}{2l_1} 2\sin\frac{k_2\pi h_2}{2l_2}\cos\frac{k_2\pi h_2}{2l_2}}{\frac{2}{h_1}\sin\frac{k_1\pi h_1}{2l_1}\frac{2}{h_2}\sin\frac{k_2\pi h_2}{2l_2}\frac{2}{3}\left(\frac{4}{h_1^2}\sin^2\frac{k_1\pi h_1}{2l_1} + \frac{4}{h_2^2}\sin^2\frac{k_2\pi h_2}{2l_2}\right)} \leq$$

$$\leq h_1 h_2 \frac{24}{l_1 l_2} \sum_{k_1=1}^{N_1-1} \sum_{k_2=1}^{N_2-1} \frac{1}{\frac{4}{h_1^2}\sin^2\frac{k_1\pi h_1}{2l_1} + \frac{4}{h_2^2}\sin^2\frac{k_2\pi h_2}{2l_2}} \leq \qquad [1.47]$$

$$\leq h_1 h_2 \frac{24}{l_1 l_2} \sum_{k_1=1}^{N_1-1} \sum_{k_2=1}^{N_2-1} \frac{1}{\frac{4k_1^2}{l_1^2} + \frac{4k_2^2}{l_2^2}} \leq h_1 h_2 \frac{6 l_1}{l_2} \sum_{k_2=1}^{N_2-1} \sum_{k_1=1}^{\infty} \frac{1}{k_1^2 + \left(\frac{k_2 l_1}{l_2}\right)^2} =$$

$$= h_1 h_2 \frac{6 l_1}{l_2} \sum_{k_2=1}^{N_2-1} \frac{\frac{\pi k_2 l_1}{l_2}\operatorname{cth}\frac{\pi k_2 l_1}{l_2} - 1}{2\left(\frac{k_2 l_1}{l_2}\right)^2} \leq h_1 h_2 3\pi \operatorname{cth}\frac{\pi l_1}{l_2} \sum_{k_2=1}^{N_2-1} \frac{1}{k_2} \leq$$

$$\leq h_1 h_2 3\pi \operatorname{cth}\frac{\pi l_1}{l_2}\left(1 + \int_1^{N_2} \frac{dx}{x}\right) = h_1 h_2 3\pi \operatorname{cth}\frac{\pi l_1}{l_2}(1 + \ln N_2) =$$

$$= h_1 h_2 3\pi \operatorname{cth}\frac{\pi l_1}{l_2} \ln\frac{e l_2}{h_2} = h_1 h_2 3\pi \operatorname{cth}\frac{\pi l_1}{l_2} \ln\frac{e l_2 \sqrt{1+(h_1/h_2)^2}}{\sqrt{h_1^2 + h_2^2}} \leq$$

$$\leq (h_1^2 + h_2^2)\frac{3}{2}\pi \operatorname{cth}\frac{\pi l_1}{l_2} \ln\frac{e l_2 \sqrt{6}}{\sqrt{h_1^2+h_2^2}} = M_2 |h|^2 \ln\frac{M_3}{|h|}$$

with $M_2 = \frac{3}{2}\pi \operatorname{cth}\frac{\pi l_1}{l_2}$ and $M_3 = \sqrt{6} e l_2$.

Now we study improvability of the estimate for $v(x)$ at the vertices of the mesh ω. For example, we consider the node (h_1, h_2). Let $U(x)$ be a solution of problem [1.35] with the right-hand side $f(x) \equiv 1$ (i.e. the problem of torsion of a prismatic bar with a rectangular section D):

$$\begin{aligned} \Delta U(x) &= -1, \quad x \in D, \\ U(x) &= 0, \quad x \in \Gamma. \end{aligned} \qquad [1.48]$$

It is known (e.g. Babenko (2002, p. 683)) that in a neighborhood of the vertex $(0,0)$ the function $U(x)$ can be presented in the form

$$U(x_1, x_2) = -\frac{x_1^2 + x_2^2}{4} - \frac{x_1 x_2}{\pi} \ln(x_1^2 + x_2^2) + \\ + \frac{x_1^2 - x_2^2}{2\pi} \operatorname{arctg} \frac{x_1^2 - x_2^2}{2 x_1 x_2} + w(x_1, x_2), \qquad [1.49]$$

where $w(x_1, x_2)$ is a regular function in the neighborhood of the point $w(x_1, x_2)$.

Applying Theorems 3.1 and 8.1 from Volkov (1965), we get that the function $w(x)$ does not influence the behavior of the function $U(x)$ in a neighborhood of the point $(0,0)$, and therefore the estimate for the function $v(x)$ in the neighborhood of the point $(0,0)$ is unimprovable and defined by the first two terms of formula [1.49].

Similar reasoning can be applied to the other vertices of the rectangle D if we take each of them as the origin and perform an appropriate linear change of variables. The theorem is proved.

The following calculations confirm the estimates obtained in Lemma 1.9. Let

$$l_1 = l_2 = 1, \quad N_1 = N_2 = N, \quad h_1 = l_1/N_1 = 1/N, \quad h_2 = l_2/N_2 = 1/N,$$

$$v(x) = \sum_{k_1=1}^{N_1-1} \sum_{k_2=1}^{N_2-1} \frac{c_{k_1 k_2}}{\lambda_{k_1 k_2} - \frac{h_1^2 + h_2^2}{12} \lambda_{k_1}^{(1)} \lambda_{k_2}^{(2)}} w_{k_1 k_2}(x), \quad x \in \bar{\omega},$$

$$c_{k_1k_2} = \frac{2}{\sqrt{l_1 l_2}} \frac{\left(1-(-1)^{k_1}\right)\left(1-(-1)^{k_2}\right)\cos\frac{k_1\pi h_1}{2l_1}\cos\frac{k_2\pi h_2}{2l_2}}{\sqrt{\lambda_{k_1}^{(1)}\lambda_{k_2}^{(2)}}} =$$

$$= \frac{2\left(1-(-1)^{k_1}\right)\left(1-(-1)^{k_2}\right)\cos\frac{k_1\pi}{2N}\cos\frac{k_2\pi}{2N}}{2N\sin\frac{k_1\pi}{2N}\sin\frac{k_2\pi}{2N}},$$

$$\lambda_{k_1k_2} = \lambda_{k_1}^{(1)} + \lambda_{k_2}^{(2)} = \frac{4}{h_1^2}\sin^2\frac{k_1\pi h_1}{2l_1} + \frac{4}{h_2^2}\sin^2\frac{k_2\pi h_2}{2l_2} = 4N^2\left(\sin^2\frac{k_1\pi}{2N} + \sin^2\frac{k_2\pi}{2N}\right),$$

$$w_{k_1k_2}(x) = w_{k_1}^{(1)}(x_1)w_{k_2}^{(2)}(x_2) = \frac{2}{\sqrt{l_1 l_2}}\sin\frac{k_1\pi x_1}{l_1}\sin\frac{k_2\pi x_2}{l_2} = \sin(k_1\pi x_1)\sin(k_2\pi x_2),$$

$$\|w_{k_1k_2}\| = 1, \quad k_\alpha = 1,\ldots,N-1, \quad \alpha = 1,2.$$

The results are presented in Table 1.1.

N	$v(h_1,h_2)$	$\overline{v(h_1,h_2)}\left((h_1^2+h_2^2)\ln\left(h_1^2+h_2^2\right)\right)^{-1}$
10	0.01308646211042133811157633	0.16725952393765521 8325540
20	0.00437472918667352612869399	0.16513654747426927 5574177
30	0.00223113044205542753312851	0.16434244730193759 9331952
40	0.00136947578792366180178461	0.16389592619183383 1035287
50	0.00093328751757113090492 6526	0.16359920743896341 7220672
100	0.00027744899528649053 1166830	0.16287583776196836 4897204
200	0.00008039402876460789 39473448	0.16235498522744190 7329565
300	0.00003859875839191985 84238664	0.16211278719241468 7115354
400	0.00002285645219204996 70499109	0.16196204579932944 8839502
500	0.00001519635979025831 62624470	0.16185530019860712 3329541
600	0.00001087544293207267 90476134	0.16177392758071777 1067540
700	0.00000819039760458596 327787628	0.16170885903356941 8727886
800	0.00000640359929847374 4423236743	0.16165505248569700 2556906
900	0.00000515220563008799 116657702	0.16160943931487873 5780303
1000	0.00000424036114786574 463001509	0.16157002461794766 8474767

Table 1.1. *Numerical calculations for Lemma 1.9*

LEMMA 1.10.– *Let $F(x)$ be an arbitrary function defined on the mesh ω and let the steps h_1 and h_2 satisfy the condition*

$$\frac{1}{\sqrt{5}} \leq \frac{h_1}{h_2} \leq \sqrt{5}. \qquad [1.50]$$

Then, for the solution $Y(x)$ of the problem

$$\Lambda Y(x) + \frac{h_1^2 + h_2^2}{12} \Lambda_1 \Lambda_2 Y(x) = -F(x), \quad x \in \omega, \qquad [1.51]$$

$$Y(x) = 0, \quad x \in \gamma,$$

the following estimate holds true:

$$|Y(x)| \leq v(x) \|F\|_{C(\omega)}, \quad x \in \omega. \qquad [1.52]$$

PROOF.– Problem [1.51] can be rewritten in the form

$$\frac{5}{3}\left(\frac{1}{h_1^2} + \frac{1}{h_2^2}\right)y(x) = \frac{1}{6}\left(\frac{5}{h_1^2} - \frac{1}{h_2^2}\right)(y(x_1+h_1,x_2) + y(x_1-h_1,x_2)) +$$

$$+ \frac{1}{6}\left(\frac{5}{h_2^2} - \frac{1}{h_1^2}\right)(y(x_1,x_2+h_2) + y(x_1,x_2-h_2)) +$$

$$+ \frac{1}{12}\left(\frac{1}{h_1^2} + \frac{1}{h_2^2}\right)(y(x_1+h_1,x_2+h_2) + y(x_1-h_1,x_2+h_2) +$$

$$+ y(x_1+h_1,x_2-h_2) + y(x_1-h_1,x_2-h_2)) + F(x), \quad x \in \omega,$$

$$V(x) = 0, \quad x \in \gamma.$$

If $1/\sqrt{5} \leq h_1/h_2 \leq \sqrt{5}$, then the assumptions of the comparison theorem (Samarskii 2001) are fulfilled and estimate [1.52] follows. The lemma is proved.

Note that in the case of a square mesh ($h_1 = h_2$), condition [1.50] is fulfilled.

1.3.2. *Accuracy with the boundary effect*

In the next lemma, we study the convergence rate of the finite-difference scheme [1.36] in the uniform mesh norm $C(\omega)$ if the solution of the differential problem [1.35] belongs to the class $W_\infty^m(D)$, $2 < m \leq 6$.

LEMMA 1.11.– *Let the solution* $u(x_1, x_2)$ *[1.35] satisfy the condition*

$$\frac{\partial^6 u}{\partial x_1^4 \partial x_2^2}, \frac{\partial^6 u}{\partial x_1^2 \partial x_2^4} \in L_\infty(D).$$

Then, for the approximation error $\psi(x)$, *the following estimate holds true*:

$$|\psi(x)| \leq \frac{7}{720} |h|^4 \left\{ \left\| \frac{\partial^6 u}{\partial x_1^4 \partial x_2^2} \right\|_{L_\infty(D)} + \left\| \frac{\partial^6 u}{\partial x_1^2 \partial x_2^4} \right\|_{L_\infty(D)} \right\}, \quad x \in \omega. \quad [1.53]$$

PROOF.– We consider the approximation error $\psi(x) = \Lambda_1 \eta_1(x) + \Lambda_2 \eta_2(x)$ in [1.37]. First, we transform the functional $\eta_1(x)$ as follows:

$$\eta_1(x) = u(x) - T_2 u(x) + \frac{h_2^2}{12} \Lambda_2 u(x) =$$

$$= \frac{1}{h_2^2} \int_{x_2 - h_2}^{x_2 + h_2} (h_2 - |x_2 - \xi|) \left[u(x_1, x_2) - u(x_1, \xi) + \frac{h_2^2}{12} \frac{\partial^2 u(x_1, \xi)}{\partial \xi^2} \right] d\xi =$$

$$= \frac{1}{h_2^2} \int_{x_2 - h_2}^{x_2 + h_2} (h_2 - |x_2 - \xi|) \left[-\frac{\partial u(x_1, x_2)}{\partial x_2}(\xi - x_2) - \frac{1}{2!} \frac{\partial^2 u(x_1, x_2)}{\partial x_2^2}(\xi - x_2)^2 - \right.$$

$$- \frac{1}{3!} \frac{\partial^3 u(x_1, x_2)}{\partial x_2^3}(\xi - x_2)^3 - \frac{1}{3!} \int_{x_2}^{\xi} (\xi - \xi_1)^3 \frac{\partial^4 u(x_1, \xi_1)}{\partial \xi_1^4} d\xi_1 +$$

$$+ \frac{h_2^2}{12} \left(\frac{\partial^2 u(x_1, x_2)}{\partial x_2^2} + \frac{\partial^3 u(x_1, x_2)}{\partial x_2^3}(\xi - x_2) + \int_{x_2}^{\xi} (\xi - \xi_1) \frac{\partial^4 u(x_1, \xi_1)}{\partial \xi_1^4} d\xi_1 \right) \Bigg] d\xi.$$

Due to the relations

$$\frac{1}{h_2^2}\int_{x_2-h_2}^{x_2+h_2}(h_2-|x_2-\xi|)(\xi-x_2)d\xi=0,\quad \frac{1}{h_2^2}\int_{x_2-h_2}^{x_2+h_2}(h_2-|x_2-\xi|)(\xi-x_2)^3 d\xi=0,$$

$$\frac{1}{h_2^2}\int_{x_2-h_2}^{x_2+h_2}(h_2-|x_2-\xi|)(\xi-x_2)^2 d\xi=\frac{h_2^2}{6},$$

we get the formula

$$\eta_1(x)=-\frac{1}{6h_2^2}\int_{x_2-h_2}^{x_2+h_2}(h_2-|x_2-\xi|)d\xi\int_{x_2}^{\xi}(\xi-\xi_1)^3\frac{\partial^4 u(x_1,\xi_1)}{\partial \xi_1^4}d\xi_1+$$

$$+\frac{1}{12h_2^2}\int_{x_2-h_2}^{x_2+h_2}(h_2-|x_2-\xi|)d\xi\int_{x_2}^{\xi}(\xi-\xi_1)\frac{\partial^4 u(x_1,\xi_1)}{\partial \xi_1^4}d\xi_1,$$

which yields the representation

$$\Lambda_1\eta_1(x)==-\frac{1}{6h_1^2 h_2^2}\int_{x_1-h_1}^{x_1+h_1}(h_1-|x_1-\xi_2|)d\xi_2\times$$

$$\times\int_{x_2-h_2}^{x_2+h_2}(h_2-|x_2-\xi|)d\xi\int_{x_2}^{\xi}(\xi-\xi_1)^3\frac{\partial^6 u(\xi_2,\xi_1)}{\partial\xi_2^2\partial\xi_1^4}d\xi_1+$$

$$+\frac{1}{12h_1^2}\int_{x_1-h_1}^{x_1+h_1}(h_1-|x_1-\xi_2|)d\xi_2\int_{x_2-h_2}^{x_2+h_2}(h_2-|x_2-\xi|)d\xi\times$$

$$\times\int_{x_2}^{\xi}(\xi-\xi_1)\frac{\partial^6 u(\xi_2,\xi_1)}{\partial\xi_2^2\partial\xi_1^4}d\xi_1,\quad x\in\omega.$$

From here, we obtain the inequality

$$|\Lambda_1\eta_1(x)|\leq\left[\frac{1}{6h_1^2 h_2^2}h_1^2\int_{x_2-h_2}^{x_2+h_2}(h_2-|x_2-\xi|)d\xi\left|\int_{x_2}^{\xi}(\xi-\xi_1)^3 d\xi_1\right|+\right.$$

$$+\frac{1}{12h_1^2}h_1^2\int_{x_2-h_2}^{x_2+h_2}(h_2-|x_2-\xi|)d\xi\left|\int_{x_2}^{\xi}(\xi-\xi_1)d\xi_1\right|\cdot\left\|\frac{\partial^6 u}{\partial x_1^2\partial x_2^4}\right\|_{L_\infty(D)}=$$

$$=\left[\frac{1}{6h_1^2h_2^2}h_1^2\int_{x_2-h_2}^{x_2+h_2}(h_2-|x_2-\xi|)\frac{(x_2-\xi)^4}{4}d\xi+\right.$$

$$\left.+\frac{1}{12h_1^2}h_1^2\int_{x_2-h_2}^{x_2+h_2}(h_2-|x_2-\xi|)\frac{(x_2-\xi)^2}{2}d\xi\right]\cdot\left\|\frac{\partial^6 u}{\partial x_1^2\partial x_2^4}\right\|_{L_\infty(D)}= \quad [1.54]$$

$$=\left(\frac{1}{6h_1^2h_2^2}\frac{h_1^2h_2^6}{60}+\frac{1}{12h_1^2}\frac{h_1^2h_2^4}{12}\right)\left\|\frac{\partial^6 u}{\partial x_1^2\partial x_2^4}\right\|_{L_\infty(D)}=\frac{7h_2^4}{720}\left\|\frac{\partial^6 u}{\partial x_1^2\partial x_2^4}\right\|_{L_\infty(D)}, \quad x\in\omega.$$

Similarly, we get the inequality

$$|\Lambda_2\eta_2(x)|\le\frac{7}{720}h_1^4\left\|\frac{\partial^6 u}{\partial x_1^4\partial x_2^2}\right\|_{L_\infty(D)}, \quad x\in\omega. \quad [1.55]$$

Inequalities [1.54] and [1.55] lead to the estimate

$$|\psi(x)|=|\Lambda_1\eta_1(x)+\Lambda_2\eta_2(x)|\le|\Lambda_1\eta_1(x)|+|\Lambda_1\eta_1(x)|\le$$

$$\le\frac{7}{720}h_2^4\left\|\frac{\partial^6 u}{\partial x_1^2\partial x_2^4}\right\|_{L_\infty(D)}+\frac{7}{720}h_1^4\left\|\frac{\partial^6 u}{\partial x_1^4\partial x_2^2}\right\|_{L_\infty(D)},$$

which completes the proof of the lemma.

We finally come to the main result.

THEOREM 1.3.– Let the solution $u(x_1,x_2)$ of problem [1.35] satisfy the condition $\frac{\partial^6 u}{\partial x_1^4\partial x_2^2}, \frac{\partial^6 u}{\partial x_1^2\partial x_2^4}\in L_\infty(D)$ and let the steps h_1, h_2 be such that $\frac{1}{\sqrt{5}}\le\frac{h_1}{h_2}\le\sqrt{5}$. Then, the accuracy of the finite-difference scheme [1.36] is characterized by the estimate

$$|z(x)|\le Mv(x)\|u\|_{W_\infty^6(D)}|h|^4, \quad x\in\omega,$$

where M is a positive constant independent of $|h| = \sqrt{h_1^2 + h_2^2}$ and the function $v(x)$ satisfies the relations $v(x) = O(|h|)$ and $v(x) = O\left(|h|^2 \ln \dfrac{1}{|h|}\right)$ near the sides and the vertices of the rectangle D respectively.

PROOF.— Applying Lemma 1.10 to problem [1.37] with $F(x) = \psi(x)$, we have

$$|z(x)| \leq v(x) \|\psi\|_{C(\omega)}, \quad x \in \omega.$$

It remains to apply Lemmas 3.9 and 3.11. The theorem is proved.

1.3.3. Conclusion

The weighted estimate obtained in Theorem 1.3 shows that the error of method [1.36] in the uniform mesh norm $C(\omega)$ is $O(|h|^5)$ near the sides and $O(|h|^6 \ln|h|^{-1})$ near the vertices of the rectangle D, whereas it is $O(|h|^4)$ in the inner nodes of the mesh ω.

REMARK 1.4.— Using methods (Samarskii et al. 1987), we can obtain the scale of estimates

$$|z(x)| \leq M v(x) \|u\|_{W_\infty^m(D)} |h|^{m-2}, \quad x \in \omega \quad (2 < m \leq 6),$$

where $\dfrac{1}{\sqrt{5}} \leq \dfrac{h_1}{h_2} \leq \sqrt{5}$, the constant M is independent of $|h| = \sqrt{h_1^2 + h_2^2}$ and the function $v(x)$ satisfies the relations $v(x) = O(|h|)$ and $v(x) = O\left(|h|^2 \ln|h|^{-1}\right)$ near the sides and vertices of the rectangle D respectively.

1.4. A finite-difference scheme for the equation with mixed derivatives

We consider the BVP

$$Lu \equiv L_1 u + L_2 u + 2L_{12} u = -f(x), \quad x \in D,$$
$$u(x) = 0, \quad x \in \Gamma, \qquad [1.56]$$

where $x = (x_1, x_2)$, $D = \{x = (x_1, x_2) : 0 < x_\alpha < l_\alpha, \alpha = 1, 2\}$ is a rectangle with the boundary $\Gamma = \partial D$,

$$L_\alpha u = k_{\alpha\alpha} \frac{\partial^2 u(x)}{\partial x_\alpha^2}, \alpha = 1, 2, \quad L_{12} u = k_{12} \frac{\partial^2 u(x)}{\partial x_1 \partial x_2},$$

and constant coefficients $k_{\alpha\beta}$ satisfy the ellipticity condition

$$k_{11}\xi_1^2 + k_{22}\xi_2^2 + 2k_{12}\xi_1\xi_2 \geq \gamma \sum_{\alpha=1}^{2} \xi_\alpha^2 \quad \forall \xi_1, \xi_2 \in \mathbb{R} \quad (\gamma = \text{const} > 0). \qquad [1.57]$$

1.4.1. *Discretization of the BVP*

We use here the sets of nodes [1.2] and additionally denote

$$\gamma_\alpha = \gamma_{-\alpha} \cup \gamma_{+\alpha}, \quad \alpha = 1, 2.$$

We recall the traditional notation for finite-difference derivatives from Samarskii (2001):

$$u_{x_1}(x) = \frac{u(x_1 + h_1, x_2) - u(x)}{h_1}, \quad x \in \omega_1^- \times \overline{\omega}_2,$$

$$u_{\bar{x}_1}(x) = \frac{u(x) - u(x_1 - h_1, x_2)}{h_1}, \quad x \in \omega_1^+ \times \overline{\omega}_2,$$

$$u_{x_1\bar{x}_1}(x) = \frac{u(x_1 + h_1, x_2) - 2u(x) + u(x_1 - h_1, x_2)}{h_1^2}, \quad x \in \omega_1 \times \overline{\omega}_2, \quad x = (x_1, x_2);$$

the finite-difference derivatives $u_{x_2}, u_{\bar{x}_2}, u_{x_2\bar{x}_2}$ are defined similarly.

We will also need the Steklov averaging operators (Samarskii et al. 1987), for example:

$$S_1^+ u(x) = \frac{1}{h_1} \int_{x_1}^{x_1+h_1} u(\xi_1, x_2) d\xi_1, \quad S_1^- u(x) = \frac{1}{h_1} \int_{x_1-h_1}^{x_1} u(\xi_1, x_2) d\xi_1,$$

$$T_1 u(x) = \frac{1}{h_1^2} \int_{x_1-h_1}^{x_1+h_1} (h_1 - |x_1 - \xi_1|) u(\xi_1, x_2) d\xi$$

(operators S_2^+, S_2^- and T_2 can be written in a similar way). Here are some of their properties and useful relations:

$$T_\alpha = S_\alpha^+ S_\alpha^- = S_\alpha^- S_\alpha^+, \quad T = T_1 T_2, \quad T_\alpha \frac{\partial^2 u}{\partial x_\alpha^2}(x) = u_{x_\alpha \bar{x}_\alpha}(x),$$

$$S_\alpha^+ \frac{\partial u}{\partial x_\alpha}(x) = u_{x_\alpha}(x), \quad S_\alpha^- \frac{\partial u}{\partial x_\alpha}(x) = u_{\bar{x}_\alpha}(x), \quad \alpha = 1, 2.$$

To construct a discrete analogue of problem [1.56], we apply the operator $T = T_1 T_2$ to both sides of the equation. Then, we get the relation (the so-called generalized balance equation)

$$k_{11}(T_2 u)_{x_1 \bar{x}_1} + k_{22}(T_1 u)_{x_2 \bar{x}_2} +$$

$$+ 2k_{12} \frac{1}{2}\left[(S_1^+ S_2^- u)_{\bar{x}_1 x_2} + (S_1^- S_2^+ u)_{x_1 \bar{x}_2} \right] = -Tf(x), \quad x \in \omega.$$

which naturally leads to the finite-difference scheme

$$\Lambda y \equiv \Lambda_1 y + \Lambda_2 y + 2\Lambda_{12} y = -Tf(x), \quad x \in \omega,$$
$$y(x) = 0, \quad x \in \gamma,$$
[1.58]

with $\Lambda_\alpha y = k_{\alpha\alpha} y_{x_\alpha \bar{x}_\alpha}$, $\alpha = 1, 2$, $\Lambda_{12} y = \frac{1}{2} k_{12}(y_{\bar{x}_1 x_2} + y_{x_1 \bar{x}_2})$.

Note that difference expression $\Lambda_{12} u$ approximates the mixed derivative $L_{12} u = k_{12} \dfrac{\partial^2 u(x)}{\partial x_1 \partial x_2}$ on a seven-point template

$$(x_1, x_2), (x_1 \pm h_1, x_2), (x_1, x_2 \pm h_2), (x_1 + h_1, x_2 - h_2), (x_1 - h_1, x_2 + h_2)$$

with the second order of accuracy on sufficiently smooth functions, namely:

$$\Lambda_{12}u = L_{12}u + O(|h|^2), \quad |h|^2 = \sqrt{h_1^2 + h_2^2}.$$

For the error $z(x) = y(x) - u(x)$, we have the problem

$$\Lambda z \equiv \Lambda_1 z + \Lambda_2 z + 2\Lambda_{12}z = -\psi(x), \quad x \in \omega,$$
$$z(x) = 0, \quad x \in \gamma,$$
[1.59]

where $\psi(x)$ is the approximation error:

$$\psi(x) = Tf(x) + \Lambda_1 u(x) + \Lambda_2 u(x) + 2\Lambda_{12}u(x) =$$
$$= \eta_{1\bar{x}_1 x_1} + \eta_{2\bar{x}_2 x_2} + 2\eta_{12\bar{x}_1 x_2},$$
[1.60]

with

$$\eta_\alpha(x) = k_{\alpha\alpha}\left(u(x) - T_{3-\alpha}u(x)\right), \quad \alpha = 1, 2,$$

$$\eta_{12} = k_{12}\left(\frac{u(x) + u(x_1 + h_1, x_2 - h_2)}{2} - S_1^+ S_2^- u(x)\right).$$

Furthermore, we will also need the notation

$$2\eta_{12\bar{x}_1 x_2} = k_{12}\left(u_{\bar{x}_1 x_2} + u_{x_1 \bar{x}_2} - 2T\frac{\partial^2 u}{\partial x_1 \partial x_2}\right).$$

1.4.2. *Properties of the finite-difference operators*

We introduce the space $\overset{0}{H}$ of mesh functions [1.22] and recall some standard notations for sums and norms:

$$(y, v)_{1,2} = \sum_{x \in \omega_1^+ \times \omega_2^+} h_1 h_2 y(x) v(x),$$

$$(y, v]_\alpha = \sum_{x \in \omega \cup \gamma_{+\alpha}} h_1 h_2 y(x) v(x), \quad \|v\|_\alpha = \sqrt{(v, v]_\alpha}, \quad \alpha = 1, 2,$$

$$|v|_{1,\omega}^2 = |v|_{W_2^1(\omega)}^2 = \sum_{\alpha=1}^{2}\|v]|_\alpha^2, \ \|v\|_{1,\omega}^2 = \|v\|_{W_2^1(\omega)}^2 = |v|_{1,\omega}^2 + \|v\|^2.$$

Note that the mesh function $\varphi(x) = Tf(x)$ is defined at the inner nodes $x \in \omega$. Setting it equal to zero at the boundary nodes $x \in \gamma$, we have $\varphi \in \overset{0}{H}$. Then, scheme [1.58] can be written as the operator equation

$$Ay \equiv A_1 y + A_2 y + 2A_{12} y = \varphi, \quad y \in \overset{0}{H}, \ \varphi \in \overset{0}{H}, \qquad [1.61]$$

where $A_1, A_2, A_{12}, A : \overset{0}{H} \to \overset{0}{H}$, $A_\alpha y = -\Lambda_\alpha y$, $\alpha = 1, 2$, $A_{12} y = -\Lambda_{12} y$.

Next, we study the properties of the operator A.

LEMMA 1.12.– *The difference operator A is symmetric and positive definite in $\overset{0}{H}$.*

PROOF.– Taking into account that $y_{\bar{x}_1} = 0$ for $x_2 = l_2$ and $y_{\bar{x}_2} = 0$ for $x_1 = l_1$, and applying summation by parts, for example:

$$\sum_{x_1 \in \omega_1} h_1 y_{x_1}(x) v(x) = -\sum_{x_1 \in \omega_1^+} h_1 y(x) v_{\bar{x}_1}(x) +$$

$$+ y(l_1, x_2) v(l_1, x_2) - y(h_1, x_2) v(0, x_2), \quad x_2 \in \omega_2,$$

we obtain the relation

$$(Ay, v) = (A_1 y + A_2 y + 2A_{12} y, v) = \left(-k_{11} y_{\bar{x}_1 x_1} - k_{22} y_{\bar{x}_2 x_2} - k_{12}(y_{\bar{x}_1 x_2} + y_{x_1 \bar{x}_2}), v\right) =$$

$$= k_{11}(y_{\bar{x}_1}, v_{\bar{x}_1}]_1 + k_{22}(y_{\bar{x}_2}, v_{\bar{x}_2}]_2 + k_{12}\left[(y_{\bar{x}_1}, v_{\bar{x}_2}) + (y_{\bar{x}_2}, v_{\bar{x}_1})\right] = (y, Av)$$

for all $y, v \in \overset{0}{H}$. It means that A is a symmetric operator.

Next, making use of condition [1.57] and the inequality (e.g. (Samarskii 2001))

$$|v|_{1,\omega}^2 \equiv \|v_{\bar{x}_1}^2]_1 + \|v_{\bar{x}_2}^2]_2 \geq \left(\frac{8}{l_1^2} + \frac{8}{l_2^2}\right)\|v\|^2 \quad \forall v \in \overset{0}{H}, \qquad [1.62]$$

we have

$$(Av,v) = k_{11}(v_{\bar{x}_1}, v_{\bar{x}_1}]_1 + k_{22}(v_{\bar{x}_2}, v_{\bar{x}_2}]_2 + k_{12}\left[(v_{\bar{x}_1}, v_{\bar{x}_2}]_2 + (v_{\bar{x}_2}, v_{\bar{x}_1}]_1\right] =$$

$$= \left(k_{11} v_{\bar{x}_1}^2 + k_{22} v_{\bar{x}_2}^2 + 2k_{12} v_{\bar{x}_1} v_{\bar{x}_2}, 1\right]_{1,2} \geq$$

$$\geq \gamma\left(\|v_{\bar{x}_1}^2\|_1 + \|v_{\bar{x}_2}^2\|_2\right) \geq \gamma\left(\frac{8}{l_1^2} + \frac{8}{l_2^2}\right)\|v\|^2 \quad \forall v \in \overset{0}{H},$$

which means that the operator A is positive definite:

$$(Av,v) \geq \gamma\left(\frac{8}{l_1^2} + \frac{8}{l_2^2}\right)\|v\|^2 \quad \forall v \in \overset{0}{H},$$

and it holds

$$(Av,v) \geq \gamma |v|_{1,\omega}^2 \quad \forall v \in \overset{0}{H}. \qquad [1.63]$$

The lemma is proved.

Now we have the following proposition.

THEOREM 1.4.– *The finite-difference problem [1.61] is uniquely solvable for an arbitrary right-hand side $\varphi(x)$, and for the solution $y(x)$, the following a priori estimate holds true:*

$$|v|_{1,\omega} \leq \frac{1}{\gamma}\left(\frac{8}{l_1^2} + \frac{8}{l_2^2}\right)^{-1/2} \|\varphi\|. \qquad [1.64]$$

PROOF.– It follows from Lemma 1.10 that the inverse operator $A^{-1} : \overset{0}{H} \to \overset{0}{H}$ exists and therefore the unique solution exists for any right-hand side $\varphi \in \overset{0}{H}$. To prove estimate [1.64], we multiply both sides of equation [1.61] by y in $\overset{0}{H}$ and then apply estimate [1.63] to the left-hand side and the Cauchy–Bunyakovsky inequality to the right-hand side. Then, we have

$$\gamma\|y\|_{1,\omega}^2 \le (Ay,y) = (\varphi,y) \le \|\varphi\|\cdot\|y\| \le \|\varphi\|\left(\frac{8}{l_1^2}+\frac{8}{l_2^2}\right)^{-1/2}|y|_{1,\omega},$$

which yields estimate [1.64] and completes the proof of the lemma.

LEMMA 1.13.– *The following estimate holds true:*

$$\|A^{-1}B_k y\| \le \frac{1}{2\left(\sqrt{k_1 k_2}-|k_{12}|\right)}\|y\|_k \quad \forall y \in H_k \quad (k=\overline{1,4}), \qquad [1.65]$$

where $\sqrt{k_1 k_2}-|k_{12}|>0$ *due to the ellipticity condition [1.57].*

PROOF.– First, we introduce the operators:

1) $B_1: H_1 \to \overset{0}{H}$, $B_1 y = -y_{x_1 x_2}$, where H_1 is a space of mesh functions defined on $\tilde{\omega} = \omega_1^+ \times \omega_2^+$;

2) $B_2: H_2 \to \overset{0}{H}$, $B_2 y = -y_{\bar{x}_1 \bar{x}_2}$, where H_2 is a space of mesh functions defined on $\tilde{\omega} = \omega_1^- \times \omega_2^-$;

3) $B_3: H_3 \to \overset{0}{H}$, $B_3 y = -y_{x_1 \bar{x}_2}$, where H_3 is a space of mesh functions defined on $\tilde{\omega} = \omega_1^+ \times \omega_2^-$;

4) $B_4: H_4 \to \overset{0}{H}$, $B_4 y = -y_{\bar{x}_1 x_2}$, where H_4 is a space of mesh functions defined on $\tilde{\omega} = \omega_1^- \times \omega_2^+$.

We also define the inner product and the associate norm in the space H_k, $k=\overline{1,4}$, by formulas

$$(y,v)_k = \sum_{x\in\tilde{\omega}} h_1 h_2 y(x)v(x), \quad \|y\|_k = \sqrt{(y,v)_k} = \left(\sum_{x\in\tilde{\omega}} h_1 h_2 y^2(x)\right)^{1/2}.$$

We find the operator $B_k^*: \overset{0}{H} \to H_k$ which is adjoint of $B_k: H_k \to \overset{0}{H}$ using the relation

$$(B_k y, v) = (y, B_k^* v)_k \quad \forall y \in H_k \quad \forall v \in \overset{0}{H}.$$

Then, we obtain $B_1^* y = -y_{\bar{x}_1 \bar{x}_2}$, $B_2^* y = -y_{x_1 x_2}$, $B_3^* y = -y_{\bar{x}_1 x_2}$, $B_4^* y = -y_{x_1 \bar{x}_2}$.

Next, we prove the inequality

$$\| Ay \| \geq 2\left(\sqrt{k_{11} k_{22}} - |k_{12}|\right) \| B_k^* y \|_k \quad \forall y \in \overset{0}{H} \quad (k = \overline{1,4}).$$

We consider $k = 1$ since for $k = 2, 3, 4$ the reasoning is the same. We have

$$\| Ay \| = \| A_1 y + A_2 y + 2 A_{12} y \| \geq \| A_1 y + A_2 y \| - \| 2 A_{12} y \|,$$

where

$$\| A_1 y + A_2 y \|^2 = (A_1 y + A_2 y, A_1 y + A_2 y) = \| A_1 y \|^2 + \| A_2 y \|^2 + 2(A_1 y, A_2 y) \geq$$

$$\geq 4(A_1 y, A_2 y) = 4 \sum_{x \in \omega} h_1 h_2 (-k_{11} y_{\bar{x}_1 x_1})(-k_{22} y_{\bar{x}_2 x_2}) =$$

$$= 4 k_{11} k_{22} \sum_{x \in \omega} h_1 h_2 y_{\bar{x}_1 x_1} y_{\bar{x}_2 x_2} = 4 k_{11} k_{22} \sum_{x \in \omega_1^+ \times \omega_2^+} h_1 h_2 y_{\bar{x}_1 \bar{x}_2}^2 = 4 k_{11} k_{22} \| B_1^* y \|_1^2,$$

$$\| 2 A_{12} y \|^2 = (2 A_{12} y, 2 A_{12} y) = \sum_{x \in \omega} h_1 h_2 \left(-k_{12}(y_{\bar{x}_1 \bar{x}_2} + y_{x_1 \bar{x}_2})\right)^2 =$$

$$= k_{12}^2 \sum_{x \in \omega} h_1 h_2 \left(y_{\bar{x}_1 x_2}^2 + y_{x_1 \bar{x}_2}^2 + 2 y_{\bar{x}_1 x_2} y_{x_1 \bar{x}_2}\right) \leq 2 k_{12}^2 \sum_{x \in \omega} h_1 h_2 \left(y_{\bar{x}_1 x_2}^2 + y_{x_1 \bar{x}_2}^2\right) \leq$$

$$\leq 4 k_{12}^2 \sum_{x \in \omega_1^+ \times \omega_2^+} h_1 h_2 y_{\bar{x}_1 \bar{x}_2}^2 = 4 k_{12}^2 \| B_1^* y \|_1^2,$$

and therefore

$$\| Ay \| \geq 2\left(\sqrt{k_{11} k_{22}} - |k_{12}|\right) \| B_1^* y \|_1 \quad \forall y \in \overset{0}{H}.$$

Using the inequality

$$\|B_k^* y\|_k \le \frac{1}{2\left(\sqrt{k_{11}k_{22}} - |k_{12}|\right)} \|Ay\| \quad \forall y \in \overset{0}{H} \quad (k = \overline{1,4})$$

and applying the main lemma from Samarskii et al. (1987, p. 54), we come to estimate [1.65] and thus complete the proof.

1.4.3. *Discrete Green's function*

We denote by $G(x,\xi)$ Green's function of the finite-difference problem [1.59]:

$$\Lambda_\xi G(x,\xi) \equiv \Lambda_{1\xi}G(x,\xi) + \Lambda_{2\xi}G(x,\xi) + 2\Lambda_{12\xi}G(x,\xi) =$$
$$= -\frac{\delta(x_1,\xi_1)\delta(x_2,\xi_2)}{h_1 h_2}, \quad \xi \in \omega, \qquad [1.66]$$
$$G(x,\xi) = 0, \quad \xi \in \gamma,$$

where $\xi = (\xi_1,\xi_2)$, $\delta(m,n)$ is the Kronecker delta symbol, and the subscript ξ means the finite-difference derivatives in the variables ξ_1 and ξ_2, for example: $\Lambda_{1\xi}G(x,\xi) \equiv k_{11}G_{\bar{\xi}_1\xi_1}(x,\xi)$.

LEMMA 1.14.– *For Green's function $G(x,\xi)$, the following estimate holds true:*

$$\|G(x,\cdot)\| \le \frac{1}{2\left(\sqrt{k_1 k_2} - |k_{12}|\right)} \rho(x), \quad x \in \omega, \qquad [1.67]$$

where $\rho(x) = \min\left\{\sqrt{x_1 x_2}, \sqrt{x_1(l_2 - x_2)}, \sqrt{x_2(l_1 - x_1)}, \sqrt{(l_1 - x_1)(l_2 - x_2)}\right\}$.

PROOF.– We introduce the Heaviside step function $H(s) = \begin{cases} 1, & s \ge 0, \\ 0, & s < 0, \end{cases}$ and rewrite problem [1.66] in the form

$$\Lambda_\xi G(x,\xi) = -\left(H(x_1 - \xi_1)H(x_2 - \xi_2)\right)_{\xi_1\xi_2}, \quad \xi \in \omega,$$
$$G(x,\xi) = 0, \quad \xi \in \gamma,$$

which means the operator equation

$$A_\xi G(x,\xi) = -B_{1\xi}\left(H(x_1 - \xi_1)H(x_2 - \xi_2)\right).$$

Applying here Lemma 1.11, we obtain

$$\| G(x,\cdot) \| \leq \| -A_\xi^{-1} B_{1\xi} H(x_1 - \cdot) H(x_2 - \cdot) \| \leq$$

$$\leq \frac{1}{2\left(\sqrt{k_1 k_2} - |k_{12}|\right)} \| H(x_1 - \cdot) H(x_2 - \cdot) \|_1 =$$

$$= \frac{1}{2\left(\sqrt{k_1 k_2} - |k_{12}|\right)} \left(\sum_{\xi \in \omega_1^+ \times \omega_2^+} h_1 h_2 H^2(x_1 - \xi_1) H^2(x_2 - \xi_2) \right)^{1/2} = \qquad [1.68]$$

$$= \frac{1}{2\left(\sqrt{k_1 k_2} - |k_{12}|\right)} \left(\sum_{\xi_1 \in \omega_1^+} h_1 H^2(x_1 - \xi_1) \right)^{1/2} \left(\sum_{\xi_2 \in \omega_2^+} h_2 H^2(x_2 - \xi_2) \right)^{1/2} =$$

$$= \frac{1}{2\left(\sqrt{k_1 k_2} - |k_{12}|\right)} \left(\sum_{\xi_1 = h_1}^{x_1} h_1 \right)^{1/2} \left(\sum_{\xi_2 = h_2}^{x_2} h_2 \right)^{1/2} = \frac{1}{2\left(\sqrt{k_1 k_2} - |k_{12}|\right)} \sqrt{x_1 x_2} \ .$$

We can also present problem [1.66] as follows:

$$\Lambda_\xi G(x,\xi) = -\left(H(\xi_1 - x_1) H(\xi_2 - x_2) \right)_{\bar{\xi}_1 \bar{\xi}_2}, \quad \xi \in \omega,$$
$$G(x,\xi) = 0, \quad \xi \in \gamma,$$

which means the operator equation

$$A_\xi G(x,\xi) = -B_{2\xi} \left(H(\xi_1 - x_1) H(\xi_2 - x_2) \right).$$

Using here Lemma 1.11, we get

$$\| G(x,\cdot) \| \leq \| -A_\xi^{-1} B_{2\xi} H(\cdot - x_1) H(\cdot - x_2) \| \leq$$

$$\leq \frac{1}{2\left(\sqrt{k_1 k_2} - |k_{12}|\right)} \| H(\cdot - x_1) H(\cdot - x_2) \|_2 = \qquad [1.69]$$

$$= \frac{1}{2\left(\sqrt{k_1 k_2} - |k_{12}|\right)} \left(\sum_{\xi \in \omega_1^- \times \omega_2^-} h_1 h_2 H^2(\xi_1 - x_1) H^2(\xi_2 - x_2) \right)^{1/2} =$$

$$= \frac{1}{2\left(\sqrt{k_1 k_2} - |k_{12}|\right)} \left(\sum_{\xi_1 \in \omega_1^-} h_1 H^2(\xi_1 - x_1)\right)^{1/2} \left(\sum_{\xi_2 \in \omega_2^-} h_2 H^2(\xi_2 - x_2)\right)^{1/2} =$$

$$= \frac{1}{2\left(\sqrt{k_1 k_2} - |k_{12}|\right)} \left(\sum_{\xi_1 = x_1}^{l_1 - h_1} h_1\right)^{1/2} \left(\sum_{\xi_2 = x_2}^{l_2 - h_2} h_2\right)^{1/2} =$$

$$= \frac{1}{2\left(\sqrt{k_1 k_2} - |k_{12}|\right)} \sqrt{(l_1 - x_1)(l_2 - x_2)}.$$

Another possible form of problem [1.66] is

$$\Lambda_\xi G(x, \xi) = \left(H(x_1 - \xi_1) H(\xi_2 - x_2)\right)_{\xi_1 \xi_2}, \quad \xi \in \omega,$$
$$G(x, \xi) = 0, \quad \xi \in \gamma,$$

which is the operator equation

$$A_\xi G(x, \xi) = B_{3\xi}\left(H(x_1 - \xi_1) H(\xi_2 - x_2)\right),$$

which due to Lemma 1.11 leads to the estimate

$$\|G(x, \cdot)\| \leq \| A_\xi^{-1} B_{3\xi} H(x_1 - \cdot) H(\cdot - x_2)\| \leq$$

$$\leq \frac{1}{2\left(\sqrt{k_1 k_2} - |k_{12}|\right)} \| H(x_1 - \cdot) H(\cdot - x_2)\|_3 = \qquad [1.70]$$

$$= \frac{1}{2\left(\sqrt{k_1 k_2} - |k_{12}|\right)} \left(\sum_{\xi \in \omega_1^+ \times \omega_2^-} h_1 h_2 H^2(x_1 - \xi_1) H^2(\xi_2 - x_2)\right)^{1/2} =$$

$$= \frac{1}{2\left(\sqrt{k_1 k_2} - |k_{12}|\right)} \left(\sum_{\xi_1 \in \omega_1^+} h_1 H^2(x_1 - \xi_1)\right)^{1/2} \left(\sum_{\xi_2 \in \omega_2^-} h_2 H^2(\xi_2 - x_2)\right)^{1/2} =$$

$$= \frac{1}{2\left(\sqrt{k_1 k_2} - |k_{12}|\right)} \left(\sum_{\xi_1 = h_1}^{x_1} h_1\right)^{1/2} \left(\sum_{\xi_2 = x_2}^{l_2 - h_2} h_2\right)^{1/2} =$$

$$= \frac{1}{2\left(\sqrt{k_1 k_2} - |k_{12}|\right)} \sqrt{x_1(l_2 - x_2)}.$$

If we present problem [1.66] as follows:

$$\Lambda_\xi G(x,\xi) = \left(H(\xi_1 - x_1)H(x_2 - \xi_2)\right)_{\bar\xi_1 \bar\xi_2}, \quad \xi \in \omega,$$
$$G(x,\xi) = 0, \quad \xi \in \gamma,$$

which implies the operator equation

$$A_\xi G(x,\xi) = B_{4\xi}\left(H(\xi_1 - x_1)H(x_2 - \xi_2)\right),$$

then again by Lemma 1.11, we come to the estimate

$$\|G(x,\cdot)\| \le \|A_\xi^{-1} B_{4\xi} H(\cdot - x_1) H(x_2 - \cdot)\| \le$$

$$\le \frac{1}{2\left(\sqrt{k_1 k_2} - |k_{12}|\right)} \|H(\cdot - x_1)H(x_2 - \cdot)\|_4 = \qquad [1.71]$$

$$= \frac{1}{2\left(\sqrt{k_1 k_2} - |k_{12}|\right)} \left(\sum_{\xi \in \omega_1^- \times \omega_2^+} h_1 h_2 H^2(\xi_1 - x_1) H^2(x_2 - \xi_2) \right)^{1/2} =$$

$$= \frac{1}{2\left(\sqrt{k_1 k_2} - |k_{12}|\right)} \left(\sum_{\xi_1 \in \omega_1^-} h_1 H^2(\xi_1 - x_1) \right)^{1/2} \left(\sum_{\xi_2 \in \omega_2^-} h_2 H^2(x_2 - \xi_2) \right)^{1/2} =$$

$$= \frac{1}{2\left(\sqrt{k_1 k_2} - |k_{12}|\right)} \left(\sum_{\xi_1 = x_1}^{l_1 - h_1} h_1 \right)^{1/2} \left(\sum_{\xi_2 = h_2}^{x_2} h_2 \right)^{1/2} = \frac{1}{2\left(\sqrt{k_1 k_2} - |k_{12}|\right)} \sqrt{(l_1 - x_1)x_2}.$$

Combining estimates [1.68]–[1.71], we arrive at inequality [1.67] and thus complete the proof of the lemma.

1.4.4. Accuracy with the boundary effect

First, we study the approximation error $\psi(x)$.

LEMMA 1.15.– *Let the solution $u(x)$ of problem [1.56] satisfy the condition $u \in W_2^4(D)$. Then, for the approximation error $\psi(x)$ the estimate holds true:*

$$\|\psi\| \leq \widetilde{M} |h|^2 |u|_{W_2^4(D)}, \qquad [1.72]$$

where $\widetilde{M} = \dfrac{8}{\sqrt{3}}(k_{11}+k_{22}) + \sqrt{1344}\,|k_{12}|$ is a positive constant independent of $u(x)$, h_1, h_2, and $|u|_{W_2^4(D)}$ is a seminorm in $W_2^4(D)$:

$$|u|_{W_2^4(D)} = \left\{ \sum_{\substack{k_1+k_2=4 \\ (k_1 \geq 0,\, k_2 \geq 0)}} \iint_D \left(\frac{\partial^{k_1+k_2} u(x_1, x_2)}{\partial x_1^{k_1} \partial x_2^{k_2}} \right)^2 dx_1 dx_2 \right\}^{1/2}.$$

PROOF.– From inequality [1.60], we have

$$\|\psi\| = \|\eta_{1\bar{x}_1 x_1} + \eta_{2\bar{x}_2 x_2} + 2\eta_{12\bar{x}_1 x_2}\| \leq \|\eta_{1\bar{x}_1 x_1}\| + \|\eta_{2\bar{x}_2 x_2}\| + \|2\eta_{12\bar{x}_1 x_2}\|. \quad [1.73]$$

Next, we will consider each of the three summands in the right-hand side. For the functional $\eta_1(x)$, we have

$$\eta_1(x) = k_{11}\left(u(x) - T_2 u(x)\right) = \frac{k_{11}}{h_2^2} \int_{x_2-h_2}^{x_2-h_2} (h_2 - |x_2 - \xi|)[u(x) - u(x_1, \xi)] d\xi =$$

$$= \frac{k_{11}}{h_2^2} \int_{x_2-h_2}^{x_2-h_2} (h_2 - |x_2 - \xi|) d\xi \int_\xi^{x_2} \frac{\partial u(x_1, \xi_1)}{\partial \xi_1} d\xi_1 =$$

$$= \frac{k_{11}}{h_2^2} \int_{x_2-h_2}^{x_2-h_2} (h_2 - |x_2 - \xi|) d\xi \int_\xi^{x_2} \left[\frac{\partial u(x_1, \xi_1)}{\partial \xi_1} - \frac{1}{2h_2} \int_{x_2-h_2}^{x_2+h_2} \frac{\partial u(x_1, \xi_2)}{\partial \xi_2} d\xi_2 \right] d\xi_1 =$$

$$= \frac{k_{11}}{2h_2^3} \int_{x_2-h_2}^{x_2-h_2} (h_2 - |x_2 - \xi|) d\xi \int_\xi^{x_2} d\xi_1 \int_{x_2-h_2}^{x_2+h_2} d\xi_2 \int_{\xi_2}^{\xi_1} \frac{\partial^2 u(x_1, \xi_3)}{\partial \xi_3^2} d\xi_3,$$

then

$$\eta_{1\bar{x}_1 x_1}(x) = \frac{k_{11}}{2h_2^3 h_1^2} \int_{x_1-h_1}^{x_1+h_1} (h_1 - |x_1 - \xi_4|) d\xi_4 \int_{x_2-h_2}^{x_2-h_2} (h_2 - |x_2 - \xi|) d\xi \times$$

$$\times \int_\xi^{x_2} d\xi_1 \int_{x_2-h_2}^{x_2+h_2} d\xi_2 \int_{\xi_2}^{\xi_1} \frac{\partial^4 u(\xi_4, \xi_3)}{\partial \xi_4^2 \partial \xi_3^2} d\xi_3 .$$

This leads to the estimate

$$|\eta_{1\bar{x}_1 x_1}(x)| \le \frac{k_{11}}{2h_2^3 h_1^2} \int_{x_1-h_1}^{x_1+h_1} (h_1 - |x_1 - \xi_4|) d\xi_4 \int_{x_2-h_2}^{x_2-h_2} (h_2 - |x_2 - \xi|) d\xi \times$$

$$\times \int_{x_2-h_2}^{x_2+h_2} d\xi_1 \int_{x_2-h_2}^{x_2+h_2} d\xi_2 \int_{x_2-h_2}^{x_2+h_2} \left| \frac{\partial^4 u(\xi_4, \xi_3)}{\partial \xi_4^2 \partial \xi_3^2} \right| d\xi_3 \le$$

$$\le \frac{k_{11} h_2^2 \cdot 2h_2 \cdot 2h_2}{2h_2^3 h_1^2} \left\{ \int_{x_1-h_1}^{x_1+h_1} (h_1 - |x_1 - \xi_4|)^2 d\xi_4 \int_{x_2-h_2}^{x_2+h_2} d\xi_3 \right\}^{1/2} \times$$

$$\times \left\{ \int_{x_1-h_1}^{x_1+h_1} d\xi_4 \int_{x_2-h_2}^{x_2+h_2} \left| \frac{\partial^4 u(\xi_4, \xi_3)}{\partial \xi_4^2 \partial \xi_3^2} \right|^2 d\xi_3 \right\}^{1/2} =$$

$$= k_{11} \frac{4}{\sqrt{3}} \sqrt{\frac{h_2^3}{h_1}} \left\{ \int_{x_1-h_1}^{x_1+h_1} d\xi_4 \int_{x_2-h_2}^{x_2+h_2} \left| \frac{\partial^4 u(\xi_4, \xi_3)}{\partial \xi_4^2 \partial \xi_3^2} \right|^2 d\xi_3 \right\}^{1/2},$$

and therefore,

$$\|\eta_{1\bar{x}_1 x_1}\| = \left(\sum_{x \in \omega} h_1 h_2 \eta_{1\bar{x}_1 x_1}^2(x) \right)^{1/2} \le \frac{8h_2^2 k_{11}}{\sqrt{3}} \left\{ \iint_D \left| \frac{\partial^4 u(x_1, x_2)}{\partial x_1^2 \partial x_2^2} \right|^2 dx_1 dx_2 \right\}^{1/2}. \quad [1.74]$$

Similarly, we can obtain the inequality

$$\|\eta_{2\bar{x}_2 x_2}\| \leq \frac{8h_1^2 k_{22}}{\sqrt{3}} \left\{ \iint_D \left|\frac{\partial^4 u(x_1, x_2)}{\partial x_1^2 \partial x_2^2}\right|^2 dx_1 dx_2 \right\}^{1/2}.$$ [1.75]

Now we move on to the third summand in [1.73]. We have

$$2\eta_{12\bar{x}_1 x_2} = k_{12}\left(u_{\bar{x}_1 x_2} + u_{x_1 \bar{x}_2} - 2T\frac{\partial^2 u}{\partial x_1 \partial x_2}\right) =$$

$$= \frac{k_{12}}{h_1^2 h_2^2} \int_{x_1 - h_1}^{x_1 + h_1} (h_1 - |x_1 - \xi_1|) d\xi_1 \int_{x_2 - h_2}^{x_2 + h_2} (h_2 - |x_2 - \xi_2|) \times$$

$$\times \left[u_{\bar{x}_1 x_2}(x) + u_{x_1 \bar{x}_2}(x) - 2\frac{\partial^2 u(\xi_1, \xi_2)}{\partial \xi_1 \partial \xi_2} \right] d\xi_2 =$$

$$= \frac{k_{12}}{h_1^3 h_2^3} \int_{x_1 - h_1}^{x_1 + h_1} (h_1 - |x_1 - \xi_1|) d\xi_1 \int_{x_2 - h_2}^{x_2 + h_2} (h_2 - |x_2 - \xi_2|) \times$$

$$\times \left[\int_{x_1}^{x_1 + h_1} d\xi_3 \int_{x_2 - h_2}^{x_2} \left(\frac{\partial^2 u(\xi_3, \xi_4)}{\partial \xi_3 \partial \xi_4} - \frac{\partial^2 u(\xi_1, \xi_2)}{\partial \xi_1 \partial \xi_2} \mp \frac{\partial^2 u(\xi_1, \xi_4)}{\partial \xi_1 \partial \xi_4}\right) d\xi_4 + \right.$$

$$\left. + \int_{x_1 - h_1}^{x_1} d\xi_5 \int_{x_2}^{x_2 + h_2} \left(\frac{\partial^2 u(\xi_5, \xi_6)}{\partial \xi_5 \partial \xi_6} - \frac{\partial^2 u(\xi_1, \xi_2)}{\partial \xi_1 \partial \xi_2} \mp \frac{\partial^2 u(\xi_1, \xi_6)}{\partial \xi_1 \partial \xi_6}\right) d\xi_6 \right] d\xi_2 =$$

$$= \frac{k_{12}}{h_1^3 h_2^3} \int_{x_1 - h_1}^{x_1 + h_1} (h_1 - |x_1 - \xi_1|) d\xi_1 \int_{x_2 - h_2}^{x_2 + h_2} (h_2 - |x_2 - \xi_2|) \times$$

$$\times \left[\int_{x_1}^{x_1 + h_1} d\xi_3 \int_{x_2 - h_2}^{x_2} d\xi_4 \int_{\xi_1}^{\xi_3} \frac{\partial^3 u(\xi_7, \xi_4)}{\partial \xi_7^2 \partial \xi_4} d\xi_7 + \int_{x_1}^{x_1 + h_1} d\xi_3 \int_{x_2 - h_2}^{x_2} d\xi_4 \int_{\xi_2}^{\xi_4} \frac{\partial^3 u(\xi_1, \xi_8)}{\partial \xi_1 \partial \xi_8^2} d\xi_8 + \right.$$

$$\int\limits_{x_1-h_1}^{x_1} d\xi_5 \int\limits_{x_2}^{x_2+h_2} d\xi_6 \int\limits_{\xi_1}^{\xi_5} \frac{\partial^3 u(\xi_9,\xi_6)}{\partial \xi_9^2 \partial \xi_6} d\xi_9 +$$

$$+ \int\limits_{x_1-h_1}^{x_1} d\xi_5 \int\limits_{x_2}^{x_2+h_2} d\xi_6 \int\limits_{\xi_2}^{\xi_6} \frac{\partial^3 u(\xi_1,\xi_{10})}{\partial \xi_1 \partial \xi_{10}^2} d\xi_{10} \Bigg] d\xi_2 =$$

$$= \frac{k_{12}}{4 h_1^4 h_2^4} \int\limits_{x_1-h_1}^{x_1+h_1} (h_1 - |x_1 - \xi_1|) d\xi_1 \int\limits_{x_2-h_2}^{x_2+h_2} (h_2 - |x_2 - \xi_2|) \times$$

$$\times \Bigg[\int\limits_{x_1}^{x_1+h_1} d\xi_3 \int\limits_{x_2-h_2}^{x_2} d\xi_4 \int\limits_{\xi_1}^{\xi_3} d\xi_7 \int\limits_{x_1-h_1}^{x_1+h_1} d\xi_{11} \int\limits_{x_2-h_2}^{x_2+h_2} d\xi_{12} \times$$

$$\times \Bigg\{ \frac{\partial^3 u(\xi_7,\xi_4)}{\partial \xi_7^2 \partial \xi_4} - \frac{\partial^3 u(\xi_{11},\xi_{12})}{\partial \xi_{11}^2 \partial \xi_{12}} \mp \frac{\partial^3 u(\xi_{11},\xi_4)}{\partial \xi_{11}^2 \partial \xi_4} \Bigg\} +$$

$$+ \int\limits_{x_1}^{x_1+h_1} d\xi_3 \int\limits_{x_2-h_2}^{x_2} d\xi_4 \int\limits_{\xi_2}^{\xi_4} d\xi_8 \int\limits_{x_1-h_1}^{x_1+h_1} d\xi_{13} \int\limits_{x_2-h_2}^{x_2+h_2} d\xi_{14} \times$$

$$\times \Bigg\{ \frac{\partial^3 u(\xi_1,\xi_8)}{\partial \xi_1 \partial \xi_8^2} - \frac{\partial^3 u(\xi_{13},\xi_{14})}{\partial \xi_{13} \partial \xi_{14}^2} \mp \frac{\partial^3 u(\xi_{13},\xi_8)}{\partial \xi_{13} \partial \xi_8^2} \Bigg\} +$$

$$+ \int\limits_{x_1-h_1}^{x_1} d\xi_5 \int\limits_{x_2}^{x_2+h_2} d\xi_6 \int\limits_{\xi_1}^{\xi_5} d\xi_9 \int\limits_{x_1-h_1}^{x_1+h_1} d\xi_{15} \int\limits_{x_2-h_2}^{x_2+h_2} d\xi_{16} \times$$

$$\times \Bigg\{ \frac{\partial^3 u(\xi_9,\xi_6)}{\partial \xi_9^2 \partial \xi_6} - \frac{\partial^3 u(\xi_{15},\xi_{16})}{\partial \xi_{15}^2 \partial \xi_{16}} \mp \frac{\partial^3 u(\xi_{15},\xi_6)}{\partial \xi_{15}^2 \partial \xi_6} \Bigg\} +$$

$$+ \int\limits_{x_1-h_1}^{x_1} d\xi_5 \int\limits_{x_2}^{x_2+h_2} d\xi_6 \int\limits_{\xi_2}^{\xi_6} d\xi_{10} \int\limits_{x_1-h_1}^{x_1+h_1} d\xi_{17} \int\limits_{x_2-h_2}^{x_2+h_2} d\xi_{18} \times$$

$$\times \left\{ \frac{\partial^3 u(\xi_1,\xi_{10})}{\partial \xi_1 \partial \xi_{10}^2} - \frac{\partial^3 u(\xi_{17},\xi_{18})}{\partial \xi_{17} \partial \xi_{18}^2} \mp \frac{\partial^3 u(\xi_{17},\xi_{10})}{\partial \xi_{17} \partial \xi_{10}^2} \right\} \right],$$

where we used the relations

$$\int_{x_1-h_1}^{x_1+h_1} (h_1-|x_1-\xi_1|) \left[\int_{x_1}^{x_1+h_1} (\xi_3-\xi_1)d\xi_3 + \int_{x_1-h_1}^{x_1} (\xi_5-\xi_1)d\xi_5 \right] d\xi_1 =$$

$$= \int_{x_1-h_1}^{x_1+h_1} (h_1-|x_1-\xi_1|) 2h_1(x_1-\xi_1)d\xi_1 = 0,$$

$$\int_{x_2-h_2}^{x_2+h_2} (h_2-|x_2-\xi_2|) \left[\int_{x_2-h_2}^{x_2} (\xi_4-\xi_2)d\xi_4 + \int_{x_2}^{x_2+h_2} (\xi_6-\xi_2)d\xi_6 \right] d\xi_2 =$$

$$= \int_{x_2-h_2}^{x_2+h_2} (h_2-|x_2-\xi_2|) 2h_2(x_2-\xi_2)d\xi_2 = 0.$$

Next, we make the following transformations:

$$2\eta_{12\overline{x}_1 x_2} = \frac{k_{12}}{4h_1^4 h_2^4} \int_{x_1-h_1}^{x_1+h_1} (h_1-|x_1-\xi_1|)d\xi_1 \int_{x_2-h_2}^{x_2+h_2} (h_2-|x_2-\xi_2|) \times$$

$$\times \left[\int_{x_1}^{x_1+h_1} d\xi_3 \int_{x_2-h_2}^{x_2} d\xi_4 \int_{\xi_1}^{\xi_3} d\xi_7 \int_{x_1-h_1}^{x_1+h_1} d\xi_{11} \int_{x_2-h_2}^{x_2+h_2} d\xi_{12} \times \right.$$

$$\times \left\{ \int_{\xi_{11}}^{\xi_7} \frac{\partial^4 u(\xi_{19},\xi_4)}{\partial \xi_{19}^3 \partial \xi_4} d\xi_{19} + \int_{\xi_{12}}^{\xi_4} \frac{\partial^4 u(\xi_{11},\xi_{20})}{\partial \xi_{11}^2 \partial \xi_{20}^2} d\xi_{20} \right\} +$$

$$+ \int_{x_1}^{x_1+h_1} d\xi_3 \int_{x_2-h_2}^{x_2} d\xi_4 \int_{\xi_2}^{\xi_4} d\xi_8 \int_{x_1-h_1}^{x_1+h_1} d\xi_{13} \int_{x_2-h_2}^{x_2+h_2} d\xi_{14} \times$$

$$\times \left\{ \int_{\xi_{13}}^{\xi_1} \frac{\partial^4 u(\xi_{21}, \xi_8)}{\partial \xi_{21}^2 \partial \xi_8^2} d\xi_{21} + \int_{\xi_{14}}^{\xi_8} \frac{\partial^4 u(\xi_{13}, \xi_{22})}{\partial \xi_{13} \partial \xi_{22}^3} d\xi_{22} \right\} +$$

$$+ \int_{x_1-h_1}^{x_1} d\xi_5 \int_{x_2}^{x_2+h_2} d\xi_6 \int_{\xi_1}^{\xi_5} d\xi_9 \int_{x_1-h_1}^{x_1+h_1} d\xi_{15} \int_{x_2-h_2}^{x_2+h_2} d\xi_{16} \times$$

$$\times \left\{ \int_{\xi_{15}}^{\xi_9} \frac{\partial^4 u(\xi_{23}, \xi_6)}{\partial \xi_{23}^3 \partial \xi_6} d\xi_{23} + \int_{\xi_{16}}^{\xi_6} \frac{\partial^4 u(\xi_{15}, \xi_{24})}{\partial \xi_{15}^2 \partial \xi_{24}^2} d\xi_{24} \right\} +$$

$$+ \int_{x_1-h_1}^{x_1} d\xi_5 \int_{x_2}^{x_2+h_2} d\xi_6 \int_{\xi_2}^{\xi_6} d\xi_{10} \int_{x_1-h_1}^{x_1+h_1} d\xi_{17} \int_{x_2-h_2}^{x_2+h_2} d\xi_{18} \times$$

$$\times \left\{ \int_{\xi_{17}}^{\xi_1} \frac{\partial^4 u(\xi_{25}, \xi_{10})}{\partial \xi_{25}^2 \partial \xi_{10}^2} d\xi_{25} + \int_{\xi_{18}}^{\xi_{10}} \frac{\partial^4 u(\xi_{17}, \xi_{26})}{\partial \xi_{17} \partial \xi_{26}^3} d\xi_{26} \right\} \Bigg] .$$

This leads to the inequality

$$|2\eta_{12\bar{x}_1 x_2}(x)| \leq \frac{|k_{12}|}{4 h_1^4 h_2^4} \int_{x_1-h_1}^{x_1+h_1} (h_1 - |x_1 - \xi_1|) d\xi_1 \int_{x_2-h_2}^{x_2+h_2} (h_2 - |x_2 - \xi_2|) \times$$

$$\times \Bigg[\int_{x_1}^{x_1+h_1} d\xi_3 \int_{x_2-h_2}^{x_2} d\xi_4 \int_{x_1-h_1}^{x_1+h_1} d\xi_7 \int_{x_1-h_1}^{x_1+h_1} d\xi_{11} \int_{x_2-h_2}^{x_2+h_2} d\xi_{12} \times$$

$$\times \left\{ \int_{x_1-h_1}^{x_1+h_1} \left| \frac{\partial^4 u(\xi_{19}, \xi_4)}{\partial \xi_{19}^3 \partial \xi_4} \right| d\xi_{19} + \int_{x_2-h_2}^{x_2+h_2} \left| \frac{\partial^4 u(\xi_{11}, \xi_{20})}{\partial \xi_{11}^2 \partial \xi_{20}^2} \right| d\xi_{20} \right\} +$$

$$+ \int_{x_1}^{x_1+h_1} d\xi_3 \int_{x_2-h_2}^{x_2} d\xi_4 \int_{x_2-h_2}^{x_2+h_2} d\xi_8 \int_{x_1-h_1}^{x_1+h_1} d\xi_{13} \int_{x_2-h_2}^{x_2+h_2} d\xi_{14} \times$$

$$\times\left\{\int_{x_1-h_1}^{x_1+h_1}\left|\frac{\partial^4 u(\xi_{21},\xi_8)}{\partial\xi_{21}^2\partial\xi_8^2}\right|d\xi_{21}+\int_{x_2-h_2}^{x_2+h_2}\left|\frac{\partial^4 u(\xi_{13},\xi_{22})}{\partial\xi_{13}\partial\xi_{22}^3}\right|d\xi_{22}\right\}+$$

$$+\int_{x_1-h_1}^{x_1}d\xi_5\int_{x_2}^{x_2+h_2}d\xi_6\int_{x_1-h_1}^{x_1+h_1}d\xi_9\int_{x_1-h_1}^{x_1+h_1}d\xi_{15}\int_{x_2-h_2}^{x_2+h_2}d\xi_{16}\times$$

$$\times\left\{\int_{x_1-h_1}^{x_1+h_1}\left|\frac{\partial^4 u(\xi_{23},\xi_6)}{\partial\xi_{23}^3\partial\xi_6}\right|d\xi_{23}+\int_{x_2-h_2}^{x_2+h_2}\left|\frac{\partial^4 u(\xi_{15},\xi_{24})}{\partial\xi_{15}^2\partial\xi_{24}^2}\right|d\xi_{24}\right\}+$$

$$+\int_{x_1-h_1}^{x_1}d\xi_5\int_{x_2}^{x_2+h_2}d\xi_6\int_{x_2-h_2}^{x_2+h_2}d\xi_{10}\int_{x_1-h_1}^{x_1+h_1}d\xi_{17}\int_{x_2-h_2}^{x_2+h_2}d\xi_{18}\times$$

$$\times\left\{\int_{x_1-h_1}^{x_1+h_1}\left|\frac{\partial^4 u(\xi_{25},\xi_{10})}{\partial\xi_{25}^2\partial\xi_{10}^2}\right|d\xi_{25}+\int_{x_2-h_2}^{x_2+h_2}\left|\frac{\partial^4 u(\xi_{17},\xi_{26})}{\partial\xi_{17}\partial\xi_{26}^3}\right|d\xi_{26}\right\}\right]\leq$$

$$\leq\frac{2\sqrt{2}\,|k_{12}|\sqrt{h_1^3}}{\sqrt{h_2}}\left\{\int_{x_2-h_2}^{x_2}\int_{x_1-h_1}^{x_1+h_1}\left(\frac{\partial^4 u(\xi_{19},\xi_4)}{\partial\xi_{19}^3\partial\xi_4}\right)^2 d\xi_{19}d\xi_4\right\}^{1/2}+$$

$$+2\,|k_{12}|\sqrt{h_1 h_2}\left\{\int_{x_1-h_1}^{x_1+h_1}\int_{x_2-h_2}^{x_2+h_2}\left(\frac{\partial^4 u(\xi_{11},\xi_{20})}{\partial\xi_{11}^2\partial\xi_{20}^2}\right)^2 d\xi_{20}d\xi_{11}\right\}^{1/2}+$$

$$+2\,|k_{12}|\sqrt{h_2 h_1}\left\{\int_{x_2-h_2}^{x_2-h_2}\int_{x_1-h_1}^{x_1+h_1}\left(\frac{\partial^4 u(\xi_{21},\xi_8)}{\partial\xi_{21}^2\partial\xi_8^2}\right)^2 d\xi_{21}d\xi_8\right\}^{1/2}+$$

$$+\frac{2\,|k_{12}|\sqrt{h_2^3}}{\sqrt{h_1}}\left\{\int_{x_1-h_1}^{x_1+h_1}\int_{x_2-h_2}^{x_2-h_2}\left(\frac{\partial^4 u(\xi_{13},\xi_{22})}{\partial\xi_{13}\partial\xi_{22}^3}\right)^2 d\xi_{22}d\xi_{13}\right\}^{1/2}+$$

$$+\frac{2\sqrt{2}\,|k_{12}|\sqrt{h_1^3}}{\sqrt{h_2}}\left\{\int_{x_2}^{x_2-h_2}\int_{x_1-h_1}^{x_1+h_1}\left(\frac{\partial^4 u(\xi_{23},\xi_6)}{\partial \xi_{23}^3 \partial \xi_6}\right)^2 d\xi_{23}d\xi_6\right\}^{1/2}+$$

$$+2\,|k_{12}|\sqrt{h_1 h_2}\left\{\int_{x_1-h_1}^{x_1+h_1}\int_{x_2-h_2}^{x_2-h_2}\left(\frac{\partial^4 u(\xi_{15},\xi_{24})}{\partial \xi_{15}^2 \partial \xi_{24}^2}\right)^2 d\xi_{24}d\xi_{15}\right\}^{1/2}+$$

$$+2\,|k_{12}|\sqrt{h_1 h_2}\left\{\int_{x_2-h_2}^{x_2-h_2}\int_{x_1-h_1}^{x_1+h_1}\left(\frac{\partial^4 u(\xi_{25},\xi_{10})}{\partial \xi_{25}^2 \partial \xi_{10}^2}\right)^2 d\xi_{25}d\xi_{10}\right\}^{1/2}+$$

$$+\frac{2\,|k_{12}|\sqrt{h_2^3}}{\sqrt{h_1}}\left\{\int_{x_1-h_1}^{x_1+h_1}\int_{x_2-h_2}^{x_2-h_2}\left(\frac{\partial^4 u(\xi_{17},\xi_{26})}{\partial \xi_{17} \partial \xi_{26}^3}\right)^2 d\xi_{26}d\xi_{17}\right\}^{1/2}\Bigg]=$$

$$=\frac{4\sqrt{2}\,|k_{12}|\sqrt{h_1^3}}{\sqrt{h_2}}\left\{\int_{x_2-h_2}^{x_2+h_2}\int_{x_1-h_1}^{x_1+h_1}\left(\frac{\partial^4 u(x_1,x_2)}{\partial x_1^3 \partial x_2}\right)^2 dx_1 dx_2\right\}^{1/2}+$$

$$+8\,|k_{12}|\sqrt{h_1 h_2}\left\{\int_{x_1-h_1}^{x_1+h_1}\int_{x_2-h_2}^{x_2+h_2}\left(\frac{\partial^4 u(x_1,x_2)}{\partial x_1^2 \partial x_2^2}\right)^2 dx_2 dx_1\right\}^{1/2}+$$

$$+\frac{4\,|k_{12}|\sqrt{h_2^3}}{\sqrt{h_1}}\left\{\int_{x_1-h_1}^{x_1+h_1}\int_{x_2-h_2}^{x_2-h_2}\left(\frac{\partial^4 u(x_1,x_2)}{\partial x_1 \partial x_2^3}\right)^2 dx_2 dx_1\right\}^{1/2}.$$

Using the elementary inequality $(a+b+c)^2 \leq 3(a^2+b^2+c^2)$, we obtain

$$\|2\eta_{12\bar{x}_1 x_2}(x)\|^2 = \sum_{x\in\omega} h_1 h_2 \left(2\eta_{12\bar{x}_1 x_2}(x)\right)^2 \leq$$

$$\leq 3k_{12}^2\left\{128 h_1^4 \iint_D \left(\frac{\partial^4 u(x_1,x_2)}{\partial x_1^3 \partial x_2}\right)^2 dx_1 dx_2 + \right.$$

$$+256h_1^2h_2^2\iint_D\left(\frac{\partial^4 u(x_1,x_2)}{\partial x_1^2\partial x_2^2}\right)^2 dx_1dx_2 + 64h_2^4\iint_D\left(\frac{\partial^4 u(x_1,x_2)}{\partial x_1\partial x_2^3}\right)^2 dx_1dx_2\Bigg\},$$

and therefore

$$\|2\eta_{12\bar{x}_1x_2}(x)\| \le |k_{12}|\sqrt{1344}\,|h|^2\,|u|_{W_2^4(D)}. \qquad [1.76]$$

Combining inequalities [1.73] and [1.74]–[1.76], we obtain the estimate

$$\|\psi\| \le \left(\frac{8}{\sqrt{3}}(k_{11}+k_{22})+\sqrt{1344}\,|k_{12}|\right)|h|^2\,|u|_{W_2^4(D)}, \qquad [1.77]$$

which is estimate [1.72]. The lemma is proved.

Lemmas 1.14 and 1.15 pave the way to the main proposition of this section.

THEOREM 1.5.– *Let the solution $u(x)$ of BVP [1.56] satisfy the condition $u \in W_2^4(D)$. Then, the accuracy of the discrete problem [1.58] is characterized by the weighted estimate*

$$\max_{x\in\omega}\left|\frac{z(x)}{\rho(x)}\right| \le M\,|h|^2\,|u|_{W_2^4(D)}, \qquad [1.78]$$

where $\rho(x)=\min\left\{\sqrt{x_1x_2},\sqrt{x_1(l_2-x_2)},\sqrt{x_2(l_1-x_1)},\sqrt{(l_1-x_1)(l_2-x_2)}\right\}$ *and M is a positive constant independent of $u(x)$, h_1, h_2:*

$$M = \frac{4(k_{11}+k_{22})+12\sqrt{7}\,|k_{12}|}{\sqrt{3}\left(\sqrt{k_1k_2}-|k_{12}|\right)},$$

$|h|^2 = h_1^2 + h_2^2$, $|u|_{W_2^4(D)}$ *is a seminorm in $W_2^4(D)$.*

PROOF.– Using Green's function $G(x,\xi)$, we can represent the solution of problem [1.59] in the form

$$z(x) = (G(x,\cdot),\psi(\cdot)) = \sum_{\xi\in\omega} h_1h_2 G(x,\xi)\psi(\xi), \quad x\in\omega.$$

Due to Lemmas 1.14 and 1.15, we obtain

$$|z(x)| = \left|(G(x,\cdot), \psi(\cdot))\right| \leq \|G(x,\cdot)\| \cdot \|\psi\| \leq$$

$$\leq \frac{1}{2\left(\sqrt{k_1 k_2} - |k_{12}|\right)} \rho(x) \widetilde{M} |h|^2 |u|_{W_2^4(D)} = M\rho(x)|h|^2 |u|_{W_2^4(D)}, \quad x \in \omega,$$

where $M = \dfrac{\widetilde{M}}{2\left(\sqrt{k_1 k_2} - |k_{12}|\right)} = \dfrac{4(k_{11} + k_{22}) + 12\sqrt{7}\,|k_{12}|}{\sqrt{3}\left(\sqrt{k_1 k_2} - |k_{12}|\right)}$.

This leads to estimate [1.78] and completes the proof.

1.4.5. Conclusion

The weighted estimate [1.78] obtained in Theorem 1.5 shows the influence of the Dirichlet boundary condition and indicates that the accuracy of the finite-discrete scheme [1.58] in the uniform mesh norm is $O(|h|^2 \sqrt{h_1})$ near the vertical sides $x_1 = 0$ and $x_1 = l_1$ of the rectangle D, $O(|h|^2 \sqrt{h_2})$ near the horizontal sides $x_2 = 0$ and $x_2 = l_2$, and $O(|h|^2 \sqrt{h_1 h_2})$ near the vertices, whereas it is $O(|h|^2)$ in the inner nodes of the mesh.

REMARK 1.5.– *It is also possible to estimate the functionals $\eta_{1\bar{x}_1 x_1}$, $\eta_{2\bar{x}_2 x_2}$, $2\eta_{12\bar{x}_1 x_2}$ in Lemma 1.15 using the Bramble–Hilbert lemma (Samarskii et al. 1987, p. 29). However, the exact form of the constant $\widetilde{M}$ in estimate [1.72] and the constant M in estimate [1.78] will be unknown.*

2

Parabolic Equations in Canonical Domains with the Dirichlet Condition on the Boundary or its Part

2.1. A standard finite-difference scheme for the one-dimensional heat equation with mixed boundary conditions

In this section, we consider the problem

$$\frac{\partial u}{\partial t} = \frac{\partial^2 u}{\partial x^2} + f(x,t), \quad (x,t) \in Q_T = (0,1) \times (0,T),$$

$$\frac{\partial u}{\partial x}(0,t) = 0, \quad u(1,t) = 0, \quad t \in (0,T), \qquad [2.1]$$

$$u(x,0) = \varphi(x), \quad x \in (0,1),$$

with the given functions $f(x,t)$, $\varphi(x)$ and the unknown solution $u(x,t)$.

2.1.1. Discretization of the problem

Using the conventional notation of the theory of finite-difference schemes (e.g. Samarskii (2001)), we introduce the following sets of nodes:

$$\omega = \{x_i = ih, \ i = 1, 2, \ldots, N-1, \ h = 1/N\}, \ N \geq 2 \text{ is an integer number,}$$

$$\omega^- = \omega \cup \{0\}, \quad \omega^+ = \omega \cup \{1\}, \quad \bar{\omega} = \omega \cup \{0\} \cup \{1\},$$

$\omega_\tau = \{t_j = j\tau, \ j = \overline{1, M-1}, \ \tau = T/M\}$, $M \geq 2$ is an integer number.

Using the operator (Samarskii et al. 1987)

$$Tv(x) = \begin{cases} \dfrac{1}{h^2} \displaystyle\int_{x-h}^{x+h} (h - |x - \xi|) v(\xi) d\xi, & x \in \omega, \\ \dfrac{2}{h^2} \displaystyle\int_0^h (h - \xi) v(\xi) d\xi, & x = 0, \end{cases} \quad [2.2]$$

and taking into account the relation $\left(T \dfrac{\partial^2 u}{\partial x^2}\right)(x) = \begin{cases} u_{\bar{x}x}, & x \in \omega, \\ \dfrac{2}{h} u_x, & x = 0, \end{cases}$ we approximate

problem [2.1] by the finite-difference scheme

$$y_{\bar{t}}(x,t) - (\Lambda y)(x,t) = (Tf)(x,t), \quad (x,t) \in \omega^- \times \omega_\tau,$$
$$y(1,t) = 0, \quad t \in \omega_\tau, \quad [2.3]$$
$$y(x,0) = \varphi(x), \quad x \in \omega^-.$$

where $\Lambda y(x) = \begin{cases} y_{\bar{x}x}, & x \in \omega, \\ \dfrac{2}{h} y_x, & x = 0. \end{cases}$

For the error $z(x,t) = y(x,t) - u(x,t)$, we have the problem

$$z_{\bar{t}}(x,t) - (\Lambda z)(x,t) = \psi(x,t), \quad (x,t) \in \omega^- \times \omega_\tau,$$
$$z(1,t) = 0, \quad t \in \omega_\tau, \quad [2.4]$$
$$z(x,0) = \varphi(x), \quad x \in \omega^-,$$

where $\psi(x,t)$ is the approximation error:

$$\psi(x,t) = (Tf)(x,t) - u_{\bar{t}}(x,t) + (\Lambda u)(x,t) = \dfrac{d(Tu)}{dt}(x,t) - u_{\bar{t}}(x,t).$$

We denote by H a set of mesh functions defined on $\bar{\omega}$ and equal to zero at the node $x = 1$. The inner product and the associate norm in H are defined by the formulas

$$(y,v) = \sum_{x \in \omega} hy(x)v(x) + \frac{h}{2} y(0)v(0),$$

$$\|v\| = \|v\|_{L_2(\omega)} = \sqrt{(v,v)} = \left(\sum_{x \in \omega} hv^2(x) + \frac{h}{2} v^2(0) \right)^{1/2}.$$

If necessary, functions defined on the mesh ω can be set equal to zero at the node $x = 1$ and any value at the node $x = 0$.

Introducing the difference operator $A: H \to H$, where $Ay = -\Lambda y$, we can rewrite problem [2.3] in the form

$$y_{\bar{t}}(\cdot, t) + Ay(\cdot, t) = Tf(\cdot, t), \quad t \in \omega_\tau,$$
$$y(\cdot, 0) = \varphi(\cdot),$$

with $y(\cdot, t), Tf(\cdot, t), \varphi(\cdot) \in H$.

It is known from Samarskii (2001) that the operator A is symmetric and positive definite in H, and therefore the discrete problem [2.3] is uniquely solvable for any $f(x,t)$ and $\varphi(x)$.

2.1.2. *Accuracy with the boundary effect*

First, we need the following two auxiliary propositions.

LEMMA 2.1.– *For the error $z(x,t)$, the estimate holds true:*

$$\sum_{\eta=\tau}^{t} \tau \left(\frac{z(x,\eta)}{1-x} \right)^2 \leq 4 \sum_{\eta=\tau}^{t} \tau \|\psi(\cdot, \eta)\|^2, \quad (x,t) \in \omega^- \times \omega_\tau. \quad [2.5]$$

PROOF.– Using Green's function $G(x, \xi) = \begin{cases} 1-\xi, & x \leq \xi, \\ 1-x, & \xi \leq x, \end{cases}$ of the finite-difference BVP

$$\Lambda y(x) = -f(x), \quad x \in \omega^-, \quad y(1) = 0,$$

we can present the solution of problem [2.4] in the form

$$z(x,t) = \big(G(x,\cdot), \psi(\cdot,t) - z_{\bar{t}}(\cdot,t)\big) =$$

$$= \sum_{\xi \in \omega} hG(x,\xi)\big(\psi(\xi,t) - z_{\bar{t}}(\xi,t)\big) + \frac{h}{2} G(x,0)\big(\psi(0,t) - z_{\bar{t}}(0,t)\big), \quad (x,t) \in \omega^{-} \times \omega_{\tau}.$$

This gives the inequality

$$|z(x,t)| \le \sum_{\xi \in \omega} hG(x,\xi)\big(|\psi(\xi,t)| + |z_{\bar{t}}(\xi,t)|\big) + \frac{h}{2} G(x,0)\big(|\psi(0,t)| + |z_{\bar{t}}(0,t)|\big) \le$$

$$\le (1-x)\left(\sum_{\xi \in \omega} h\big(|\psi(\xi,t)| + |z_{\bar{t}}(\xi,t)|\big) + \frac{h}{2}\big(|\psi(0,t)| + |z_{\bar{t}}(0,t)|\big)\right) \le$$

$$\le (1-x)\big(\|\psi(\cdot,t)\| + \|z_{\bar{t}}(\cdot,t)\|\big), \quad (x,t) \in \omega^{-} \times \omega_{\tau},$$

and then we have

$$\frac{|z(x,t)|}{1-x} \le \|\psi(\cdot,t)\| + \|z_{\bar{t}}(\cdot,t)\|, \quad (x,t) \in \omega^{-} \times \omega_{\tau}. \qquad [2.6]$$

We consider now the second summand in the right-hand side of [2.6]. To this end, we find the squared norm in H of both sides of equation [2.4]:

$$\|z_{\bar{t}}(\cdot,t)\|^2 + 2\big(z_{\bar{t}}(\cdot,t), -\Lambda z(\cdot,t)\big) + \|\Lambda z(\cdot,t)\|^2 = \|\psi(\cdot,t)\|^2, \quad t \in \omega_{\tau}.$$

Due to the transformations

$$\sum_{\eta=\tau}^{t} \tau 2\big(z_{\bar{t}}(\cdot,\eta), -\Lambda z(\cdot,\eta)\big) = -2\sum_{\eta=\tau}^{t} \tau \left(\sum_{\xi \in \omega} h z_{\bar{\eta}}(\xi,\eta) z_{\bar{\xi}\xi}(\xi,\eta) + \frac{h}{2} z_{\bar{\eta}}(0,\eta) \frac{2}{h} z_{\xi}(0,\eta)\right) =$$

$$= -2\sum_{\eta=\tau}^{t} \tau \left(-\sum_{\xi \in \omega^{+}} h z_{\bar{\eta}\bar{\xi}}(\xi,\eta) z_{\bar{\xi}}(\xi,\eta) + z_{\bar{\eta}}(1,\eta) z_{\bar{\xi}}(1,\eta) - \right.$$

$$\left. - z_{\bar{\eta}}(0,\eta) z_{\bar{\xi}}(h,\eta) + \frac{h}{2} z_{\bar{\eta}}(0,\eta) \frac{2}{h} z_{\xi}(0,\eta)\right) = 2\sum_{\eta=\tau}^{t} \tau \sum_{\xi \in \omega^{+}} h z_{\bar{\eta}\bar{\xi}}(\xi,\eta) z_{\bar{\xi}}(\xi,\eta) =$$

$$= 2\sum_{\eta=\tau}^{t}\tau\sum_{\xi\in\omega^+} h\frac{z_{\bar{\xi}}(\xi,\eta)-z_{\bar{\xi}}(\xi,\eta-\tau)}{\tau}\times$$

$$\times\frac{1}{2}\Big[\big(z_{\bar{\xi}}(\xi,\eta)-z_{\bar{\xi}}(\xi,\eta-\tau)\big)+\big(z_{\bar{\xi}}(\xi,\eta-\tau)+z_{\bar{\xi}}(\xi,\eta)\big)\Big]=$$

$$=\tau\sum_{\eta=\tau}^{t}\tau\sum_{\xi\in\omega^+} hz_{\bar{\eta}\bar{\xi}}^2(\xi,\eta)+\sum_{\xi\in\omega^+} h\sum_{\eta=\tau}^{t}\big(z_{\bar{\xi}}^2(\xi,\eta)-z_{\bar{\xi}}^2(\xi,\eta-\tau)\big)=$$

$$=\tau\sum_{\eta=\tau}^{t}\tau\sum_{\xi\in\omega^+} hz_{\bar{\eta}\bar{\xi}}^2(\xi,\eta)+\sum_{\xi\in\omega^+} hz_{\bar{\xi}}^2(\xi,t)\geq 0,$$

we obtain the estimate

$$\sum_{\eta=\tau}^{t}\tau\,\|z_{\bar{\eta}}(\cdot,\eta)\|^2 \leq \sum_{\eta=\tau}^{t}\tau\,\|\psi(\cdot,\eta)\|^2, \quad t\in\omega_\tau.$$

Then, inequality [2.6] yields

$$\sum_{\eta=\tau}^{t}\tau\left(\frac{z(x,\eta)}{1-x}\right)^2 \leq \sum_{\eta=\tau}^{t}\tau\big(\|\psi(\cdot,\eta)\|+\|z_{\bar{t}}(\cdot,\eta)\|\big)^2 \leq$$

$$\leq 2\sum_{\eta=\tau}^{t}\tau\big(\|\psi(\cdot,\eta)\|^2+\|z_{\bar{t}}(\cdot,\eta)\|^2\big)\leq 4\sum_{\eta=\tau}^{t}\tau\,\|\psi(\cdot,\eta)\|^2$$

and proves the lemma.

LEMMA 2.2.– *Let the solution $u(x,t)$ of problem [2.1] satisfy the conditions $\frac{\partial^2 u}{\partial t^2}, \frac{\partial^2 u}{\partial x \partial t} \in L_2(Q_T)$. Then, for the approximation error $\psi(x,t)$, the following estimate holds true:*

$$\sum_{\eta=\tau}^{t}\tau\,\|\psi(\cdot,\eta)\|^2 \leq 8\tau^2\int_0^1 d\xi\int_0^t\left(\frac{\partial^2 u(\xi,\eta_1)}{\partial\eta_1^2}\right)^2 d\eta_1 + 8h^2\int_0^1 d\xi_1\int_0^t\left(\frac{\partial^2 u(\xi_1,\eta)}{\partial\xi_1\partial\eta}\right)^2 d\eta. \quad [2.7]$$

PROOF.– Using the averaging operator [2.2], we have

$$\psi(x,t) = \frac{d}{dt}T(u(\cdot,t)) - u_{\bar{t}}(x,t) =$$

$$= \frac{1}{\tau h^2} \int_{x-h}^{x+h} (h-|x-\xi|)d\xi \int_{t-\tau}^{t} \frac{\partial u(\xi,t)}{\partial t} d\eta - \frac{1}{\tau} \int_{t-\tau}^{t} \frac{\partial u(x,\eta)}{\partial \eta} d\eta =$$

$$= \frac{1}{\tau h^2} \int_{x-h}^{x+h} (h-|x-\xi|)d\xi \int_{t-\tau}^{t} \left(\frac{\partial u(\xi,t)}{\partial t} - \frac{\partial u(x,\eta)}{\partial \eta} \pm \frac{\partial u(\xi,\eta)}{\partial \eta} \right) d\eta =$$

$$= \frac{1}{\tau h^2} \int_{x-h}^{x+h} (h-|x-\xi|)d\xi \int_{t-\tau}^{t} d\eta \int_{\eta}^{t} \frac{\partial^2 u(\xi,\eta_1)}{\partial \eta_1^2} d\eta_1 +$$

$$+ \frac{1}{\tau h^2} \int_{x-h}^{x+h} (h-|x-\xi|)d\xi \int_{t-\tau}^{t} d\eta \int_{x}^{\xi} \frac{\partial^2 u(\xi_1,\eta)}{\partial \xi_1 \partial \eta} d\xi_1, \quad x \in \omega,$$

which gives the estimate

$$|\psi(x,t)| \leq \frac{1}{\tau h^2} \int_{x-h}^{x+h} (h-|x-\xi|)d\xi \int_{t-\tau}^{t} d\eta \int_{t-\tau}^{t} \left| \frac{\partial^2 u(\xi,\eta_1)}{\partial \eta_1^2} \right| d\eta_1 +$$

$$+ \frac{1}{\tau h^2} \int_{x-h}^{x+h} (h-|x-\xi|)d\xi \int_{t-\tau}^{t} d\eta \int_{x-h}^{x+h} \left| \frac{\partial^2 u(\xi_1,\eta)}{\partial \xi_1 \partial \eta} \right| d\xi_1 \leq$$

$$\leq \frac{1}{\tau h^2} h\tau\sqrt{2h\tau} \left(\int_{x-h}^{x+h} d\xi \int_{t-\tau}^{t} \left(\frac{\partial^2 u(\xi,\eta_1)}{\partial \eta_1^2} \right)^2 d\eta_1 \right)^{1/2} +$$

$$+ \frac{1}{\tau h^2} h^2\sqrt{2h\tau} \left(\int_{t-\tau}^{t} d\eta \int_{x-h}^{x+h} \left(\frac{\partial^2 u(\xi_1,\eta)}{\partial \xi_1 \partial \eta} \right)^2 d\xi_1 \right)^{1/2} =$$

$$= \frac{\sqrt{2\tau}}{\sqrt{h}} \left(\int_{x-h}^{x+h} d\xi \int_{t-\tau}^{t} \left(\frac{\partial^2 u(\xi,\eta_1)}{\partial \eta_1^2} \right)^2 d\eta_1 \right)^{1/2} + \frac{\sqrt{2h}}{\sqrt{\tau}} \left(\int_{x-h}^{x+h} d\xi_1 \int_{t-\tau}^{t} \left(\frac{\partial^2 u(\xi_1,\eta)}{\partial \xi_1 \partial \eta} \right)^2 d\eta \right)^{1/2}.$$

Next, at the node $x=0$ due to [2.2], we have

$$\psi(0,t) = \frac{d}{dt} T(u(\cdot,t)) - u_{\bar{t}}(0,t) = \frac{2}{h^2} \int_0^h (h-\xi) \frac{\partial u(\xi,t)}{\partial t} d\xi - \frac{u(0,t)-u(0,t-\tau)}{\tau} =$$

$$= \frac{2}{\tau h^2} \int_0^h (h-\xi) d\xi \int_{t-\tau}^{t} \left(\frac{\partial u(\xi,t)}{\partial t} - \frac{\partial u(\xi,\eta)}{\partial \eta} \pm \frac{\partial u(\xi,\eta)}{\partial \eta} \right) d\eta =$$

$$= \frac{2}{\tau h^2} \int_0^h (h-\xi) d\xi \int_{t-\tau}^{t} d\eta \int_{\eta}^{t} \frac{\partial^2 u(\xi,\eta_1)}{\partial \eta_1^2} d\eta_1 + \frac{2}{\tau h^2} \int_0^h (h-\xi) d\xi \int_{t-\tau}^{t} d\eta \int_0^{\xi} \frac{\partial^2 u(\xi_1,\eta)}{\partial \xi_1 \partial \eta} d\xi_1,$$

which leads to the estimate

$$|\psi(0,t)| \leq \frac{2}{\tau h^2} \int_0^h (h-\xi) d\xi \int_{t-\tau}^{t} d\eta \int_{t-\tau}^{t} \left| \frac{\partial^2 u(\xi,\eta_1)}{\partial \eta_1^2} \right| d\eta_1 +$$

$$+ \frac{2}{\tau h^2} \int_0^h (h-\xi) d\xi \int_{t-\tau}^{t} d\eta \int_0^h \left| \frac{\partial^2 u(\xi_1,\eta)}{\partial \xi_1 \partial \eta} \right| d\xi_1 \leq$$

$$\leq \frac{2}{\tau h^2} h\tau \sqrt{h\tau} \left(\int_0^h d\xi \int_{t-\tau}^{t} \left(\frac{\partial^2 u(\xi,\eta_1)}{\partial \eta_1^2} \right)^2 d\eta_1 \right)^{1/2} +$$

$$+ \frac{2}{\tau h^2} \frac{h^2}{2} \sqrt{h\tau} \left(\int_{t-\tau}^{t} d\eta \int_0^h \left(\frac{\partial^2 u(\xi_1,\eta)}{\partial \xi_1 \partial \eta} \right)^2 d\xi_1 \right)^{1/2} =$$

$$= \frac{2\sqrt{\tau}}{\sqrt{h}} \left(\int_0^h d\xi \int_{t-\tau}^{t} \left(\frac{\partial^2 u(\xi,\eta_1)}{\partial \eta_1^2} \right)^2 d\eta_1 \right)^{1/2} + \frac{\sqrt{h}}{\sqrt{\tau}} \left(\int_0^h d\xi_1 \int_{t-\tau}^{t} \left(\frac{\partial^2 u(\xi_1,\eta)}{\partial \xi_1 \partial \eta} \right)^2 d\eta \right)^{1/2}.$$

Now, we can find

$$\|\psi(\cdot,t)\|^2 = \sum_{x\in\omega} h\psi^2(x,t) + \frac{h}{2}\psi^2(0,t) \leq$$

$$\leq 4\tau \sum_{x\in\omega} \int_{x-h}^{x+h} d\xi \int_{t-\tau}^{t} \left(\frac{\partial^2 u(\xi,\eta_1)}{\partial \eta_1^2}\right)^2 d\eta_1 + \frac{4h^2}{\tau} \sum_{x\in\omega} \int_{x-h}^{x+h} d\xi_1 \int_{t-\tau}^{t} \left(\frac{\partial^2 u(\xi_1,\eta)}{\partial \xi_1 \partial \eta}\right)^2 d\eta +$$

$$+ 4\tau \int_{0}^{h} d\xi \int_{t-\tau}^{t} \left(\frac{\partial^2 u(\xi,\eta_1)}{\partial \eta_1^2}\right)^2 d\eta_1 + \frac{h^2}{\tau} \int_{0}^{h} d\xi_1 \int_{t-\tau}^{t} \left(\frac{\partial^2 u(\xi_1,\eta)}{\partial \xi_1 \partial \eta}\right)^2 d\eta \leq$$

$$\leq 8\tau \int_{0}^{1} d\xi \int_{t-\tau}^{t} \left(\frac{\partial^2 u(\xi,\eta_1)}{\partial \eta_1^2}\right)^2 d\eta_1 + \frac{8h^2}{\tau} \int_{0}^{1} d\xi_1 \int_{t-\tau}^{t} \left(\frac{\partial^2 u(\xi_1,\eta)}{\partial \xi_1 \partial \eta}\right)^2 d\eta,$$

and thus complete the proof of the lemma.

Applying Lemmas 2.1 and 2.2, we finally come to the proposition.

THEOREM 2.1.– Let the solution of problem [2.1] satisfy the conditions $\frac{\partial^2 u}{\partial t^2}, \frac{\partial^2 u}{\partial x \partial t} \in L_2(Q_T)$. Then, the accuracy of the finite-difference scheme is characterized by the weighted estimate

$$\left(\sum_{\eta=\tau}^{t} \tau\left(\frac{z(x,\eta)}{1-x}\right)^2\right)^{1/2} \leq 4\sqrt{2} \left\{\tau\left(\int_{0}^{t} \left\|\frac{\partial^2 u}{\partial \eta^2}(\cdot,\eta)\right\|^2 d\eta\right)^{1/2} + h\left(\int_{0}^{t} \left\|\frac{\partial^2 u}{\partial x \partial \eta}(\cdot,\eta)\right\|^2 d\eta\right)^{1/2}\right\},$$

$$(x,t) \in \omega^- \times \omega_\tau.$$

PROOF.– From the estimates [2.5] and [2.7], we have

$$\left(\sum_{\eta=\tau}^{t} \tau\left(\frac{z(x,\eta)}{1-x}\right)^2\right)^{1/2} \leq$$

$$\leq 2\left\{8\tau^2 \int_{0}^{1} d\xi \int_{0}^{t} \left(\frac{\partial^2 u(\xi,\eta_1)}{\partial \eta_1^2}\right)^2 d\eta_1 + 8h^2 \int_{0}^{1} d\xi_1 \int_{0}^{t} \left(\frac{\partial^2 u(\xi_1,\eta)}{\partial \xi_1 \partial \eta}\right)^2 d\eta\right\}^{1/2} \leq$$

$$\leq 4\sqrt{2}\tau\left(\int_0^1 d\xi\int_0^t\left(\frac{\partial^2 u(\xi,\eta_1)}{\partial\eta_1^2}\right)^2 d\eta_1\right)^{1/2} + 4\sqrt{2}h\left(\int_0^1 d\xi_1\int_0^t\left(\frac{\partial^2 u(\xi_1,\eta)}{\partial\xi_1\partial\eta}\right)^2 d\eta\right)^{1/2}.$$

Here, we used the inequality $\sqrt{a+b} \leq \sqrt{a}+\sqrt{b}$. The theorem is proved.

2.1.3. Accuracy with the initial effect

We begin with the following supporting proposition.

LEMMA 2.3.– *For the error $z(x,t)$, the following estimate holds true:*

$$\left\|\frac{z(\cdot,t)}{[(2+\ln^2 t)t]^{1/2}}\right\| \leq 2^{\pi/\sqrt{2}}\left(\sum_{\eta=\tau}^{t}\tau\|\psi(\cdot,\eta)\|^2\right)^{1/2}, \quad t \in \omega_\tau.$$

PROOF.– Multiplying equation [2.4] by $z(\cdot,t)$ in H, we have

$$\sum_{x\in\omega}hz_{\bar{t}}(x,t)z(x,t) + \frac{h}{2}z_{\bar{t}}(0,t)z(0,t) - \sum_{x\in\omega}hz_{\bar{x}x}(x,t)z(x,t) -$$

$$-\frac{h}{2}\cdot\frac{2}{h}z_x(0,t)z(0,t) = \sum_{x\in\omega}h\psi(x,t)z(x,t) + \frac{h}{2}\psi(0,t)z(0,t).$$

For the first two summands in the left-hand side, we find

$$\sum_{x\in\omega}hz_{\bar{t}}(x,t)z(x,t) + \frac{h}{2}z_{\bar{t}}(0,t)z(0,t) =$$

$$= \sum_{x\in\omega}h\frac{z(x,t)-z(x,t-\tau)}{\tau}\cdot\frac{z(x,t)-z(x,t-\tau)+z(x,t)+z(x,t-\tau)}{2} +$$

$$+\frac{h}{2}\frac{z(0,t)-z(0,t-\tau)}{\tau}\cdot\frac{z(0,t)-z(0,t-\tau)+z(0,t)+z(0,t-\tau)}{2} =$$

$$= \sum_{x\in\omega}\frac{h\tau}{2}z_{\bar{t}}^2(x,t) + \sum_{x\in\omega}\frac{h}{2\tau}\left(z^2(x,t)-z^2(x,t-\tau)\right) +$$

$$+\frac{h\tau}{4}z_{\bar{t}}^2(0,t)+\frac{h}{4\tau}\left(z^2(0,t)-z^2(0,t-\tau)\right),$$

and for the last two summands, we have

$$-\sum_{x\in\omega}hz_{\bar{x}x}(x,t)z(x,t)-z_x(0,t)z(0,t)=$$

$$=\sum_{x\in\omega^-}hz_{\bar{x}}^2(x,t)-z_{\bar{x}}(1,t)z(1,t)+z_{\bar{x}}(h,t)z(0,t)-z_x(0,t)z(0,t)=\sum_{x\in\omega^-}hz_{\bar{x}}^2(x,t).$$

After some easy transformations, we obtain the inequality

$$\sum_{x\in\omega}\frac{h\tau}{2}z_{\bar{t}}^2(x,t)+\sum_{x\in\omega}\frac{h}{2\tau}\left(z^2(x,t)-z^2(x,t-\tau)\right)+\frac{h\tau}{4}z_{\bar{t}}^2(0,t)+$$

$$+\frac{h}{4\tau}\left(z^2(0,t)-z^2(0,t-\tau)\right)+\sum_{x\in\omega^-}hz_{\bar{x}}^2(x,t)=\left(\psi(\cdot,t),z(\cdot,t)\right).$$

Multiplying both sides of it by τ and making summation from $\eta=\tau$ to $\eta=t$, we have

$$\frac{\tau}{2}\sum_{\eta=\tau}^{t}\tau\|z_{\bar{\eta}}(\cdot,\eta)\|^2+\frac{1}{2}\sum_{x\in\omega}h\left(z^2(x,t)-z^2(x,0)\right)+\frac{h}{4}\left(z^2(0,t)-z^2(0,0)\right)+$$

$$+\sum_{\eta=\tau}^{t}\tau\sum_{x\in\omega^-}hz_{\bar{x}}^2(x,\eta)=\sum_{\eta=\tau}^{t}\tau\left(\psi(\cdot,\eta),z(\cdot,\eta)\right),$$

i.e.

$$\frac{\tau}{2}\sum_{\eta=\tau}^{t}\tau\|z_{\bar{\eta}}(\cdot,\eta)\|^2+\frac{1}{2}\|z(\cdot,t)\|^2+\sum_{\eta=\tau}^{t}\tau\sum_{x\in\omega^-}hz_{\bar{x}}^2(x,\eta)=$$

$$=\sum_{\eta=\tau}^{t}\tau\left(\psi(\cdot,\eta)\sqrt{(2+\ln^2\eta)\eta},\frac{z(\cdot,\eta)}{\sqrt{(2+\ln^2\eta)\eta}}\right).$$

Taking into account the relation $(y,v) \le \|y\| \cdot \|v\| \le \frac{1}{2}\left(\|y\|^2 + \|v\|^2\right)$, we have

$$\frac{1}{2}\|z(\cdot,t)\|^2 \le \frac{1}{2}\sum_{\eta=\tau}^{t}\tau\|\psi(\cdot,\eta)\|^2 (2+\ln^2\eta)\eta + \frac{1}{2}\sum_{\eta=\tau}^{t}\tau\frac{\|z(\cdot,\eta)\|^2}{(2+\ln^2\eta)\eta}.$$

Dividing by $(2+\ln^2 t)t$, we come to the inequality

$$\frac{\|z(\cdot,t)\|^2}{(2+\ln^2 t)t} \le \sum_{\eta=\tau}^{t}\tau\|\psi(\cdot,\eta)\|^2 \frac{(2+\ln^2\eta)\eta}{(2+\ln^2 t)t} + \sum_{\eta=\tau}^{t}\tau\frac{\|z(\cdot,\eta)\|^2}{(2+\ln^2\eta)\eta(2+\ln^2 t)t} \le$$

$$\le \frac{\|z(\cdot,t)\|^2}{(2+\ln^2 t)t} \le \sum_{\eta=\tau}^{t}\tau\|\psi(\cdot,\eta)\|^2 + \sum_{\eta=\tau}^{t}\tau\frac{1}{(2+\ln^2\eta)\eta}\frac{\|z(\cdot,\eta)\|^2}{(2+\ln^2\eta)\eta}.$$

Denoting here $Z(t) = \dfrac{\|z(\cdot,t)\|^2}{(2+\ln^2 t)t}$ and $g(t) = \dfrac{1}{(2+\ln^2 t)t}$, we get the inequality

$$Z(t) \le \sum_{\eta=\tau}^{t}\tau\|\psi(\cdot,\eta)\|^2 + \sum_{\eta=\tau}^{t}\tau g(\eta)Z(\eta). \qquad [2.8]$$

To solve [2.8], we consider the auxiliary equation

$$v(t) = \sum_{\eta=\tau}^{t}\tau\|\psi(\cdot,\eta)\|^2 + \sum_{\eta=\tau}^{t}\tau g(\eta)v(\eta), \quad v(0) = 0,$$

From which we obtain

$$v(t) - v(t-\tau) = \tau\|\psi(\cdot,t)\|^2 + \tau g(t)v(t)$$

and therefore

$$v(t) = \frac{v(t-\tau)}{1-\tau g(t)} + \frac{\tau}{1-\tau g(t)}\|\psi(\cdot,t)\|^2 \le 4^{\tau g(t)}\left(v(t-\tau) + \tau\|\psi(\cdot,t)\|^2\right) \le$$

$$\le 4^{\tau g(t)}\left[4^{\tau g(t-\tau)}\left(v(t-\tau) + \tau\|\psi(\cdot,t-\tau)\|^2\right) + \tau\|\psi(\cdot,t)\|^2\right] \le$$

$$\leq 4^{\tau g(t)+\tau g(t-\tau)}\left[v(t-2\tau)+\tau\|\psi(\cdot,t-\tau)\|^2+\tau\|\psi(\cdot,t)\|^2\right]\leq\cdots\leq$$

$$\leq 4^{\tau\sum_{\eta=\tau}^{t}g(\eta)}\sum_{\eta=\tau}^{t}\tau\|\psi(\cdot,\eta)\|^2 \leq 4^{\int_0^t g(\eta)d\eta}\sum_{\eta=\tau}^{t}\tau\|\psi(\cdot,\eta)\|^2 =$$

$$= 4^{\int_0^t \frac{d\eta}{(2+\ln^2\eta)\eta}}\sum_{\eta=\tau}^{t}\tau\|\psi(\cdot,\eta)\|^2 = 4^{\int_0^t \frac{d\eta}{(2+\ln^2\eta)\eta}}\sum_{\eta=\tau}^{t}\tau\|\psi(\cdot,\eta)\|^2 =$$

$$= 4^{\frac{1}{\sqrt{2}}\left(\operatorname{arctg}\frac{\ln t}{\sqrt{2}}+\frac{\pi}{2}\right)}\sum_{\eta=\tau}^{t}\tau\|\psi(\cdot,\eta)\|^2 \leq 2^{\pi\sqrt{2}}\sum_{\eta=\tau}^{t}\tau\|\psi(\cdot,\eta)\|^2.$$

We take into account that the function $\varphi(x)=(1-x)^{-1/x}=e^{-\ln(1-x)/x}$ monotonically increases on the interval $(0,1)$ and some useful inequalities hold true:

$$e < e^{-\ln(1-x)/x} \leq e^{\ln 4} = 4 \quad (0 < x \leq 1/2),$$

$$0 < \tau g(\eta) = \frac{\tau}{(2+\ln^2\eta)\eta} \leq \frac{1}{2+\ln^2\eta} \leq \frac{1}{2} \quad (\eta \geq \tau),$$

$$\left(1-\tau g(\eta)\right)^{-1}=\left[\left(1-\tau g(\eta)\right)^{-\frac{1}{\tau g(\eta)}}\right]^{\tau g(\eta)} \leq 4^{\tau g(\eta)} \quad (\eta \geq \tau).$$

Then, estimate [2.8] yields the estimate

$$Z(t) \leq 2^{\pi\sqrt{2}}\sum_{\eta=\tau}^{t}\tau\|\psi(\cdot,\eta)\|^2$$

and therefore completes the proof.

Finally, we arrive at another main result of this section.

THEOREM 2.2.– *Let the solution of problem [2.1] satisfy the conditions* $\dfrac{\partial^2 u}{\partial t^2}, \dfrac{\partial^2 u}{\partial x \partial t} \in L_2(Q_T)$. *Then, the accuracy of the finite-difference scheme [2.3] is characterized by the weighted estimate*

$$\left\| \frac{z(\cdot,t)}{\left[(2+\ln^2 t)t\right]^{1/2}} \right\| \leq 2^{\pi/\sqrt{2}} 2\sqrt{2} \left[\tau \left(\int_0^1 d\xi \int_0^t \left(\frac{\partial^2 u(\xi,\eta_1)}{\partial \eta_1^2} \right)^2 d\eta_1 \right)^{1/2} + \right.$$

$$\left. + h \left(\int_0^1 d\xi_1 \int_0^t \left(\frac{\partial^2 u(\xi_1,\eta)}{\partial \xi_1 \partial \eta} \right)^2 d\eta \right)^{1/2} \right], \quad t \in \omega_\tau.$$

PROOF.– Applying Lemmas 2.2 and 2.3 and using the inequality $\sqrt{a+b} \leq \sqrt{a}+\sqrt{b}$, we obtain

$$\left\| \frac{z(\cdot,t)}{\left[(2+\ln^2 t)t\right]^{1/2}} \right\| \leq 2^{\pi/\sqrt{2}} \left(\sum_{\eta=\tau}^t \tau \|\psi(\cdot,\eta)\|^2 \right)^{1/2} \leq$$

$$\leq 2^{\pi/\sqrt{2}} \left[2\sqrt{2}\tau \left(\int_0^1 d\xi \int_0^t \left(\frac{\partial^2 u(\xi,\eta_1)}{\partial \eta_1^2} \right)^2 d\eta_1 \right)^{1/2} + 2\sqrt{2}h \left(\int_0^1 d\xi_1 \int_0^t \left(\frac{\partial^2 u(\xi_1,\eta)}{\partial \xi_1 \partial \eta} \right)^2 d\eta \right)^{1/2} \right].$$

The theorem is proved.

2.1.4. *Conclusion*

The weighted a priori estimate obtained in Theorem 2.1 shows that the accuracy of scheme [2.3] is $O(h(\tau+h))$ near the right side $x=1$ of the space–time rectangle $Q_T = (0,1) \times (0,T)$, while it is $O(\tau+h)$ further away from it.

The weighted a priori estimate obtained in Theorem 2.2 shows that the accuracy of scheme [2.3] is $O\left(\sqrt{(2+\ln^2 \tau)\tau}\,(\tau+h)\right)$ near the bottom side $t=0$ of the space–time rectangle.

2.2. A standard finite-difference scheme for the two-dimensional heat equation with mixed boundary conditions

We consider the problem

$$\frac{\partial u(x,t)}{\partial t} = \Delta u(x,t) + f(x,t), \quad (x,t) \in D_T = D \times (0,T),$$
$$\frac{\partial u(x,t)}{\partial x_1} = 0, \quad (x,t) \in \Gamma_{-1} \times (0,T), \qquad [2.9]$$
$$u(x,t) = 0, \quad (x,t) \in (\Gamma \setminus \Gamma_{-1}) \times (0,T),$$
$$u(x,0) = \varphi(x), \quad x \in D,$$

where $x = (x_1, x_2)$, $\Delta = \dfrac{\partial^2}{\partial x_1^2} + \dfrac{\partial^2}{\partial x_2^2}$ is the Laplace operator in a Cartesian coordinate system, $D = \{x = (x_1, x_2): 0 < x_\alpha < l_\alpha, \alpha = 1,2\}$ is a rectangle, $\Gamma = \partial D$ is a boundary of D, $\Gamma_{-1} = \{x = (x_1, x_2): x_1 = 0, 0 < x_2 < l_2\}$ is the left side of D, $f(x,t)$ and $\varphi(x)$ are given functions, $u(x,t)$ is an unknown solution.

2.2.1. *Discretization of the differential problem and properties of the finite-difference operators*

Using the traditional notation of the theory of finite-difference schemes (Samarskii 2001), we introduce the sets of nodes:

$$\omega_\alpha = \{i_\alpha h_\alpha, \; i_\alpha = 1, \ldots, N_\alpha - 1, \; h_\alpha = l_\alpha / N_\alpha\}, \; N_\alpha \geq 2 \text{ is an integer number,}$$

$$\omega_\alpha^- = \omega_\alpha \cup \{0\}, \quad \omega_\alpha^+ = \omega_\alpha \cup \{1\}, \quad \bar{\omega}_\alpha = \omega_\alpha \cup \{0\} \cup \{1\},$$

$$\omega = \omega_1 \times \omega_2, \quad \bar{\omega} = \bar{\omega}_1 \times \bar{\omega}_2, \quad \gamma = \bar{\omega} \setminus \omega, \qquad [2.10]$$

$$\gamma_{-\alpha} = \{x_\alpha = 0, \; x_{3-\alpha} \in \omega_{3-\alpha}\}, \quad \gamma_{+\alpha} = \{x_\alpha = l_\alpha, \; x_{3-\alpha} \in \omega_{3-\alpha}\}, \quad \alpha = 1,2,$$

$$\omega_\tau = \{t_j = j\tau, \; j = 1, 2, \ldots, M-1, \; \tau = T/M\}, \; M \geq 2 \text{ is an integer number.}$$

Applying the operators (Samarskii et al. 1987)

$$T_2 v(x_1, x_2) = \frac{1}{h_2^2} \int_{x_2-h_2}^{x_2+h_2} (h_2 - |x_2 - \xi_2|) v(x_1, \xi_2) d\xi_2, \quad x \in \omega \cup \gamma_{-1},$$

$$T_1 v(x_1, x_2) = \begin{cases} \dfrac{1}{h_1^2} \displaystyle\int_{x_1-h_1}^{x_1+h_1} (h_1 - |x_1 - \xi_1|) v(\xi_1, x_2) d\xi_1, & x \in \omega, \\ \dfrac{2}{h_1^2} \displaystyle\int_0^{h_1} (h_1 - \xi_1) v(\xi_1, x_2) d\xi_1, & x \in \gamma_{-1}, \end{cases} \quad [2.10]$$

we approximate problem [2.9] by the finite-difference scheme

$$\begin{aligned} y_{\bar{t}}(x,t) - \Lambda y(x,t) &= T_1 T_2 f(x,t), & (x,t) &\in (\omega \cup \gamma_{-1}) \times \omega_\tau, \\ y(x,t) &= 0, & (x,t) &\in (\gamma \setminus \gamma_{-1}) \times \omega_\tau, \\ y(x,0) &= \varphi(x), & x &\in \omega, \end{aligned} \quad [2.11]$$

with $\Lambda = \Lambda_1 + \Lambda_2$, $\Lambda_1 y = \begin{cases} y_{\bar{x}_1 x_1}, & x \in \omega, \\ \dfrac{2}{h_1} y_{x_1}, & x \in \gamma_{-1}, \end{cases}$ $\Lambda_2 y = y_{\bar{x}_2 x_2}, \quad x \in \omega \cup \gamma_{-1}.$

For the error $z(x) = y(x) - u(x)$, we have the problem

$$\begin{aligned} z_{\bar{t}}(x,t) - \Lambda z(x,t) &= \psi(x,t), & (x,t) &\in (\omega \cup \gamma_{-1}) \times \omega_\tau, \\ z(x,t) &= 0, & (x,t) &\in (\gamma \setminus \gamma_{-1}) \times \omega_\tau, \\ z(x,0) &= 0, & x &\in \omega, \end{aligned} \quad [2.12]$$

where $\psi = T_1 T_2 f - u_{\bar{t}} + \Lambda u = \Lambda_1 \eta_1 + \Lambda_2 \eta_2 + \eta_3$ is the approximation error:

$$\eta_\alpha(x,t) = u(x,t) - T_{3-\alpha} u(x,t), \ \alpha = 1, 2, \ \eta_3(x,t) = \frac{\partial T u(x,t)}{\partial t} - u_{\bar{t}}(x,t).$$

Let H be a set of mesh functions defined on $\bar{\omega}$ and equal to zero on $\gamma \setminus \gamma_{-1}$. We introduce the inner product and the associate norm in H by the formulas

$$(y,v) = \sum_{x \in \omega} h_1 h_2 y(x) v(x) + \frac{h_1}{2} \sum_{x \in \gamma_{-1}} h_2 y(x) v(x),$$

$$\|v\|=\|v\|_{L_2(\omega)}=\sqrt{(v,v)}=\left(\sum_{x\in\omega}h_1h_2v^2(x)+\frac{h_1}{2}\sum_{x\in\gamma_{-1}}h_2v^2(x)\right)^{1/2}.$$

We consider the finite-difference operators acting in H:

$$A_\alpha, A: H \to H, \quad A_\alpha = -\Lambda_\alpha, \quad \alpha=1,2, \quad A = A_1 + A_2 = -\Lambda.$$

(If necessary, functions defined on ω are set equal to zero for $x \in \gamma \setminus \gamma_{-1}$ and equal to arbitrary values for $x \in \gamma_{-1}$.)

As in Lemma 1.1, we can verify that the operator A is symmetric and positive definite in H. Therefore, there exists the inverse operator $A^{-1}: H \to H$.

We begin with the key supporting proposition.

LEMMA 2.4.– *The following estimate holds true:*

$$\|A^{-1}B_k v\| \le \frac{1}{\sqrt{2}}\|v\|_k \quad \forall v \in H_k \quad (k=1,2). \tag{2.13}$$

PROOF.– We find

$$\|Ay\|^2 = (Ay, Ay) = \sum_{x\in\omega} h_1 h_2 (-y_{\bar{x}_1 x_1} - y_{\bar{x}_2 x_2})^2 + \frac{h_1}{2}\sum_{x\in\gamma_{-1}} h_2 \left(-\frac{2}{h_1}y_{x_1} - y_{\bar{x}_2 x_2}\right)^2 =$$

$$= \sum_{x\in\omega} h_1 h_2 y_{\bar{x}_1 x_1}^2 + \frac{h_1}{2}\sum_{x\in\gamma_{-1}} h_2 \left(\frac{2}{h_1}y_{x_1}\right)^2 + \sum_{x\in\omega} h_1 h_2 y_{\bar{x}_2 x_2}^2 +$$

$$+ \frac{h_1}{2}\sum_{x\in\gamma_{-1}} h_2 y_{\bar{x}_2 x_2}^2 + 2\sum_{x\in\omega_1^- \times \omega_2^-} h_1 h_2 y_{\bar{x}_1 x_2}^2 = \|A_1 y\|^2 + \|A_2 y\|^2 + 2\|B_1^* y\|_1^2$$

and similarly

$$\|Ay\|^2 = \sum_{x\in\omega} h_1 h_2 y_{\bar{x}_1 x_1}^2 + \frac{h_1}{2}\sum_{x\in\gamma_{-1}} h_2 \left(\frac{2}{h_1}y_{x_1}\right)^2 + \sum_{x\in\omega} h_1 h_2 y_{\bar{x}_2 x_2}^2 +$$

$$+\frac{h_1}{2}\sum_{x\in\gamma_{-1}}h_2 y_{\bar{x}_2 x_2}^2 + 2\sum_{x\in\omega_1^-\times\omega_2^+}h_1 h_2 y_{\bar{x}_1\bar{x}_2}^2 = \|A_1 y\|^2 + \|A_2 y\|^2 + 2\|B_2^* y\|_2^2,$$

where B_1^* and B_2^* are the same operators as in Lemma 1.2.

Repeating the considerations of Lemma 1.2, we define the operators B_1 and B_2 which are the adjoint operators of B_1^* and B_2^* respectively.

Applying the main lemma from Samarskii et al. (1987, p. 54) to the operators A, B_1, B_2, we arrive at estimate [2.13]. The lemma is proved.

2.2.2. Discrete Green's function

Let $G(x,\xi)$ be Green's function of the finite-difference problem [2.12]:

$$-G_{\bar{\xi}_1\xi_1}(x,\xi) - G_{\bar{\xi}_2\xi_2}(x,\xi) = \frac{\delta(x_1,\xi_1)\delta(x_2,\xi_2)}{h_1 h_2}, \quad \xi\in\omega,$$

$$-\frac{2}{h_1}G_{\xi_1}(x,\xi) - G_{\bar{\xi}_2\xi_2}(x,\xi) = \frac{2}{h_1}\frac{\delta(x_1,\xi_1)\delta(x_2,\xi_2)}{h_2}, \quad \xi\in\gamma_{-1}, \qquad [2.14]$$

$$G(x,\xi) = 0, \quad \xi\in\gamma\setminus\gamma_{-1},$$

where $\delta(m,n)$ is the Kronecker delta symbol and $\xi = (\xi_1,\xi_2)$.

LEMMA 2.5.– *For Green's function $G(x,\xi)$, the following estimate holds true:*

$$\|G(x,\cdot)\| \leq \frac{1}{\sqrt{2}}\rho(x), \quad x\in\omega,$$

where $\rho(x) = \min\left\{\sqrt{(l_1-x_1)(l_2-x_2)},\ \sqrt{(l_1-x_1)x_2}\right\}$.

PROOF.– Our reasoning and transformations are basically the same as in Lemma 1.3, which allows us to omit some details.

Using the Heaviside step function $H(s) = \begin{cases} 1, & s \geq 0, \\ 0, & s < 0, \end{cases}$ we rewrite problem [2.14] as follows:

$$-G_{\bar{\xi}_1\xi_1}(x,\xi) - G_{\bar{\xi}_2\xi_2}(x,\xi) = \left(H(\xi_1 - x_1)H(\xi_2 - x_2)\right)_{\bar{\xi}_1\bar{\xi}_2}, \quad \xi \in \omega,$$

$$-\frac{2}{h_1}G_{\xi_1}(x,\xi) - G_{\bar{\xi}_2\xi_2}(x,\xi) = \frac{2}{h_1}\left(H(\xi_1 - x_1)H(\xi_2 - x_2)\right)_{\bar{\xi}_2}, \quad \xi \in \gamma_{-1},$$

$$G(x,\xi) = 0, \quad \xi \in \gamma \setminus \gamma_{-1},$$

or in the operator form:

$$A_\xi G(x,\xi) = -B_{1\xi}\left(H(\xi_1 - x_1)H(\xi_2 - x_2)\right).$$

Using estimate [2.13], we find

$$\|G(x,\cdot)\| = \|-A_\xi^{-1} B_{1\xi}\left(H(\cdot - x_1)H(\cdot - x_2)\right)\| \leq \frac{1}{\sqrt{2}}\sqrt{(l_1 - x_1)(l_2 - x_2)}. \quad [2.15]$$

Next, representing problem [2.14] in this way:

$$-G_{\bar{\xi}_1\xi_1}(x,\xi) - G_{\bar{\xi}_2\xi_2}(x,\xi) = -\left(H(\xi_1 - x_1)H(x_2 - \xi_2)\right)_{\bar{\xi}_1\bar{\xi}_2}, \quad \xi \in \omega,$$

$$-\frac{2}{h_1}G_{\xi_1}(x,\xi) - G_{\bar{\xi}_2\xi_2}(x,\xi) = -\frac{2}{h_1}\left(H(\xi_1 - x_1)H(x_2 - \xi_2)\right)_{\bar{\xi}_2}, \quad \xi \in \gamma_{-1},$$

$$G(x,\xi) = 0, \quad \xi \in \gamma \setminus \gamma_{-1},$$

or as the operator equation

$$A_\xi G(x,\xi) = B_{2\xi}\left(H(\xi_1 - x_1)H(x_2 - \xi_2)\right),$$

we apply estimate [2.13] and obtain

$$\|G(x,\cdot)\| = \|A_\xi^{-1} B_{2\xi}\left(H(\cdot - x_1)H(x_2 - \cdot)\right)\| \leq \frac{1}{\sqrt{2}}\sqrt{(l_1 - x_1)x_2}. \quad [2.16]$$

Estimates [2.15] and [2.16] prove the lemma.

2.2.3. Accuracy with the boundary effect

To prove the main theorem, we first need some auxiliary results.

LEMMA 2.6.– *For the error* $z(x,t)$, *the following estimate holds true:*

$$\sum_{\eta=\tau}^{t} \tau \frac{|z(x,\eta)|^2}{\rho^2(x)} \leq 2\sum_{\eta=\tau}^{t} \tau \|\psi(\cdot,\eta)\|^2 . \qquad [2.17]$$

PROOF.– Making use of Green's function $G(x,\xi)$, we can represent the solution $z(x,t)$ of problem [2.12] in the form:

$$z(x,t) = \big(G(x,\cdot), \psi(\cdot,t) - z_{\overline{t}}(\cdot,t)\big), \quad (x,t) \in (\omega \cup \gamma_{-1}) \times \omega_\tau .$$

Applying here Lemma 2.5, we have

$$|z(x,t)| = \big|\big(G(x,\cdot), \psi(\cdot,t) - z_{\overline{t}}(\cdot,t)\big)\big| \leq \|G(x,\cdot)\| \cdot \|\psi(\cdot,t) - z_{\overline{t}}(\cdot,t)\| \leq$$

$$\leq \|G(x,\cdot)\| \cdot \big(\|\psi(\cdot,t)\| + \|z_{\overline{t}}(\cdot,t)\|\big) \leq \frac{1}{\sqrt{2}} \rho(x) \big(\|\psi(\cdot,t)\| + \|z_{\overline{t}}(\cdot,t)\|\big).$$

After squaring and summing, we obtain

$$\frac{|z(x,t)|^2}{\rho^2(x)} \leq \frac{1}{2} \big(\|\psi(\cdot,t)\| + \|z_{\overline{t}}(\cdot,t)\|\big)^2 \leq \|\psi(\cdot,t)\|^2 + \|z_{\overline{t}}(\cdot,t)\|^2 ,$$

$$\sum_{\eta=\tau}^{t} \tau \frac{|z(x,\eta)|^2}{\rho^2(x)} \leq \sum_{\eta=\tau}^{t} \tau \big(\|\psi(\cdot,\eta)\|^2 + \|z_{\overline{t}}(\cdot,\eta)\|^2\big). \qquad [2.18]$$

Similarly to the one-dimensional case (see Lemma 2.1), we obtain from equation [2.12] the relation

$$\|z_{\overline{t}}(\cdot,t)\|^2 - 2\big(z_{\overline{t}}(\cdot,t), \Lambda z(\cdot,t)\big) + \|\Lambda z(\cdot,t)\|^2 = \|\psi(\cdot,t)\|^2 .$$

For the second summand in the left-hand side, we have

$$2\sum_{\eta=\tau}^{t} \tau \big(z_{\overline{\eta}}(\cdot,\eta), -\Lambda z(\cdot,\eta)\big) =$$

$$= 2\sum_{\eta=\tau}^{t} \tau \Bigg[\sum_{\xi \in \omega} h_1 h_2 z_{\overline{\eta}}(\xi,\eta)\big(-z_{\overline{\xi}_1 \xi_1}(\xi,\eta) - z_{\overline{\xi}_2 \xi_2}(\xi,\eta)\big) +$$

$$+\frac{h_1}{2}\sum_{\xi\in\gamma_{-1}}h_2 z_{\bar{\eta}}(\xi,\eta)\left(-\frac{2}{h_1}z_{\xi_1}(\xi,\eta)-z_{\bar{\xi}_2\xi_2}(\xi,\eta)\right)\right]=$$

$$=\tau\sum_{\xi\in\omega_1^+\times\omega_2}h_1 h_2\sum_{\eta=\tau}^{t}\tau z_{\bar{\xi}_1\bar{\eta}}^2(\xi,\eta)+\sum_{\xi\in\omega_1^+\times\omega_2}h_1 h_2 z_{\bar{\xi}_1}^2(\xi,t)+$$

$$+\tau\sum_{\xi\in\omega_1\times\omega_2^+}h_1 h_2\sum_{\eta=\tau}^{t}\tau z_{\bar{\xi}_2\bar{\eta}}^2(\xi,\eta)+\sum_{\xi\in\omega_1\times\omega_2^+}h_1 h_2 z_{\bar{\xi}_2}^2(\xi,t)+$$

$$+\frac{h_1}{2}\tau\sum_{\substack{\xi_2\in\omega_2^+ \\ (\xi_1=0)}}h_2\sum_{\eta=\tau}^{t}\tau z_{\bar{\xi}_2\bar{\eta}}^2(\xi,\eta)+\frac{h_1}{2}\sum_{\substack{\xi_2\in\omega_2^+ \\ (\xi_1=0)}}h_2 z_{\bar{\xi}_2}^2(\xi,t)\geq 0,$$

and therefore, $\|z_{\bar{t}}(\cdot,\eta)\|\leq\|\psi(\cdot,\eta)\|$.

Then, inequality [2.18] leads to the estimate

$$\sum_{\eta=\tau}^{t}\tau\frac{|z(x,\eta)|^2}{\rho^2(x)}\leq\sum_{\eta=\tau}^{t}\tau\left(\|\psi(\cdot,\eta)\|^2+\|z_{\bar{t}}(\cdot,\eta)\|^2\right)\leq 2\sum_{\eta=\tau}^{t}\tau\|\psi(\cdot,\eta)\|^2,$$

and proves the lemma.

LEMMA 2.7.– *Let the solution* $u(x_1,x_2,t)$ *of problem [2.9] satisfy the conditions*

$$\frac{\partial^4 u}{\partial x_1^2\partial x_2\partial t},\frac{\partial^4 u}{\partial x_1^2\partial x_2^2},\frac{\partial^4 u}{\partial x_1\partial x_2^2\partial t},\frac{\partial^3 u}{\partial x_1 x_2^2},\frac{\partial^3 u}{x_1^2\partial t},\frac{\partial^3 u}{x_2^2\partial t},\frac{\partial^3 u}{\partial x_1\partial x_2\partial t},\frac{\partial^2 u}{\partial t^2},\frac{\partial^2 u}{\partial x_1\partial t}\in L_2(D_T).$$

Then, for the approximation error $\psi(x,t)$, *the following estimate holds true*:

$$\sum_{\eta=\tau}^{t}\tau\|\psi(\cdot,\eta)\|^2\leq M_1\left(\tau^2 h_2^2+h_2^4+\tau^2 h_1^2+h_1^4+\tau^2+h_1^2 h_2^2+h_1^2\right),$$

where M_1 *is a positive constant containing the norms of the above derivatives of the solution* $u(x,t)$.

PROOF.– For the squared norm of the approximation error $\psi = \Lambda_1\eta_1 + \Lambda_2\eta_2 + \eta_3$, we get

$$\|\psi(\cdot,t)\|^2 = \|\Lambda_1\eta_1(\cdot,t) + \Lambda_2\eta_1(\cdot,t) + \eta_3(\cdot,t)\|^2 \le$$

$$\le 3\left(\|\Lambda_1\eta_1(\cdot,t)\|^2 + \|\Lambda_2\eta_1(\cdot,t)\|^2 + \|\eta_3(\cdot,t)\|^2\right),$$

then

$$\sum_{\eta=\tau}^{t}\tau\|\psi(\cdot,\eta)\|^2 \le 3\sum_{\eta=\tau}^{t}\tau\left(\|\Lambda_1\eta_1(\cdot,t)\|^2 + \|\Lambda_2\eta_1(\cdot,t)\|^2 + \|\eta_3(\cdot,t)\|^2\right). \quad [2.19]$$

Now we consider each summand in the right-hand side of [2.19]. For the summand $\|\Lambda_1\eta_1(\cdot,t)\|^2$ at the node $x \in \omega \cup \gamma_{-1}$, we have

$$\eta_1(x,t) = u(x,t) - T_2 u(x,t) =$$

$$= \frac{1}{\tau h_2^3} \int_{t-\tau}^{t} d\eta \int_{x_2-\frac{h_2}{2}}^{x_2+\frac{h_2}{2}} d\xi_2 \int_{\eta}^{t} d\eta_1 \int_{x_2-h_2}^{x_2+h_2} (h_2 - |x_2 - \xi|) d\xi \int_{\xi}^{x_2} \frac{\partial^2 u(x_1,\xi_1,\eta_1)}{\partial \eta_1 \partial \xi_1} d\xi_1 +$$

$$+ \frac{1}{\tau h_2^3} \int_{t-\tau}^{t} d\eta \int_{x_2-\frac{h_2}{2}}^{x_2+\frac{h_2}{2}} d\xi_2 \int_{x_2-h_2}^{x_2+h_2} (h_2 - |x_2 - \xi|) d\xi \int_{\xi}^{x_2} d\xi_1 \int_{\xi_2}^{\xi_1} \frac{\partial^2 u(x_1,\xi_3,\eta)}{\partial \xi_3^2} d\xi_3,$$

which due to the relation $\left(T_1 \dfrac{\partial^2 u}{\partial x_1^2}\right)(x,t) = u_{\bar{x}_1 x_1}(x,t), x \in \omega$, finally gives the representation

$$\eta_{1\bar{x}_1 x_1}(x,t) = \frac{1}{h_1^2 \tau h_2^3} \int_{x_1-h_1}^{x_1+h_1} (h_1 - |x_1 - \xi_4|) d\xi_4 \int_{t-\tau}^{t} d\eta \times$$

$$\times \int_{x_2-\frac{h_2}{2}}^{x_2+\frac{h_2}{2}} d\xi_2 \int_{\eta}^{t} d\eta_1 \int_{x_2-h_2}^{x_2+h_2} (h_2-|x_2-\xi|)d\xi \int_{\xi}^{x_2} \frac{\partial^4 u(\xi_4,\xi_1,\eta_1)}{\partial \xi_4^2 \partial \eta_1 \partial \xi_1} d\xi_1 +$$

$$+ \frac{1}{h_1^2 \tau h_2^3} \int_{x_1-h_1}^{x_1+h_1} (h_1-|x_1-\xi_4|)d\xi_4 \int_{t-\tau}^{t} d\eta \int_{x_2-\frac{h_2}{2}}^{x_2+\frac{h_2}{2}} d\xi_2 \times$$

$$\times \int_{x_2-h_2}^{x_2+h_2} (h_2-|x_2-\xi|)d\xi \int_{\xi}^{x_2} d\xi_1 \int_{\xi_2}^{\xi_1} \frac{\partial^2 u(x_1,\xi_3,\eta)}{\partial \xi_3^2} d\xi_3 \ .$$

This yields the estimate

$$\left|\eta_{1\bar{x}_1 x_1}(x,t)\right| \leq 4\sqrt{\frac{\tau h_2}{h_1}} \left(\int_{x_1-h_1}^{x_1+h_1} d\xi_4 \int_{t-\tau}^{t} d\eta_1 \int_{x_2-h_2}^{x_2+h_2} \left(\frac{\partial^4 u(\xi_4,\xi_1,\eta_1)}{\partial \xi_4^2 \partial \eta_1 \partial \xi_1} \right)^2 d\xi_1 \right)^{1/2} +$$

$$+ 8\sqrt{\frac{h_2^3}{h_1 \tau}} \left(\int_{x_1-h_1}^{x_1+h_1} d\xi_4 \int_{t-\tau}^{t} d\eta \int_{x_2-h_2}^{x_2+h_2} \left(\frac{\partial^4 u(\xi_4,\xi_3,\eta)}{\partial \xi_4^2 \partial \xi_3^2} \right)^2 d\xi_3 \right)^{1/2},$$

which due to the elementary inequality $(a+b)^2 \leq 2(a^2+b^2)$ leads to the following:

$$\left|\eta_{1\bar{x}_1 x_1}(x,t)\right|^2 \leq 2 \left[16\frac{\tau h_2}{h_1} \int_{x_1-h_1}^{x_1+h_1} d\xi_4 \int_{t-\tau}^{t} d\eta_1 \int_{x_2-h_2}^{x_2+h_2} \left(\frac{\partial^4 u(\xi_4,\xi_1,\eta_1)}{\partial \xi_4^2 \partial \eta_1 \partial \xi_1} \right)^2 d\xi_1 + \right.$$

$$\left. + 64\frac{h_2^3}{h_1 \tau} \int_{x_1-h_1}^{x_1+h_1} d\xi_4 \int_{t-\tau}^{t} d\eta \int_{x_2-h_2}^{x_2+h_2} \left(\frac{\partial^4 u(\xi_4,\xi_3,\eta)}{\partial \xi_4^2 \partial \xi_3^2} \right)^2 d\xi_3 \right], \quad x \in \omega. \qquad [2.20]$$

Next, applying the relation $\left(T_1 \frac{\partial^2 u}{\partial x_1^2} \right)(x,t) = \frac{2}{h_1} u_{x_1}(x,t)$, $x \in \gamma_{-1}$, we have

$$\frac{2}{h_1}\eta_{1x_1}(x,t) = \frac{2}{h_1^2 \tau h_2^3}\int_0^{h_1}(h_1-\xi_4)d\xi_4 \int_{t-\tau}^t d\eta \times$$

$$\times \int_{x_2-\frac{h_2}{2}}^{x_2+\frac{h_2}{2}} d\xi_2 \int_\eta^t d\eta_1 \int_{x_2-h_2}^{x_2+h_2}(h_2-|x_2-\xi|)d\xi \int_\xi^{x_2} \frac{\partial^4 u(\xi_4,\xi_1,\eta_1)}{\partial \xi_4^2 \partial \eta_1 \partial \xi_1} d\xi_1 +$$

$$+\frac{2}{h_1^2 \tau h_2^3}\int_0^{h_1}(h_1-\xi_4)d\xi_4 \int_{t-\tau}^t d\eta \int_{x_2-\frac{h_2}{2}}^{x_2+\frac{h_2}{2}} d\xi_2 \int_{x_2-h_2}^{x_2+h_2}(h_2-|x_2-\xi|)d\xi \times$$

$$\times \int_{x_2-h_2}^{x_2+h_2}(h_2-|x_2-\xi|)d\xi \int_\xi^{x_2} d\xi_1 \int_{\xi_2}^{\xi_1}\frac{\partial^2 u(x_1,\xi_3,\eta)}{\partial \xi_3^2} d\xi_3,$$

which yields the inequality

$$\left|\frac{2}{h_1}\eta_{1x_1}(x,t)\right| \le \frac{4\sqrt{2\tau h_2}}{\sqrt{h_1}}\left(\int_0^{h_1} d\xi_4 \int_{t-\tau}^t d\eta_1 \int_{x_2-h_2}^{x_2+h_2}\left(\frac{\partial^4 u(\xi_4,\xi_1,\eta_1)}{\partial \xi_4^2 \partial \eta_1 \partial \xi_1}\right)^2 d\xi_1\right)^{1/2} +$$

$$+\frac{8\sqrt{2h_2^3}}{\sqrt{h_1 \tau}}\left(\int_0^{h_1} d\xi_4 \int_{t-\tau}^t d\eta \int_{x_2-h_2}^{x_2+h_2}\left(\frac{\partial^4 u(\xi_4,\xi_3,\eta)}{\partial \xi_4^2 \partial \xi_3^2}\right)^2 d\xi_3\right)^{1/2},$$

and therefore

$$\left|\frac{2}{h_1}\eta_{1x_1}(x,t)\right|^2 \le 2\left[32\frac{\tau h_2}{h_1}\int_0^{h_1} d\xi_4 \int_{t-\tau}^t d\eta_1 \int_{x_2-h_2}^{x_2+h_2}\left(\frac{\partial^4 u(\xi_4,\xi_1,\eta_1)}{\partial \xi_4^2 \partial \eta_1 \partial \xi_1}\right)^2 d\xi_1 +\right.$$

$$\left.+128\frac{h_2^3}{h_1 \tau}\int_0^{h_1} d\xi_4 \int_{t-\tau}^t d\eta \int_{x_2-h_2}^{x_2+h_2}\left(\frac{\partial^4 u(\xi_4,\xi_3,\eta)}{\partial \xi_4^2 \partial \xi_3^2}\right)^2 d\xi_3\right], \quad x \in \gamma_{-1}. \quad [2.21]$$

Combining [2.20] and [2.21], we obtain the estimate

$$\sum_{\eta=\tau}^{t} \tau \| \Lambda_1 \eta_1(\cdot,\eta) \|^2 =$$

$$= \sum_{\eta=\tau}^{t} \tau \left(\sum_{x \in \omega} h_1 h_2 \left(\eta_{1\bar{x}_1 x_1}(x,\eta) \right)^2 + \frac{h_1}{2} \sum_{x \in \gamma_{-1}} h_2 \left(\frac{2}{h_1} \eta_{1 x_1}(x,\eta) \right)^2 \right) \leq \quad [2.22]$$

$$\leq 128 \left[\tau^2 h_2^2 \int_0^t d\eta \iint_D \left(\frac{\partial^4 u(\xi_1,\xi_2,\eta)}{\partial \xi_1^2 \partial \xi_2 \partial \eta} \right)^2 d\xi_1 d\xi_2 + 4 h_2^4 \int_0^t d\eta \iint_D \left(\frac{\partial^4 u(\xi_1,\xi_2,\eta)}{\partial \xi_1^2 \partial \xi_2^2} \right)^2 d\xi_1 d\xi_2 \right].$$

Next, we consider the summand $\| A_2 \eta_2(\cdot,t) \|^2$ in formula [2.19]. Similarly to [2.20], at the node $x \in \omega$, we have

$$\left| \eta_{2\bar{x}_2 x_2}(x,t) \right|^2 \leq 2 \left[16 \frac{\tau h_1}{h_2} \int_{x_1-h_1}^{x_1+h_1} d\xi_1 \int_{t-\tau}^{t} d\eta \int_{x_2-h_2}^{x_2+h_2} \left(\frac{\partial^4 u(\xi_1,\xi_2,\eta)}{\partial \xi_2^2 \partial \eta \partial \xi_1} \right)^2 d\xi_2 + \right.$$

$$\left. + 64 \frac{h_1^3}{h_2 \tau} \int_{x_1-h_1}^{x_1+h_1} d\xi_1 \int_{t-\tau}^{t} d\eta \int_{x_2-h_2}^{x_2+h_2} \left(\frac{\partial^4 u(\xi_1,\xi_2,\eta)}{\partial \xi_1^2 \partial \xi_2^2} \right)^2 d\xi_2 \right], \quad x \in \omega. \quad [2.23]$$

At the node $x \in \gamma_{-1}$, we have

$$\eta_2(x,t) = u(x,t) - T_1 u(x,t) = -\frac{2}{h_1^2} \int_0^{h_1} (h_1 - \xi) \left(u(\xi,x_2,t) - u(0,x_2,t) \right) d\xi =$$

$$= -\frac{2}{h_1^2} \int_0^{h_1} (h_1 - \xi) d\xi \int_0^{\xi} \frac{\partial u(\xi_1,x_2,t)}{\partial \xi_1} d\xi_1 =$$

$$= -\frac{2}{\tau h_1^3} \int_0^{h_1} (h_1 - \xi) d\xi \int_0^{\xi} d\xi_1 \int_{t-\tau}^{t} d\eta \int_0^{h_1} \left(\frac{\partial u(\xi_1,x_2,t)}{\partial \xi_1} - \frac{\partial u(\xi_2,x_2,\eta)}{\partial \xi_2} \right) d\xi_2 -$$

$$-\frac{2}{\tau h_1^3}\int_0^{h_1}(h_1-\xi)d\xi\int_0^\xi d\xi_1\int_{t-\tau}^t d\eta\int_0^{h_1}\frac{\partial u(\xi_2,x_2,\eta)}{\partial \xi_2}d\xi_2 =$$

$$=-\frac{2}{\tau h_1^3}\int_0^{h_1}(h_1-\xi)d\xi\int_0^\xi d\xi_1\int_{t-\tau}^t d\eta\int_0^{h_1}d\xi_2\int_\eta^t\frac{\partial^2 u(\xi_1,x_2,\eta_1)}{\partial \eta_1\partial \xi_1}d\eta_1 -$$

$$-\frac{2}{\tau h_1^3}\int_0^{h_1}(h_1-\xi)d\xi\int_0^\xi d\xi_1\int_{t-\tau}^t d\eta\int_0^{h_1}d\xi_2\int_{\xi_2}^{\xi_1}\frac{\partial^2 u(\xi_3,x_2,\eta)}{\partial \xi_3^2}d\eta_1 -$$

$$-\frac{2}{\tau h_1^3}\int_0^{h_1}(h_1-\xi)d\xi\int_0^\xi d\xi_1\int_{t-\tau}^t d\eta\int_0^{h_1}\frac{\partial u(\xi_2,x_2,\eta)}{\partial \xi_2}d\xi_2 .$$

Applying here the relation $\left(T_2\frac{\partial^2 u}{\partial x_2^2}\right)(x,t) = u_{\bar{x}_2 x_2}(x,t)$, $x\in \omega\cup\gamma_{-1}$, we obtain the representation

$$\eta_{2\bar{x}_2 x_2}(x,t) = -\frac{2}{h_2^2\tau h_1^3}\int_{x_2-h_2}^{x_2+h_2}(h_2-|x_2-\xi_4|)d\xi_4 \times$$

$$\times \int_0^{h_1}(h_1-\xi)d\xi\int_0^\xi d\xi_1\int_{t-\tau}^t d\eta\int_0^{h_1}d\xi_2\int_\eta^t\frac{\partial^4 u(\xi_1,\xi_4,\eta_1)}{\partial \xi_4^2\partial \eta_1\partial \xi_1}d\eta_1 -$$

$$-\frac{2}{h_2^2\tau h_1^3}\int_{x_2-h_2}^{x_2+h_2}(h_2-|x_2-\xi_4|)d\xi_4\int_0^{h_1}(h_1-\xi)d\xi\int_0^\xi d\xi_1\int_{t-\tau}^t d\eta\int_0^{h_1}d\xi_2\int_{\xi_2}^{\xi_1}\frac{\partial^4 u(\xi_3,\xi_4,\eta)}{\partial \xi_4^2\partial \xi_3^2}d\eta_1 -$$

$$-\frac{2}{h_2^2\tau h_1^3}\int_{x_2-h_2}^{x_2+h_2}(h_2-|x_2-\xi_4|)d\xi_4\int_0^{h_1}(h_1-\xi)d\xi\int_0^\xi d\xi_1\int_{t-\tau}^t d\eta\int_0^{h_1}\frac{\partial^3 u(\xi_2,\xi_4,\eta)}{\partial \xi_4^2\partial \xi_2}d\xi_2 ,$$

which gives the inequality

$$|\eta_{2\bar{x}_2 x_2}(x,t)| \leq \frac{\sqrt{2\tau h_1}}{\sqrt{h_2}} \left(\int_{x_2-h_2}^{x_2+h_2} d\xi_4 \int_0^{h_1} d\xi_1 \int_{t-\tau}^t \left(\frac{\partial^4 u(\xi_1,\xi_4,\eta_1)}{\partial \xi_4^2 \partial \eta_1 \partial \xi_1} \right)^2 d\eta_1 \right)^{1/2} +$$

$$+ \frac{\sqrt{2h_1^3}}{3\sqrt{\tau h_2}} \left(\int_{x_2-h_2}^{x_2+h_2} d\xi_4 \int_{t-\tau}^t d\eta \int_0^{h_1} \left(\frac{\partial^4 u(\xi_3,\xi_4,\eta)}{\partial \xi_4^2 \partial \xi_3^2} \right)^2 d\xi_3 \right)^{1/2} +$$

$$+ \frac{\sqrt{2h_1}}{3\sqrt{\tau h_2}} \left(\int_{x_2-h_2}^{x_2+h_2} d\xi_4 \int_{t-\tau}^t d\eta \int_0^{h_1} \left(\frac{\partial^3 u(\xi_2,\xi_4,\eta)}{\partial \xi_4^2 \partial \xi_2} \right)^2 d\xi_2 \right)^{1/2}.$$

Using the relation $(a+b+c)^2 \leq 3(a^2+b^2+c^2)$, we have

$$|\eta_{2\bar{x}_2 x_2}(x,t)|^2 \leq 3 \left[\frac{2\tau h_1}{h_2} \int_{x_2-h_2}^{x_2+h_2} d\xi_2 \int_0^{h_1} d\xi_1 \int_{t-\tau}^t \left(\frac{\partial^4 u(\xi_1,\xi_2,\eta)}{\partial \xi_2^2 \partial \eta \partial \xi_1} \right)^2 d\eta +$$

$$+ \frac{2h_1^3}{9\tau h_2} \int_{x_2-h_2}^{x_2+h_2} d\xi_2 \int_{t-\tau}^t d\eta \int_0^{h_1} \left(\frac{\partial^4 u(\xi_1,\xi_2,\eta)}{\partial \xi_1^2 \partial \xi_2^2} \right)^2 d\xi_1 + \qquad [2.24]$$

$$+ \frac{2h_1}{9\tau h_2} \int_{x_2-h_2}^{x_2+h_2} d\xi_2 \int_{t-\tau}^t d\eta \int_0^{h_1} \left(\frac{\partial^3 u(\xi_1,\xi_2,\eta)}{\partial \xi_2^2 \partial \xi_1} \right)^2 d\xi_1 \right], \quad x \in \gamma_{-1}.$$

Combining [2.23] and [2.24], we come to the estimate

$$\sum_{\eta=\tau}^t \tau \|\Lambda_2 \eta_2(\cdot,\eta)\|^2 =$$

$$= \sum_{\eta=\tau}^t \tau \left(\sum_{x \in \omega} h_1 h_2 \left(\eta_{2\bar{x}_2 x_2}(x,\eta) \right)^2 + \frac{h_1}{2} \sum_{x \in \gamma_{-1}} h_2 \left(\eta_{2\bar{x}_2 x_2}(x,\eta) \right)^2 \right) = \qquad [2.25]$$

$$\leq 128 \left[\tau^2 h_1^2 \int_0^t d\eta \iint_D \left(\frac{\partial^4 u(\xi_1,\xi_2,\eta)}{\partial \xi_1 \partial \xi_2^2 \partial \eta} \right)^2 d\xi_1 d\xi_2 + \right.$$

$$+4h_1^4\int_0^t d\eta\iint_D\left(\frac{\partial^4 u(\xi_1,\xi_2,\eta)}{\partial\xi_1^2\partial\xi_2^2}\right)^2 d\xi_1 d\xi_2\Bigg]+\frac{2h_1^2}{3}\int_0^t d\eta\iint_{D_h}\left(\frac{\partial^3 u(\xi_1,\xi_2,\eta)}{\partial\xi_2^2\partial\xi_1}\right)^2 d\xi_1 d\xi_2,$$

where $D_h=\{x=(x_1,x_2): 0\le x_1\le h_1, 0\le x_2\le l_2\}$ is a strip near the left side of D.

At last, we study the summand $\|\eta_3(\cdot,t)\|^2$ in [2.19]. At the node $x\in\omega$, we have

$$\eta_3(x,t)=\frac{\partial(Tu)}{\partial t}(x,t)-u_{\bar{t}}(x,t)=\frac{1}{\tau h_1^2 h_2^2}\int_{x_1-h_1}^{x_1+h_1}(h_1-|x_1-\xi_1|)d\xi_1\times$$

$$\times\int_{x_2-h_2}^{x_2+h_2}(h_2-|x_2-\xi_2|)d\xi_2\int_{t-\tau}^t d\eta\int_\eta^t\frac{\partial^2 u(\xi_1,\xi_2,\eta_1)}{\partial\eta_1^2}d\eta_1 +$$

$$+\frac{1}{\tau h_1^3 h_2^2}\int_{x_1-h_1}^{x_1+h_1}(h_1-|x_1-\xi_1|)d\xi_1\int_{x_2-h_2}^{x_2+h_2}(h_2-|x_2-\xi_2|)d\xi_2\int_{t-\tau}^t d\eta\int_{x_1}^{\xi_1}d\xi_3\times$$

$$\times\int_{x_1-\frac{h_1}{2}}^{x_1+\frac{h_1}{2}}d\xi_5\int_{\xi_5}^{\xi_3}\frac{\partial^3 u(\xi_8,\xi_2,\eta)}{\partial\xi_8^2\partial\eta}d\xi_8 +$$

$$+\frac{1}{\tau h_1^3 h_2^3}\int_{x_1-h_1}^{x_1+h_1}(h_1-|x_1-\xi_1|)d\xi_1\int_{x_2-h_2}^{x_2+h_2}(h_2-|x_2-\xi_2|)d\xi_2\int_{t-\tau}^t d\eta\int_{x_2}^{\xi_2}d\xi_4\times$$

$$\times\int_{x_1-\frac{h_1}{2}}^{x_1+\frac{h_1}{2}}d\xi_6\int_{x_2-\frac{h_2}{2}}^{x_2+\frac{h_2}{2}}d\xi_7\int_{x_6}^{x_1}\frac{\partial^3 u(\xi_9,\xi_4,\eta)}{\partial\xi_9\partial\xi_4\partial\eta}d\xi_9 +$$

$$+\frac{1}{\tau h_1^3 h_2^3}\int_{x_1-h_1}^{x_1+h_1}(h_1-|x_1-\xi_1|)d\xi_1\int_{x_2-h_2}^{x_2+h_2}(h_2-|x_2-\xi_2|)d\xi_2\int_{t-\tau}^t d\eta\int_{x_2}^{\xi_2}d\xi_4\times$$

$$\times \int_{x_1-\frac{h_1}{2}}^{x_1+\frac{h_1}{2}} d\xi_6 \int_{x_2-\frac{h_2}{2}}^{x_2+\frac{h_2}{2}} d\xi_7 \int_{\xi_7}^{\xi_4} \frac{\partial^3 u(\xi_6,\xi_{10},\eta)}{\partial \xi_{10}^2 \partial \eta} d\xi_{10},$$

which yields the inequality

$$|\eta_3(x,t)| \leq \frac{4\sqrt{\tau}}{3\sqrt{h_1 h_2}} \left(\int_{x_1-h_1}^{x_1+h_1} d\xi_1 \int_{x_2-h_2}^{x_2+h_2} d\xi_2 \int_{t-\tau}^{t} \left(\frac{\partial^2 u(\xi_1,\xi_2,\eta_1)}{\partial \eta_1^2} \right)^2 d\eta_1 \right)^{1/2} +$$

$$+ \frac{8\sqrt{h_1^3}}{\sqrt{\tau h_2}} \left(\int_{x_2-h_2}^{x_2+h_2} d\xi_2 \int_{t-\tau}^{t} d\eta \int_{x_1-h_1}^{x_1+h_1} \left(\frac{\partial^3 u(\xi_8,\xi_2,\eta)}{\partial \xi_8^2 \partial \eta} \right)^2 d\xi_8 \right)^{1/2} +$$

$$+ \frac{4\sqrt{2h_1 h_2}}{\sqrt{\tau}} \left(\int_{t-\tau}^{t} d\eta \int_{x_2-h_2}^{x_2+h_2} d\xi_4 \int_{x_1-\frac{h_1}{2}}^{x_1+\frac{h_1}{2}} \left(\frac{\partial^3 u(\xi_9,\xi_4,\eta)}{\partial \xi_9 \partial \xi_4 \partial \eta} \right)^2 d\xi_9 \right)^{1/2} +$$

$$+ \frac{8\sqrt{2h_2^3}}{\sqrt{\tau h_1}} \left(\int_{t-\tau}^{t} d\eta \int_{x_1-\frac{h_1}{2}}^{x_1+\frac{h_1}{2}} d\xi_6 \int_{x_2-h_2}^{x_2+h_2} \left(\frac{\partial^3 u(\xi_6,\xi_{10},\eta)}{\partial \xi_{10}^2 \partial \eta} \right)^2 d\xi_{10} \right)^{1/2}.$$

Using the relation $(a+b+c+d)^2 \leq 4(a^2+b^2+c^2+d^2)$, we then have

$$|\eta_3(x,t)|^2 \leq 4 \left[\frac{16\tau}{9h_1 h_2} \int_{x_1-h_1}^{x_1+h_1} d\xi_1 \int_{x_2-h_2}^{x_2+h_2} d\xi_2 \int_{t-\tau}^{t} \left(\frac{\partial^2 u(\xi_1,\xi_2,\eta_1)}{\partial \eta_1^2} \right)^2 d\eta_1 + \right.$$

$$\left. + \frac{64 h_1^3}{\tau h_2} \int_{x_2-h_2}^{x_2+h_2} d\xi_2 \int_{t-\tau}^{t} d\eta \int_{x_1-h_1}^{x_1+h_1} \left(\frac{\partial^3 u(\xi_8,\xi_2,\eta)}{\partial \xi_8^2 \partial \eta} \right)^2 d\xi_8 + \right. \qquad [2.26]$$

$$+\frac{32h_1h_2}{\tau}\int_{t-\tau}^{t}d\eta\int_{x_2-h_2}^{x_2+h_2}d\xi_4\int_{x_1-\frac{h_1}{2}}^{x_1+\frac{h_1}{2}}\left(\frac{\partial^3 u(\xi_9,\xi_4,\eta)}{\partial\xi_9\partial\xi_4\partial\eta}\right)^2 d\xi_9 +$$

$$+\frac{128h_2^3}{\tau h_1}\int_{t-\tau}^{t}d\eta\int_{x_1-\frac{h_1}{2}}^{x_1+\frac{h_1}{2}}d\xi_6\int_{x_2-h_2}^{x_2+h_2}\left(\frac{\partial^3 u(\xi_6,\xi_{10},\eta)}{\partial\xi_{10}^2\partial\eta}\right)^2 d\xi_{10}\Biggr], \quad x\in\omega.$$

At the node $x\in\gamma_{-1}$, we have the representation

$$\eta_3(x,t) = \frac{d(Tu)}{dt}(x,t) - u_{\bar{t}}(x,t) =$$

$$= \frac{2}{h_1^2 h_2^2}\int_0^{h_1}(h_1-\xi_1)d\xi_1\int_{x_2-h_2}^{x_2+h_2}(h_2-|x_2-\xi_2|)\frac{\partial u(\xi_1,\xi_2,t)}{\partial t}d\xi_2 -$$

$$-\frac{u(0,x_2,t)-u(0,x_2,t-\tau)}{\tau} =$$

$$= \frac{2}{h_1^2 h_2^2}\int_0^{h_1}(h_1-\xi_1)d\xi_1\int_{x_2-h_2}^{x_2+h_2}(h_2-|x_2-\xi_2|)d\xi_2\int_{t-\tau}^{t}\left[\frac{\partial u(\xi_1,\xi_2,t)}{\partial t}-\frac{\partial u(0,x_2,\eta)}{\partial\eta}\right]d\eta =$$

$$= \frac{2}{h_1^2 h_2^2}\int_0^{h_1}(h_1-\xi_1)d\xi_1\int_{x_2-h_2}^{x_2+h_2}(h_2-|x_2-\xi_2|)d\xi_2\int_{t-\tau}^{t}d\eta\int_{\eta}^{t}\frac{\partial^2 u(\xi_1,\xi_2,\eta_1)}{\partial\eta_1^2}d\eta_1 +$$

$$+\frac{2}{h_1^2 h_2^2}\int_0^{h_1}(h_1-\xi_1)d\xi_1\int_{x_2-h_2}^{x_2+h_2}(h_2-|x_2-\xi_2|)d\xi_2\int_{t-\tau}^{t}d\eta\int_0^{\xi_1}\frac{\partial^2 u(\xi_3,\xi_2,\eta)}{\partial\xi_3\partial\eta}d\xi_3 +$$

$$+\frac{2}{h_1^2 h_2^2}\int_0^{h_1}(h_1-\xi_1)d\xi_1\int_{x_2-h_2}^{x_2+h_2}(h_2-|x_2-\xi_2|)d\xi_2\int_{t-\tau}^{t}d\eta\int_{x_2}^{\xi_2}\frac{\partial^2 u(0,\xi_4,\eta)}{\partial\xi_4\partial\eta}d\xi_4 =$$

$$= \frac{2}{h_1^2 h_2^2} \int_0^{h_1} (h_1 - \xi_1) d\xi_1 \int_{x_2-h_2}^{x_2+h_2} (h_2 - |x_2 - \xi_2|) d\xi_2 \int_{t-\tau}^{t} d\eta \int_{\eta}^{t} \frac{\partial^2 u(\xi_1, \xi_2, \eta_1)}{\partial \eta_1^2} d\eta_1 +$$

$$+ \frac{2}{h_1^2 h_2^2} \int_0^{h_1} (h_1 - \xi_1) d\xi_1 \int_{x_2-h_2}^{x_2+h_2} (h_2 - |x_2 - \xi_2|) d\xi_2 \int_{t-\tau}^{t} d\eta \int_0^{\xi_1} \frac{\partial^2 u(\xi_3, \xi_2, \eta)}{\partial \xi_3 \partial \eta} d\xi_3 +$$

$$+ \frac{2}{h_1^2 h_2^2} \int_0^{h_1} (h_1 - \xi_1) d\xi_1 \int_{x_2-h_2}^{x_2+h_2} (h_2 - |x_2 - \xi_2|) d\xi_2 \int_{t-\tau}^{t} d\eta \times$$

$$\times \int_{x_2}^{\xi_2} \left[\frac{\partial^2 u(0, \xi_4, \eta)}{\partial \xi_4 \partial \eta} - \frac{1}{h_1 h_2} \int_0^{h_1} d\xi_5 \int_{x_2 - \frac{h_2}{2}}^{x_2 + \frac{h_2}{2}} \frac{\partial^2 u(\xi_5, \xi_6, \eta)}{\partial \xi_6 \partial \eta} d\xi_6 \right] d\xi_4 =$$

$$= \frac{2}{h_1^2 h_2^2} \int_0^{h_1} (h_1 - \xi_1) d\xi_1 \int_{x_2-h_2}^{x_2+h_2} (h_2 - |x_2 - \xi_2|) d\xi_2 \int_{t-\tau}^{t} d\eta \int_{\eta}^{t} \frac{\partial^2 u(\xi_1, \xi_2, \eta_1)}{\partial \eta_1^2} d\eta_1 +$$

$$+ \frac{2}{h_1^2 h_2^2} \int_0^{h_1} (h_1 - \xi_1) d\xi_1 \int_{x_2-h_2}^{x_2+h_2} (h_2 - |x_2 - \xi_2|) d\xi_2 \int_{t-\tau}^{t} d\eta \int_0^{\xi_1} \frac{\partial^2 u(\xi_3, \xi_2, \eta)}{\partial \xi_3 \partial \eta} d\xi_3 +$$

$$+ \frac{2}{h_1^3 h_2^3} \int_0^{h_1} (h_1 - \xi_1) d\xi_1 \int_{x_2-h_2}^{x_2+h_2} (h_2 - |x_2 - \xi_2|) d\xi_2 \int_{t-\tau}^{t} d\eta \times$$

$$\times \int_{x_2}^{\xi_2} d\xi_4 \int_0^{h_1} d\xi_5 \int_{x_2 - \frac{h_2}{2}}^{x_2 + \frac{h_2}{2}} d\xi_6 \int_{\xi_5}^{0} \frac{\partial^3 u(\xi_7, \xi_4, \eta)}{\partial \xi_7 \partial \xi_4 \partial \eta} d\xi_7 +$$

$$+ \frac{2}{h_1^3 h_2^3} \int_0^{h_1} (h_1 - \xi_1) d\xi_1 \int_{x_2-h_2}^{x_2+h_2} (h_2 - |x_2 - \xi_2|) d\xi_2 \int_{t-\tau}^{t} d\eta \times$$

$$\times \int_{x_2}^{\xi_2} d\xi_4 \int_0^{h_1} d\xi_5 \int_{x_2-\frac{h_2}{2}}^{x_2+\frac{h_2}{2}} d\xi_6 \int_{\xi_6}^{\xi_4} d\xi_8 \frac{\partial^3 u(\xi_5,\xi_8,\eta)}{\partial \xi_8^2 \partial \eta} d\xi_8 \; ,$$

which brings the estimate

$$|\eta_3(x,t)| \le \frac{4\sqrt{2\tau}}{3\sqrt{h_1 h_2}} \left(\int_0^{h_1} d\xi_1 \int_{x_2-h_2}^{x_2+h_2} d\xi_2 \int_{t-\tau}^{t} \left(\frac{\partial^2 u(\xi_1,\xi_2,\eta_1)}{\partial \eta_1^2} \right)^2 d\eta_1 \right)^{1/2} +$$

$$+ \frac{\sqrt{2h_1}}{\sqrt{\tau h_2}} \left(\int_{x_2-h_2}^{x_2+h_2} d\xi_2 \int_{t-\tau}^{t} d\eta \int_0^{h_1} \left(\frac{\partial^2 u(\xi_3,\xi_2,\eta)}{\partial \xi_3 \partial \eta} \right)^2 d\xi_3 \right)^{1/2} +$$

$$+ \frac{\sqrt{2h_1 h_2}}{\sqrt{\tau}} \left(\int_{t-\tau}^{t} d\eta \int_{x_2-h_2}^{x_2+h_2} d\xi_4 \int_0^{h_1} \left(\frac{\partial^3 u(\xi_7,\xi_4,\eta)}{\partial \xi_7 \partial \xi_4 \partial \eta} \right)^2 d\xi_7 \right)^{1/2} +$$

$$+ \frac{\sqrt{2h_2^3}}{\sqrt{\tau h_1}} \left(\int_{t-\tau}^{t} d\eta \int_0^{h_1} d\xi_5 \int_{x_2-h_2}^{x_2+h_2} \left(\frac{\partial^3 u(\xi_5,\xi_8,\eta)}{\partial \xi_8^2 \partial \eta} \right)^2 d\xi_8 \right)^{1/2} , \; x \in \gamma_{-1},$$

and therefore

$$|\eta_3(x,t)|^2 \le 4 \left[\frac{32\tau}{9h_1 h_2} \int_0^{h_1} d\xi_1 \int_{x_2-h_2}^{x_2+h_2} d\xi_2 \int_{t-\tau}^{t} \left(\frac{\partial^2 u(\xi_1,\xi_2,\eta_1)}{\partial \eta_1^2} \right)^2 d\eta_1 + \right.$$

$$+ \frac{2h_1}{\tau h_2} \int_{x_2-h_2}^{x_2+h_2} d\xi_2 \int_{t-\tau}^{t} d\eta \int_0^{h_1} \left(\frac{\partial^2 u(\xi_3,\xi_2,\eta)}{\partial \xi_3 \partial \eta} \right)^2 d\xi_3 + \qquad [2.27]$$

$$+ \frac{2h_1 h_2}{\tau} \int_{t-\tau}^{t} d\eta \int_{x_2-h_2}^{x_2+h_2} d\xi_4 \int_0^{h_1} \left(\frac{\partial^3 u(\xi_7,\xi_4,\eta)}{\partial \xi_7 \partial \xi_4 \partial \eta} \right)^2 d\xi_7 +$$

$$+\frac{2h_2^3}{\tau h_1}\int_{t-\tau}^{t}d\eta\int_{0}^{h_1}d\xi_5\int_{x_2-h_2}^{x_2+h_2}\left(\frac{\partial^3 u(\xi_5,\xi_8,\eta)}{\partial \xi_8^2 \partial \eta}\right)^2 d\xi_8\Bigg], \quad x\in \gamma_{-1}.$$

Inequalities [2.26] and [2.27] are combined in the estimate

$$\sum_{\eta=\tau}^{t}\tau\|\eta_3(\cdot,\eta)\|^2 = \sum_{\eta=\tau}^{t}\tau\left(\sum_{x\in\omega}h_1 h_2\left(\eta_3(x,\eta)\right)^2 + \frac{h_1}{2}\sum_{x\in\gamma_{-1}}h_2\left(\eta_3(x,\eta)\right)^2\right)\leq$$

$$\leq \frac{4\cdot 4\cdot 16}{9}\tau^2 \int_0^t d\eta \iint_D \left(\frac{\partial^2 u(\xi_1,\xi_2,\eta)}{\partial \eta^2}\right)^2 d\xi_1 d\xi_2 +$$

$$+4\cdot 4\cdot 64 h_1^4 \int_0^t d\eta \iint_D \left(\frac{\partial^3 u(\xi_1,\xi_2,\eta)}{\partial \xi_1^2 \partial \eta}\right)^2 d\xi_1 d\xi_2 + \qquad [2.28]$$

$$+2\cdot 4\cdot 32 h_1^2 h_2^2 \int_0^t d\eta \iint_D \left(\frac{\partial^3 u(\xi_1,\xi_2,\eta)}{\partial \xi_1 \partial \xi_2 \partial \eta}\right)^2 d\xi_1 d\xi_2 +$$

$$+8\cdot 128 h_2^4 \int_0^t d\eta \iint_D \left(\frac{\partial^3 u(\xi_1,\xi_2,\eta)}{\partial \xi_2^2 \partial \eta}\right)^2 d\xi_1 d\xi_2 + 8 h_1^2 \int_0^t d\eta \iint_{D_h}\left(\frac{\partial^2 u(\xi_1,\xi_2,\eta)}{\partial \xi_1 \partial \eta}\right)^2 d\xi_1 d\xi_2,$$

where $D_h = \{x=(x_1,x_2): 0\leq x_1 \leq h_1, 0\leq x_2 \leq l_2\}$ is a strip of width h_1.

Estimates [2.19], [2.22], [2.25] and [2.28] prove the lemma.

Lemmas 2.6 and 2.7 bring us to the following main result of this section.

THEOREM 2.3.– *Let the solution* $u(x_1,x_2,t)$ *of problem [2.9] satisfy the conditions*

$$\frac{\partial^4 u}{\partial x_1^2 \partial x_2 \partial t}, \frac{\partial^4 u}{\partial x_1^2 \partial x_2^2}, \frac{\partial^4 u}{\partial x_1 \partial x_2^2 \partial t}, \frac{\partial^3 u}{\partial x_1 x_2^2}, \frac{\partial^3 u}{\partial x_1^2 \partial t}, \frac{\partial^3 u}{\partial x_2^2 \partial t}, \frac{\partial^3 u}{\partial x_1 \partial x_2 \partial t}, \frac{\partial^2 u}{\partial t^2}, \frac{\partial^2 u}{\partial x_1 \partial t} \in L_2(D_T).$$

Then, the accuracy of the finite-difference scheme [2.11] is characterized by the weighted estimate

$$\left(\sum_{\eta=\tau}^{t}\tau\frac{|z(x,\eta)|^2}{\rho^2(x)}\right)^{1/2} \leq M\left(\tau^2 h_2^2 + h_2^4 + \tau^2 h_1^2 + h_1^4 + \tau^2 + h_1^2 h_2^2 + h_1^2\right)^{1/2}, \quad [2.29]$$

$$(x,t)\in(\omega\cup\gamma_{-1})\times\omega_\tau,$$

where $\rho(x) = \min\left\{\sqrt{(l_1-x_1)(l_2-x_2)}, \sqrt{(l_1-x_1)x_2}\right\}$ *and M is a positive constant expressed in terms of norms of the above derivatives of the solution* $u(x,t)$.

2.2.4. Conclusion

Thus, the weighted a priori estimate obtained in Theorem 2.3 indicates that the accuracy of the finite-difference method [2.11] is higher near those faces of the space–time parallelepiped $D_T = D \times (0,T)$, where the Dirichlet boundary condition is specified, than further away from them. More precisely, the error in the weighted norm is $O\left(\sqrt{h_1}\,(h_1 + h_2^2 + \tau)\right)$ near the face $x_1 = 1$ and $O\left(\sqrt{h_2}\,(h_1 + h_2^2 + \tau)\right)$ near the faces $x_2 = 0$ and $x_2 = 1$, while it is $O(h_1 + h_2^2 + \tau)$ in the inner nodes of the mesh $(\omega\cup\gamma_{-1})\times\omega_\tau$.

REMARK 2.1.– *Let the conditions of Theorem 2.3 be fulfilled and, in addition,* $\dfrac{\partial^4 u}{\partial x_1 \partial x_2^3} \in L_2(D_T)$. *We apply the reasoning from Samarskii et al. (1987, p. 161) to the summands*

$$h_1^2 \int_0^t d\eta \iint_{D_h} \left(\frac{\partial^3 u(\xi_1,\xi_2,\eta)}{\partial \xi_2^2 \partial \xi_1}\right)^2 d\xi_1 d\xi_2 \quad \text{and} \quad h_1^2 \int_0^t d\eta \iint_{D_h} \left(\frac{\partial^2 u(\xi_1,\xi_2,\eta)}{\partial \xi_1 \partial \eta}\right)^2 d\xi_1 d\xi_2$$

in inequalities [2.25] and [2.28] respectively. Then, instead of [2.29], we obtain the estimate

$$\left(\sum_{\eta=\tau}^{t}\tau\frac{|z(x,\eta)|^2}{\rho^2(x)}\right)^{1/2} \leq M\left(\tau^2 h_2^2 + h_2^4 + \tau^2 h_1^2 + h_1^4 + \tau^2 + h_1^2 h_2^2 + h_1^3\right)^{1/2}.$$

2.3. A standard finite-difference scheme for the two-dimensional heat equation with the Dirichlet boundary condition

We consider the problem

$$\frac{\partial u(x,t)}{\partial t} = \Delta u(x,t) + f(x,t), \quad (x,t) \in D_T = D \times (0,T),$$
$$u(x,t) = 0, \quad (x,t) \in \Gamma \times (0,T), \qquad [2.30]$$
$$u(x,0) = u_0(x), \quad x \in D,$$

where $x = (x_1, x_2)$, $\Delta = \dfrac{\partial^2}{\partial x_1^2} + \dfrac{\partial^2}{\partial x_2^2}$ is the Laplace operator in a Cartesian coordinate system, $D = \{x = (x_1, x_2): 0 < x_\alpha < 1, \alpha = 1,2\}$ is a unit square, $\Gamma = \partial D$ is a boundary of D, $f(x,t)$ and $u_0(x)$ are given functions, $u(x,t)$ is an unknown solution.

2.3.1. *Discretization of the differential problem*

As usual, we first introduce some typical sets of nodes (Samarskii 2001):

$$\omega_\alpha = \{i_\alpha h, \; i_\alpha = 1, \ldots, N_\alpha - 1, \; h = 1/N\}, \; N \geq 2 \text{ is an integer number,}$$

$$\bar{\omega}_\alpha = \omega_\alpha \cup \{0\} \cup \{1\}, \quad \omega = \omega_1 \times \omega_2, \quad \bar{\omega} = \bar{\omega}_1 \times \bar{\omega}_2, \quad \gamma = \bar{\omega} \setminus \omega, \quad \alpha = 1,2, [2.31]$$

$$\omega_\tau = \{t_j = j\tau, \; j = 1,2,\ldots, M-1, \; \tau = T/M\}, \; M \geq 2 \text{ is an integer number,}$$

and define the averaging operators (Samarskii et al. 1987) for $h_1 = h_2 = h$:

$$T_1 v(x) = \frac{1}{h_1^2} \int_{x_1 - h_1}^{x_1 + h_1} (h_1 - |x_1 - \xi|) v(\xi, x_2) d\xi, \quad x \in \omega,$$

$$T_2 v(x) = \frac{1}{h_2^2} \int_{x_2 - h_2}^{x_2 + h_2} (h_2 - |x_2 - \xi|) v(x_1, \xi) d\xi, \quad x \in \omega.$$

Then, we approximate problem [2.30] by the finite-difference scheme

$$y_{\bar{t}}(x,t) = \Lambda y(x,t) + T_1 T_2 f(x,t), \quad (x,t) \in \omega \times \omega_\tau,$$
$$y(x,t) = 0, \quad (x,t) \in \gamma \times \omega_\tau,$$
$$y(x,0) = u_0(x), \quad x \in \omega,$$

[2.32]

where $\Lambda = \Lambda_1 + \Lambda_2$, $\Lambda_\alpha y(x) = y_{\bar{x}_\alpha x_\alpha}(x)$, $x \in \omega$, $\alpha = 1, 2$.

For the error $z(x,t) = y(x,t) - u(x,t)$, we obtain the problem

$$z_{\bar{t}}(x,t) = (\Lambda z)(x,t) + \psi(x,t), \quad (x,t) \in \omega \times \omega_\tau,$$
$$z(x,t) = 0, \quad (x,t) \in \gamma \times \omega_\tau,$$
$$z(x,0) = 0, \quad x \in \omega,$$

[2.33]

where $\psi(x,t)$ is the approximation error:

$$\psi(x,t) = \Lambda u(x,t) + T_1 T_2 f(x,t) - u_{\bar{t}}(x,t) =$$
$$= \Lambda_1 \eta_1(x,t) + \Lambda_2 \eta_2(x,t) + \eta_3(x,t)$$

[2.34]

with $\eta_\alpha(x,t) = u(x,t) - T_{3-\alpha} u(x,t)$, $\alpha = 1, 2$, and

$$\eta_3(x,t) = \frac{\partial(Tu)}{\partial t}(x,t) - u_{\bar{t}}(x,t), \quad (x,t) \in \omega \times \omega_\tau.$$

Next, we denote by $\overset{0}{H}$ the space of mesh functions defined on $\bar{\omega}$ and equal to zero on γ with the inner product and the associate norm

$$(y,v) = \sum_{x \in \omega} h^2 y(x) v(x), \quad \|v\| = \|v\|_{L_2(\omega)} = \sqrt{(v,v)} = \left(\sum_{x \in \omega} h^2 v^2(x) \right)^{1/2}.$$

We will also need the norms

$$\|v\|_{C(\omega)} = \max_{x \in \omega} |v(x)|, \quad \|v\|_{C(\omega \times \omega_\tau)} = \max_{(x,t) \in \omega \times \omega_\tau} |v(x,t)|.$$

2.3.2. Accuracy with the boundary effect

We begin with three auxiliary statements.

LEMMA 2.8.– *For the error $z(x,t)$, the following estimate holds true:*

$$|z(x,t)| \le v(x,t) \cdot \max_{t \in \omega_\tau} \|\psi(\cdot,t)\|_{C(\omega)}, \quad (x,t) \in \omega \times \omega_\tau, \qquad [2.35]$$

where $\psi(\cdot,t)$ is the approximation error [2.34],

$$v(x,t) = \sum_{k_1=1}^{N-1} \sum_{k_2=1}^{N-1} \left(1 - \left(\frac{1}{1+\tau\lambda_{k_1 k_2}}\right)^{t/\tau}\right) \frac{c_{k_1 k_2}}{\lambda_{k_1 k_2}} w_{k_1 k_2}(x), \quad (x,t) \in \omega \times \omega_\tau,$$

$$w_{k_1 k_2}(x) = 2\sin k_1 \pi x_1 \cdot \sin k_2 \pi x_2, \quad \|w_{k_1 k_2}\| = 1,$$

$$\lambda_{k_1 k_2} = \frac{4}{h^2} \sin^2 \frac{k_1 \pi h}{2} + \frac{4}{h^2} \sin^2 \frac{k_2 \pi h}{2},$$

$$c_{k_1 k_2} = h^2 \frac{\left(1-(-1)^{k_1}\right)\left(1-(-1)^{k_2}\right)\cos\frac{k_1 \pi h}{2}\cos\frac{k_2 \pi h}{2}}{2\sin\frac{k_1 \pi h}{2} \sin\frac{k_2 \pi h}{2}}, \quad k_1, k_2 = 1, \ldots, N-1.$$

PROOF.– It is well known (e.g. Samarskii (2001)) that the solutions of the spectral finite-difference problem

$$\Lambda w(x) + \lambda w(x) = 0, \quad x \in \omega,$$
$$w(x) = 0, \quad x \in \gamma, \qquad [2.36]$$

are given by the formulas

$$w_{k_1 k_2}(x) = 2\sin k_1 \pi x_1 \cdot \sin k_2 \pi x_2, \quad \|w_{k_1 k_2}\| = 1,$$
$$\lambda_{k_1 k_2} = \frac{4}{h^2}\sin^2\frac{k_1 \pi h}{2} + \frac{4}{h^2}\sin^2\frac{k_2 \pi h}{2}, \quad k_1, k_2 = 1, \ldots, N-1. \qquad [2.37]$$

Next, we consider the auxiliary problem

$$v_{\bar t}(x,t) = \Lambda v(x,t) + 1, \quad (x,t) \in \omega \times \omega_\tau,$$
$$v(x,t) = 0, \quad (x,t) \in \gamma \times \omega_\tau, \qquad [2.38]$$
$$v(x,0) = 0, \quad x \in \omega,$$

and try to find its solution in the form of the sum

$$v(x,t) = \sum_{k_1=1}^{N-1}\sum_{k_2=1}^{N-1} v_{k_1 k_2}(t) w_{k_1 k_2}(x).\qquad [2.39]$$

The function $v(x,t)$ obviously satisfies the boundary condition [2.38]. To make $v(x,t)$ satisfy the initial condition [2.38], we set $v_{k_1 k_2}(0) = 0$, $k_1, k_2 = 1,\ldots, N-1$.

Next, we use the representation

$$1 = \sum_{k_1=1}^{N-1}\sum_{k_2=1}^{N-1} c_{k_1 k_2} w_{k_1 k_2}(x) \qquad [2.40]$$

with

$$c_{k_1 k_2} = (1, w_{k_1 k_2}) = \sum_{x\in\omega} h^2 w_{k_1 k_2}(x) = \sum_{x\in\omega} h^2 2\sin k_1\pi x_1 \cdot \sin k_2\pi x_2 = \qquad [2.41]$$

$$= 2h^2 \frac{\sin\dfrac{(N-1)k_1\pi h}{2}\sin\dfrac{Nk_1\pi h}{2}}{\sin\dfrac{k_1\pi h}{2}} \cdot \frac{\sin\dfrac{(N-1)k_2\pi h}{2}\sin\dfrac{Nk_2\pi h}{2}}{\sin\dfrac{k_2\pi h}{2}} =$$

$$= h^2 \frac{\left(1-(-1)^{k_1}\right)\left(1-(-1)^{k_2}\right)\cos\dfrac{k_1\pi h}{2}\cos\dfrac{k_2\pi h}{2}}{2\sin\dfrac{k_1\pi h}{2}\sin\dfrac{k_2\pi h}{2}}.$$

Substituting [2.39] and [2.40] in equation [2.38], we obtain

$$\sum_{k_1=1}^{N-1}\sum_{k_2=1}^{N-1}\left[\frac{v_{k_1 k_2}(t) - v_{k_1 k_2}(t-\tau)}{\tau} + \lambda_{k_1 k_2} v_{k_1 k_2}(t) - c_{k_1 k_2}\right] w_{k_1 k_2}(x) = 0,$$

which implies the relation

$$\frac{v_{k_1 k_2}(t) - v_{k_1 k_2}(t-\tau)}{\tau} + \lambda_{k_1 k_2} v_{k_1 k_2}(t) - c_{k_1 k_2} = 0, \quad k_1, k_2 = 1,\ldots, N-1,$$

i.e.

$$v_{k_1k_2}(t) = \frac{1}{1+\tau\lambda_{k_1k_2}}v_{k_1k_2}(t-\tau) + \frac{\tau}{1+\tau\lambda_{k_1k_2}}c_{k_1k_2}, \quad t \in \omega_\tau,$$

$$v_{k_1k_2}(0) = 0, \quad k_1, k_2 = 1, \ldots, N-1.$$

It is easy to find the solution of this recurrence relation:

$$v_{k_1k_2}(t) = \left(1 - \left(\frac{1}{1+\tau\lambda_{k_1k_2}}\right)^{t/\tau}\right)\frac{c_{k_1k_2}}{\lambda_{k_1k_2}}, \quad t \in \omega_\tau.$$

Therefore

$$v(x,t) = \sum_{k_1=1}^{N-1}\sum_{k_2=1}^{N-1}\left(1 - \left(\frac{1}{1+\tau\lambda_{k_1k_2}}\right)^{t/\tau}\right)\frac{c_{k_1k_2}}{\lambda_{k_1k_2}}w_{k_1k_2}(x), \quad (x,t) \in \omega \times \omega_\tau, \quad [2.42]$$

with $w_{k_1k_2}(x)$, $\lambda_{k_1k_2}$, $c_{k_1k_2}$ defined in formulas [2.37] and [2.41].

Applying the comparison theorem (Samarskii 2001) to problems [2.33] and [2.38], we complete the proof of the lemma.

LEMMA 2.9.– *Let the solution $u(x,t)$ of problem [2.30] satisfy the condition $u \in W_\infty^4(D_T)$. Then, for the approximation error $\psi(x,t)$, the following estimate holds true:*

$$|\psi(x,t)| \leq M(\tau + h^2), \quad (x,t) \in \omega \times \omega_\tau, \quad [2.43]$$

where M is a positive constant independent of h and τ.

PROOF.– We will study each summand of the approximation error [2.34]. For the functional $\eta_1(x,t)$, we have

$$\eta_1(x,t) = u(x,t) - \frac{1}{h^2}\int_{x_2-h}^{x_2+h}(h - |x_2 - \xi|)u(x_1,\xi,t)d\xi =$$

$$= \frac{1}{h^2}\int_{x_2-h}^{x_2+h}(h - |x_2 - \xi|)\left(u(x_1,x_2,t) - u(x_1,\xi,t)\right)d\xi =$$

$$= \frac{1}{h_2^2} \int_{x_2-h_2}^{x_2+h_2} (h_2-|x_2-\xi|)d\xi \int_{\xi}^{x_2} \frac{\partial u(x_1,\xi_1,t)}{\partial \xi_1} d\xi_1 =$$

$$= \frac{1}{h^2} \int_{x_2-h}^{x_2+h} (h-|x_2-\xi|)d\xi \int_{\xi}^{x_2} d\xi_1 \left(\frac{\partial u(x_1,\xi_1,t)}{\partial \xi_1} - \frac{1}{2h} \int_{x_2-h}^{x_2+h} \frac{\partial u(x_1,\xi_2,t)}{\partial \xi_2} d\xi_2 \right) =$$

$$= \frac{1}{2h^3} \int_{x_2-h}^{x_2+h} (h-|x_2-\xi|)d\xi \int_{\xi}^{x_2} d\xi_1 \int_{x_2-h}^{x_2+h} \left(\frac{\partial u(x_1,\xi_1,t)}{\partial \xi_1} - \frac{\partial u(x_1,\xi_2,t)}{\partial \xi_2} \right) d\xi_2 =$$

$$= \frac{1}{2h^3} \int_{x_2-h}^{x_2+h} (h-|x_2-\xi|)d\xi \int_{\xi}^{x_2} d\xi_1 \int_{x_2-h}^{x_2+h} d\xi_2 \int_{\xi_2}^{\xi_1} \frac{\partial^2 u(x_1,\xi_3,t)}{\partial \xi_3^2} d\xi_3 ,$$

which due to the relation $\left(T_1 \frac{\partial^2 u}{\partial x_1^2} \right)(x,t) = u_{\bar{x}_1 x_1}(x,t)$, $(x,t) \in \omega \times \omega_\tau$, gives the representation

$$\eta_{1\bar{x}_1 x_1}(x,t) = = \frac{1}{2h^5} \int_{x_1-h}^{x_1+h} (h-|x_1-\xi_4|)d\xi_4 \times$$

$$\times \int_{x_2-h}^{x_2+h} (h-|x_2-\xi|)d\xi \int_{\xi}^{x_2} d\xi_1 \int_{x_2-h}^{x_2+h} d\xi_2 \int_{\xi_2}^{\xi_1} \frac{\partial^4 u(\xi_4,\xi_3,t)}{\partial \xi_4^2 \partial \xi_3^2} d\xi_3 .$$

This leads to the estimate

$$|\eta_{1\bar{x}_1 x_1}(x,t)| \le 4h^2 \left\| \frac{\partial^4 u(x_1,x_2,t)}{\partial x_1^2 \partial x_2^2} \right\|_{L_\infty(D_T)} . \qquad [2.44]$$

In a similar way, we come to the inequality

$$|\eta_{2\bar{x}_2 x_2}(x,t)| \le 4h^2 \left\| \frac{\partial^4 u(x_1,x_2,t)}{\partial x_1^2 \partial x_2^2} \right\|_{L_\infty(D_T)} . \qquad [2.45]$$

Now, we consider the summand $\eta_3(x,t)$:

$$\eta_3(x,t) = \frac{\partial(Tu)}{\partial t}(x,t) - u_{\bar{t}}(x,t) = \frac{1}{\tau h^4}\int\limits_{x_1-h}^{x_1+h}(h-|x_1-\xi_1|)d\xi_1\times$$

$$\times \int\limits_{x_2-h}^{x_2+h}(h-|x_2-\xi_2|)d\xi_2 \int\limits_{t-\tau}^{t}\left(\frac{\partial u(\xi_1,\xi_2,t)}{\partial t} - \frac{\partial u(x_1,x_2,\eta)}{\partial \eta}\right)d\eta =$$

$$= \frac{1}{\tau h^4}\int\limits_{x_1-h}^{x_1+h}(h-|x_1-\xi_1|)d\xi_1 \int\limits_{x_2-h}^{x_2+h}(h-|x_2-\xi_2|)d\xi_2 \times$$

$$\times \int\limits_{t-\tau}^{t}d\eta\left(\int\limits_{\eta}^{t}\frac{\partial^2 u(\xi_1,\xi_2,\eta_1)}{\partial \eta_1^2}d\eta_1 + \int\limits_{x_1}^{\xi_1}\frac{\partial^2 u(\xi_3,\xi_2,\eta)}{\partial \xi_3 \partial \eta}d\xi_3 + \int\limits_{x_2}^{\xi_2}\frac{\partial^2 u(x_1,\xi_4,\eta)}{\partial \xi_4 \partial \eta}d\xi_4\right) =$$

$$= \frac{1}{\tau h^4}\int\limits_{x_1-h}^{x_1+h}(h-|x_1-\xi_1|)d\xi_1 \int\limits_{x_2-h}^{x_2+h}(h-|x_2-\xi_2|)d\xi_2 \int\limits_{t-\tau}^{t}d\eta\int\limits_{\eta}^{t}\frac{\partial^2 u(\xi_1,\xi_2,\eta_1)}{\partial \eta_1^2}d\eta_1 +$$

$$+ \frac{1}{\tau h^4}\int\limits_{x_1-h}^{x_1+h}(h-|x_1-\xi_1|)d\xi_1 \int\limits_{x_2-h}^{x_2+h}(h-|x_2-\xi_2|)d\xi_2 \times$$

$$\times \int\limits_{t-\tau}^{t}d\eta\int\limits_{x_1}^{\xi_1}d\xi_3\left(\frac{\partial^2 u(\xi_3,\xi_2,\eta)}{\partial \xi_3 \partial \eta} - \frac{1}{2h}\int\limits_{x_1-h}^{x_1+h}\frac{\partial^2 u(\xi_5,\xi_2,\eta)}{\partial \xi_5 \partial \eta}d\xi_5\right) +$$

$$+ \frac{1}{\tau h^4}\int\limits_{x_1-h}^{x_1+h}(h-|x_1-\xi_1|)d\xi_1 \int\limits_{x_2-h}^{x_2+h}(h-|x_2-\xi_2|)d\xi_2 \times$$

$$\times \int\limits_{t-\tau}^{t}d\eta\int\limits_{x_2}^{\xi_2}d\xi_4\left(\frac{\partial^2 u(x_1,\xi_4,\eta)}{\partial \xi_4 \partial \eta} - \frac{1}{4h^2}\int\limits_{x_1-h}^{x_1+h}d\xi_6\int\limits_{x_2-h}^{x_2+h}\frac{\partial^2 u(\xi_6,\xi_7,\eta)}{\partial \xi_5 \partial \eta}d\xi_7\right) =$$

$$= \frac{1}{\tau h^4} \int_{x_1-h}^{x_1+h} (h-|x_1-\xi_1|)d\xi_1 \int_{x_2-h}^{x_2+h} (h-|x_2-\xi_2|)d\xi_2 \cdot \int_{t-\tau}^{t} d\eta \int_{\eta}^{t} \frac{\partial^2 u(\xi_1,\xi_2,\eta_1)}{\partial \eta_1^2} d\eta_1 +$$

$$+ \frac{1}{2\tau h^5} \int_{x_1-h}^{x_1+h} (h-|x_1-\xi_1|)d\xi_1 \int_{x_2-h}^{x_2+h} (h-|x_2-\xi_2|)d\xi_2 \times$$

$$\times \int_{t-\tau}^{t} d\eta \int_{x_1}^{\xi_1} d\xi_3 \int_{x_1-h}^{x_1+h} d\xi_5 \int_{\xi_5}^{\xi_3} \frac{\partial^3 u(\xi_8,\xi_2,\eta)}{\partial \xi_8^2 \partial \eta} d\xi_8 +$$

$$+ \frac{1}{4\tau h^6} \int_{x_1-h}^{x_1+h} (h-|x_1-\xi_1|)d\xi_1 \int_{x_2-h}^{x_2+h} (h-|x_2-\xi_2|)d\xi_2 \times$$

$$\times \int_{t-\tau}^{t} d\eta \int_{x_2}^{\xi_2} d\xi_4 \int_{x_1-h}^{x_1+h} d\xi_6 \int_{x_2-h}^{x_2+h} d\xi_7 \left(\int_{\xi_6}^{x_1} \frac{\partial^3 u(\xi_9,\xi_4,\eta)}{\partial \xi_9 \partial \xi_4 \partial \eta} d\xi_9 + \int_{\xi_7}^{\xi_4} \frac{\partial^3 u(\xi_6,\xi_{10},\eta)}{\partial \xi_{10}^2 \partial \eta} d\xi_{10} \right),$$

which brings us to the estimate

$$|\eta_3(x,t)| \leq \tau \left\| \frac{\partial^2 u(x_1,x_2,t)}{\partial t^2} \right\|_{L_\infty(D_T)} + \qquad [2.46]$$

$$+ 4h^2 \left(\left\| \frac{\partial^3 u(x_1,x_2,t)}{\partial x_1^2 \partial t} \right\|_{L_\infty(D_T)} + \left\| \frac{\partial^3 u(x_1,x_2,t)}{\partial x_2^2 \partial t} \right\|_{L_\infty(D_T)} + \left\| \frac{\partial^3 u(x_1,x_2,t)}{\partial x_1 \partial x_2 \partial t} \right\|_{L_\infty(D_T)} \right).$$

The statement of the lemma follows now from estimates [2.44], [2.45], [2.46].

In the next proposition, we study the function $v(x,t)$ from estimate [2.35].

LEMMA 2.10.– *The function $v(x,t)$ uniformly in $t \in \omega_\tau$ satisfies the relations $v(x,t) = O(h)$ and $v(x,t) = O(h^2 \ln h^{-1})$ near the sides and the vertices of the square D respectively.*

PROOF.– At the node $(x,t) \in \omega \times \omega_\tau$, we have

$$|v(x,t)| = \left| \sum_{k_1=1}^{N-1} \sum_{k_2=1}^{N-1} \left(1 - \left(\frac{1}{1+\tau\lambda_{k_1 k_2}}\right)^{t/\tau}\right) \times \right.$$

$$\left. \times \frac{h^4 \left(1-(-1)^{k_1}\right)\left(1-(-1)^{k_2}\right) \cos\frac{k_1 \pi h}{2} \cos\frac{k_2 \pi h}{2}}{4 \sin\frac{k_1 \pi h}{2} \sin\frac{k_2 \pi h}{2} \left(\sin^2\frac{k_1 \pi h}{2} + \sin^2\frac{k_2 \pi h}{2}\right)} \sin k_1 \pi x_1 \cdot \sin k_2 \pi x_2 \right| \le$$

$$\le \sum_{k_1=1}^{N-1} \sum_{k_2=1}^{N-1} \frac{1}{k_1 k_2 \left(k_1^2 + k_2^2\right)} \le \sum_{k_1=1}^{N-1} \sum_{k_2=1}^{N-1} \frac{1}{2 k_1^2 k_2^2} < \frac{1}{2} \left(\sum_{k=1}^{\infty} \frac{1}{k^2}\right)^2 = \frac{\pi^4}{72}.$$

Next, we consider the function $v(x,t)$ at the nodes close to the boundary: (h, x_2, t), $(1-h, x_2, t)$, (x_1, h, t), $(x_1, 1-h, t)$. For example, at (h, x_2, t), we have

$$|v(h, x_2, t)| = \left| \sum_{k_1=1}^{N-1} \sum_{k_2=1}^{N-1} \left(1 - \left(\frac{1}{1+\tau\lambda_{k_1 k_2}}\right)^{t/\tau}\right) \times \right.$$

$$\left. \times \frac{h^4 \left(1-(-1)^{k_1}\right)\left(1-(-1)^{k_2}\right) \cos\frac{k_1 \pi h}{2} \cos\frac{k_2 \pi h}{2}}{4 \sin\frac{k_1 \pi h}{2} \sin\frac{k_2 \pi h}{2} \left(\sin^2\frac{k_1 \pi h}{2} + \sin^2\frac{k_2 \pi h}{2}\right)} \sin k_1 \pi h \cdot \sin k_2 \pi x_2 \right| \le$$

$$\le 2h \sum_{k_1=1}^{N-1} \sum_{k_2=1}^{N-1} \frac{1}{k_2 \left(k_1^2 + k_2^2\right)} < 2h \sum_{k_2=1}^{N-1} \frac{1}{k_2} \sum_{k_1=1}^{\infty} \frac{1}{k_1^2 + k_2^2} =$$

$$= 2h \sum_{k_2=1}^{N-1} \frac{1}{k_2} \frac{\pi k_2 \operatorname{cth}(\pi k_2) - 1}{2 k_2^2} < 2h \sum_{k_2=1}^{\infty} \frac{\pi k_2 \operatorname{cth}(\pi k_2) - 1}{2 k_2^3} < 4h.$$

Finally, we study $v(x,t)$ near the vertices of the square D. For example, at the node (h, h, t), we obtain

$$|v(h,h,t)| = \left| \sum_{k_1=1}^{N-1} \sum_{k_2=1}^{N-1} \left(1 - \left(\frac{1}{1+\tau\lambda_{k_1 k_2}}\right)^{t/\tau}\right) \times \right.$$

$$\times \frac{h^4\left(1-(-1)^{k_1}\right)\left(1-(-1)^{k_2}\right)\cos\dfrac{k_1\pi h}{2}\cos\dfrac{k_2\pi h}{2}}{4\sin\dfrac{k_1\pi h}{2}\sin\dfrac{k_2\pi h}{2}\left(\sin^2\dfrac{k_1\pi h}{2}+\sin^2\dfrac{k_2\pi h}{2}\right)}\sin k_1\pi h\cdot\sin k_2\pi h \Bigg| \leq$$

$$\leq 4h^2 \sum_{k_1=1}^{N-1}\sum_{k_2=1}^{N-1}\frac{1}{k_1^2+k_2^2} < 4h^2\sum_{k_1=1}^{N-1}\frac{\pi k_1\,\text{cth}(\pi k_1)-1}{2k_1^2} < 2h^2\pi\,\text{cth}\,\pi\sum_{k_1=1}^{N-1}\frac{1}{k_1} <$$

$$< 2h^2\pi\,\text{cth}\,\pi\left(1+\sum_{k_1=2}^{N}\frac{1}{k_1}\right)\leq 2h^2\pi\,\text{cth}\,\pi\left(1+\int_1^N\frac{dx}{x}\right)=h^2\pi\,\text{cth}\,\pi(1+\ln N)<$$

$$< 2\pi\,\text{cth}\,\pi\,h^2\ln\frac{1}{h}.$$

The lemma is proved.

Lemmas 2.8–2.10 bring us to the main result of this section.

THEOREM 2.4.– Let the solution $u(x,t)$ of problem [2.30] satisfy the condition $u\in W_\infty^4(D_T)$. Then, the accuracy of the finite-difference scheme [2.32] is characterized by the estimate which takes into account the influence of the boundary condition:

$$|z(x,t)|\leq M v(x,t)(\tau+h^2),\quad (x,t)\in\omega\times\omega_\tau,\qquad [2.47]$$

with a positive constant M independent of h and τ and the function $v(x,t)$ uniformly in $t\in\omega_\tau$ satisfying the relations $v(x,t)=O(h)$ and $v(x,t)=O(h^2\ln h^{-1})$ near the sides and the vertices of the square D respectively.

2.3.3. Accuracy with the initial effect

Now, we study the behavior of the solution $y(x,t)$ near the bottom side of the space–time parallelepiped D_T.

THEOREM 2.5.– Let the solution $u(x,t)$ of problem [2.30] satisfy the condition $u\in W_\infty^4(D_T)$. Then, the accuracy of the finite-difference scheme [2.32] is characterized by the estimate, which takes into account the influence of the initial condition:

$$\|z(\cdot,\tau)\| \leq M\tau(\tau+h^2),$$

where M is a positive constant independent of h and τ.

PROOF.– We consider the function $v(x,t)$ at $t = \tau$:

$$v(x,\tau) = \sum_{k_1=1}^{N-1}\sum_{k_2=1}^{N-1} \frac{\tau}{1+\tau\lambda_{k_1 k_2}} \frac{c_{k_1 k_2}}{\lambda_{k_1 k_2}} w_{k_1 k_2}(x), \quad x \in \omega.$$

Due to the relation

$$\lambda_{k_1 k_2} = \frac{4}{h^2}\sin^2\frac{k_1\pi h}{2} + \frac{4}{h^2}\sin^2\frac{k_2\pi h}{2} \geq 4(k_1^2+k_1^2) \geq 8$$

we obtain the inequality

$$\|v(\cdot,\tau)\|^2 = \sum_{k_1=1}^{N-1}\sum_{k_2=1}^{N-1}\left(\frac{\tau}{1+\tau\lambda_{k_1 k_2}}\frac{c_{k_1 k_2}}{\lambda_{k_1 k_2}}\right)^2 < \tau^2 \sum_{k_1=1}^{N-1}\sum_{k_2=1}^{N-1}\left(\frac{c_{k_1 k_2}}{\lambda_{k_1 k_2}}\right)^2 \leq$$

$$\leq \frac{\tau^2}{64}\sum_{k_1=1}^{N-1}\sum_{k_2=1}^{N-1}c_{k_1 k_2}^2 = \frac{\tau^2}{64}\|1\|^2 = \frac{\tau^2}{64}\sum_{k_1=1}^{N-1}\sum_{k_2=1}^{N-1}h^2 = \frac{\tau^2}{64}(1-h)^2 < \frac{\tau^2}{64},$$

which together with estimate [2.47] proves the theorem.

2.3.4. Conclusion

The weighted a priori estimate obtained in Theorem 2.4 indicates that in the uniform mesh norm $C(\omega \times \omega_\tau)$ the accuracy of scheme [2.32] is $O(h(\tau+h^2))$ and $O(h^2\ln h^{-1}(\tau+h^2))$ near the side faces and the side edges of the space–time parallelepiped $D_T = D\times(0,T)$ respectively, while it is $O(\tau+h^2)$ further away from them.

The weighted a priori estimate obtained in Theorem 2.5 shows that the accuracy of scheme [2.32] in the mesh norm $L_2(\omega)$ near the bottom side $t = 0$ of the space–time rectangle $D_T = D \times (0,T)$ is $O\left(\tau(\tau + h^2)\right)$.

3

Differential Equations with Fractional Derivatives

3.1. BVP for a differential equation with constant coefficients and a fractional derivative of order ½

3.1.1. *A weighted estimate for the exact solution*

We begin with the following BVP:

$$u''(x) - \frac{\alpha}{\sqrt{\pi}} \frac{d}{dx} \int_0^x \frac{u(y)}{\sqrt{x-y}} dy = -f(x), \quad x \in (0,1),$$ [3.1]

$$u(0) = 0, \ u(1) = 0.$$

The second term on the left-hand side of the equation, up to a constant factor α, is the Riemann–Liouville fractional derivative of order $1/2$.

Throughout this section, we use the norms

$$\| w \|_{k,\infty} = \max_{0 \leq x \leq 1} |w^{(k)}(x)|, \ k = 1, 2, \ldots, \ \| w \|_\infty = \max_{0 \leq x \leq 1} |w(x)|.$$

THEOREM 3.1.– *Let the parameter α in equation [3.1] satisfy the condition*

$$\frac{2\alpha}{\sqrt{\pi}} K < 1,$$ [3.2]

where $K = \max\limits_{0 \le x \le 1} \varphi(x) = 0,3046916809...$ and

$$\varphi(x) = \frac{1}{x(1-x)}\left\{x\int_x^1 y(1-y)^{3/2}dy + (1-x)\int_0^x y\sqrt{1-y}\,dy\right\}.$$

Then, for the solution $u(x)$, the following estimate holds true:

$$\left\|\frac{u}{x(1-x)}\right\|_\infty \le \frac{1}{2}\left(1 - \frac{2\alpha}{\sqrt{\pi}}K\right)^{-1}\|f\|_\infty.$$

PROOF.– From problem [3.1], we obtain the Fredholm integral equation of the second kind

$$u(x) + \frac{2\alpha}{\sqrt{\pi}}\left\{x\int_0^1 \sqrt{1-y}\,u(y)dy - \int_0^x \sqrt{x-y}\,u(y)dy\right\} = \int_0^1 G(x,\xi)f(\xi)d\xi, \quad x \in (0,1),$$

which can be rewritten in a slightly different way:

$$u(x) + \frac{2\alpha}{\sqrt{\pi}}\left\{x\int_x^1 \sqrt{1-y}\,u(y)dy + \int_0^x\left(x\sqrt{1-y} - \sqrt{x-y}\right)u(y)dy\right\} =$$

$$= \int_0^1 G(x,\xi)f(\xi)d\xi, \quad x \in (0,1),$$ [3.3]

with Green's function $G(x,\xi) = \begin{cases} x(1-\xi) & \text{for } x \le \xi, \\ \xi(1-x) & \text{for } \xi \le x, \end{cases}$ of the BVP

$$u''(x) = -f(x), x \in (0,1), \; u(0) = 0, \; u(1) = 0.$$

Substituting $v(x) = \dfrac{u(x)}{x(1-x)}$ in [3.3], we obtain the equation

$$v(x) + \frac{2\alpha}{\sqrt{\pi}x(1-x)}\left\{x\int_x^1 y(1-y)^{3/2}v(y)dy + \right.$$ [3.4]

Differential Equations with Fractional Derivatives 117

$$+\int_0^x y(1-y)\left(x\sqrt{1-y}-\sqrt{x-y}\right)v(y)dy\bigg\} = \int_0^1 \frac{G(x,\xi)}{x(1-x)}f(\xi)d\xi, \quad x \in (0,1).$$

For the expression in curly brackets, we have

$$\left| x\int_x^1 y(1-y)^{3/2}v(y)dy + \int_0^x y(1-y)\left(x\sqrt{1-y}-\sqrt{x-y}\right)v(y)dy \right| \le$$

$$\le \left\{ x\int_x^1 y(1-y)^{3/2} dy + \int_0^x y(1-y)\left|x\sqrt{1-y}-\sqrt{x-y}\right| dy \right\} \cdot \|v\|_\infty =$$

$$= \left\{ x\int_x^1 y(1-y)^{3/2} dy + (1-x)\int_0^x y(1-y)\frac{|y(1+x)-x|}{x\sqrt{1-y}+\sqrt{x-y}} dy \right\} \cdot \|v\|_\infty \le$$

$$\le \left\{ x\int_x^1 y(1-y)^{3/2} dy + (1-x)\int_0^x y\sqrt{1-y}\, dy \right\} \cdot \|v\|_\infty ,$$

and for the right-hand side of the equation, we have

$$\left| \int_0^1 \frac{G(x,\xi)}{x(1-x)}f(\xi)d\xi \right| \le \int_0^1 \frac{G(x,\xi)}{x(1-x)}d\xi \|f\|_\infty = \left(\frac{1}{x}\int_0^x \xi d\xi + \frac{1}{1-x}\int_x^1 (1-\xi)d\xi \right),$$

$$\|f\|_\infty = \frac{1}{2}\|f\|_\infty .$$

Then, [3.4] leads to the estimate

$$\|v\|_\infty \le \frac{2\alpha}{\sqrt{\pi}} K \|v\|_\infty + \frac{1}{2}\|f\|_\infty , \qquad [3.5]$$

where

$$K = \max_{0 \le x \le 1} \varphi(x) = 0.3046916809... ,$$

$$\varphi(x) = \frac{1}{x(1-x)}\left\{ x\int_x^1 y(1-y)^{3/2} dy + (1-x)\int_0^x y\sqrt{1-y}\, dy \right\}.$$

Since α satisfies inequality [3.2], then [3.5] gives the weighted estimate

$$\left\| \frac{u}{x(1-x)} \right\|_\infty \leq \frac{1}{2}\left(1 - \frac{2\alpha}{\sqrt{\pi}} K\right)^{-1} \| f \|_\infty,$$

which proves the theorem.

3.1.2. Weighted estimates for approximate solutions

To solve equation [3.3], we will use the mesh method.

We introduce a mesh set $\bar{\omega} = \{x_j = jh, \ j = 0,1,\ldots,N, \ h = 1/N\}$ and put $x = x_j$ in the integral equation [3.3]:

$$u(x_j) + \frac{2\alpha}{\sqrt{\pi}} \left\{ x_j \sum_{k=j+1}^{N} \int_{x_{k-1}}^{x_k} \sqrt{1-y}\, u(y)dy \right.$$

$$\left. + \sum_{k=1}^{j} \int_{x_{k-1}}^{x_k} \left(x_j\sqrt{1-y} - \sqrt{x_j - y}\right) u(y)dy \right\} = \int_0^1 G(x_j,\xi) f(\xi) d\xi, \quad j = 1,2,\ldots,N-1.$$

From here we obtain the mesh scheme

$$u^h(x_j) + \frac{2\alpha}{\sqrt{\pi}} \left\{ x_j \sum_{k=j+1}^{N} u^h(x_k) \int_{x_{k-1}}^{x_k} \sqrt{1-y}\, dy \right. \qquad [3.6]$$

$$\left. + \sum_{k=1}^{j} u^h(x_k) \int_{x_{k-1}}^{x_k} \left(x_j\sqrt{1-y} - \sqrt{x_j - y}\right) dy \right\} = \int_0^1 G(x_j,\xi) f(\xi) d\xi,$$

$$j = 1,2,\ldots,N-1, \ u^h(x_N) = 0.$$

It is easy to see that the error $z(x) = u^h(x) - u(x)$ is a solution of the following discrete problem:

$$z(x_j) + \frac{2\alpha}{\sqrt{\pi}} \left\{ x_j \sum_{k=j+1}^{N} z(x_k) \int_{x_{k-1}}^{x_k} \sqrt{1-y}\, dy \right. \qquad [3.7]$$

$$+ \sum_{k=1}^{j} z(x_k) \int_{x_{k-1}}^{x_k} \left(x_j \sqrt{1-y} - \sqrt{x_j - y} \right) dy \bigg\} = \psi(x_j),$$

$$j = 1, 2, \ldots, N-1, \quad z(x_N) = 0,$$

where $\psi(x_j)$ is the approximation error:

$$\psi(x_j) = \frac{2\alpha}{\sqrt{\pi}} \Bigg\{ x_j \sum_{k=j+1}^{N} \int_{x_{k-1}}^{x_k} \sqrt{1-y} \left(u(y) - u(x_k) \right) dy$$

$$+ \sum_{k=1}^{j} \int_{x_{k-1}}^{x_k} \left(x_j \sqrt{1-y} - \sqrt{x_j - y} \right) \left(u(y) - u(x_k) \right) dy \Bigg\}.$$

[3.8]

Substituting $Z(x) = \dfrac{z(x)}{x(1-x)}$ in [3.7], we obtain the discrete problem

$$Z(x_j) + \frac{2\alpha}{\sqrt{\pi} \, x_j (1 - x_j)} \Bigg\{ x_j \sum_{k=j+1}^{N} Z(x_k) x_k (1 - x_k) \int_{x_{k-1}}^{x_k} \sqrt{1-y} \, dy$$

$$+ \sum_{k=1}^{j} Z(x_k) x_k (1 - x_k) \int_{x_{k-1}}^{x_k} \left(x_j \sqrt{1-y} - \sqrt{x_j - y} \right) dy \Bigg\} = \frac{\psi(x_j)}{x_j (1 - x_j)},$$

[3.9]

$$j = 1, 2, \ldots, N-1, \quad Z(x_N) = 0.$$

Next, we introduce the notation $\| w \|_{\infty, \bar{\omega}} = \max\limits_{x \in \bar{\omega}} |w(x)|$.

We will now study the accuracy of the discrete problem [3.6] in some weighted norm.

THEOREM 3.2.– *Let the solution* $u(x)$ *of problem [3.3] satisfy the condition* $u \in C^1_{[0,1]}$ *and let the parameter* α *of problem [3.1] be such that* $1 - \dfrac{\alpha}{\sqrt{\pi}} > 0$. *Then, the accuracy of the mesh scheme [3.6] is characterized by the weighted estimate, which takes into account the boundary effect:*

$$\left\| \frac{u - u^h}{x(1-x)} \right\|_{\infty, \bar{\omega}} \leq M h \| u \|_{1,\infty},$$

where $M = \left(1 - \dfrac{\alpha}{\sqrt{\pi}}\right)^{-1} \dfrac{4\alpha}{\sqrt{\pi}}$ is a positive constant independent of h and $u(x)$.

PROOF.– From [3.9], we have

$$|Z(x_j)| \leq \frac{2\alpha}{\sqrt{\pi} x_j(1-x_j)} \left\{ x_j \int_{x_j}^{1} \sqrt{1-y}\, dy + \int_0^{x_j} \left| x_j\sqrt{1-y} - \sqrt{x_j - y} \right| dy \right\} \frac{1}{4} \| Z \|_{\infty, \bar{\omega}} +$$

$$+ \frac{2\alpha}{\sqrt{\pi} x_j(1-x_j)} \left\{ x_j \int_{x_j}^{1} \sqrt{1-y}\, dy + \int_0^{x_j} \left| x_j\sqrt{1-y} - \sqrt{x_j - y} \right| dy \right\} h \| u \|_{1,\infty} \leq$$

$$\leq \frac{2\alpha}{\sqrt{\pi} x_j(1-x_j)} \left\{ x_j \int_{x_j}^{1} \sqrt{1-y}\, dy + (1-x_j)\int_0^{x_j} \frac{|y(1+x_j) - x_j|}{x_j\sqrt{1-y} + \sqrt{x_j-y}} dy \right\} \frac{1}{4} \| Z \|_{\infty, \bar{\omega}} +$$

$$+ \frac{2\alpha}{\sqrt{\pi} x_j(1-x_j)} \left\{ x_j \int_{x_j}^{1} \sqrt{1-y}\, dy + (1-x_j)\int_0^{x_j} \frac{|y(1+x_j) - x_j|}{x_j\sqrt{1-y} + \sqrt{x_j-y}} dy \right\} h \| u \|_{1,\infty} \leq$$

$$\leq \frac{2\alpha}{\sqrt{\pi} x_j(1-x_j)} \left\{ x_j \int_{x_j}^{1} \sqrt{1-y}\, dy + (1-x_j)\int_0^{x_j} \frac{1}{\sqrt{1-y}} dy \right\} \frac{1}{4} \| Z \|_{\infty, \bar{\omega}} +$$

$$+ \frac{2\alpha}{\sqrt{\pi} x_j(1-x_j)} \left\{ x_j \int_{x_j}^{1} \sqrt{1-y}\, dy + (1-x_j)\int_0^{x_j} \frac{1}{\sqrt{1-y}} dy \right\} h \| u \|_{1,\infty}, \quad [3.10]$$

$$j = 1, 2, \ldots, N-1.$$

Next, we find

$$\max_{0 \leq x \leq 1} \frac{x \int_x^1 \sqrt{1-y}\, dy + (1-x)\int_0^x \frac{1}{\sqrt{1-y}}\, dy}{x(1-x)} = \quad [3.11]$$

$$= \max_{0 \le x \le 1}\left(\frac{2}{3}\sqrt{1-x} + \frac{2}{1+\sqrt{1-x}}\right) = 2.$$

Then, [3.10] implies the inequality

$$\|Z\|_{\infty,\bar{\omega}} \le \frac{\alpha}{\sqrt{\pi}}\|Z\|_{\infty,\bar{\omega}} + \frac{4\alpha}{\sqrt{\pi}}h\|u\|_{1,\infty}.$$

Due to the condition $1 - \frac{\alpha}{\sqrt{\pi}} > 0$, we finally get

$$\|Z\|_{\infty,\bar{\omega}} = \left\|\frac{z(x)}{x(1-x)}\right\|_{\infty,\bar{\omega}} = \left\|\frac{u - u^h}{x(1-x)}\right\|_{\infty,\bar{\omega}} \le h\left(1 - \frac{\alpha}{\sqrt{\pi}}\right)^{-1}\frac{4\alpha}{\sqrt{\pi}}\|u\|_{1,\infty}.$$

The theorem is proved.

We will now construct and study the mesh scheme of the second order of approximation:

$$u^h(x_j) + \frac{2\alpha}{\sqrt{\pi}}\left\{x_j \sum_{k=j+1}^{N}\int_{x_{k-1}}^{x_k}\sqrt{1-y}\,L_k(y;u^h)dy + \right. \qquad [3.12]$$

$$\left. + \sum_{k=1}^{j}\int_{x_{k-1}}^{x_k}\left(x_j\sqrt{1-y} - \sqrt{x_j - y}\right)L_k(y;u^h)dy\right\} = \int_{0}^{1}G(x_j,\xi)f(\xi)d\xi,$$

$$j = 1, 2, \ldots, N-1, \quad u^h(x_N) = 0,$$

where $L_k(y; w)$ is the Lagrange interpolating polynomial of the first degree for the function $w(x)$:

$$L_k(y; w) = \frac{y - x_{k-1}}{x_k - x_{k-1}}w(x_k) + \frac{x_k - y}{x_k - x_{k-1}}w(x_{k-1}).$$

For the error $z(x) = u^h(x) - u(x)$, we have the mesh scheme

$$z(x_j) + \frac{2\alpha}{\sqrt{\pi}}\left\{x_j \sum_{k=j+1}^{N}\int_{x_{k-1}}^{x_k}\sqrt{1-y}\,L_k(y;z)dy + \right.$$

$$+\sum_{k=1}^{j} \int_{x_{k-1}}^{x_k} \left(x_j\sqrt{1-y} - \sqrt{x_j - y}\right) L_k(y;z)dy \Bigg\} = \psi(x_j),$$

$$j = 1, 2, \ldots, N-1, \quad z(x_N) = 0,$$

where $\psi(x_j)$ is the approximation error defined in [3.8].

Then, the function $Z(x) = \dfrac{z(x)}{x(1-x)}$ is a solution of the mesh problem

$$Z(x_j) + \frac{2\alpha}{\sqrt{\pi} x_j (1-x_j)} \Bigg\{ x_j \sum_{k=j+1}^{N} \int_{x_{k-1}}^{x_k} \sqrt{1-y}\, L_k\left(y; x(1-x)Z\right) dy +$$

$$+ \sum_{k=1}^{j} \int_{x_{k-1}}^{x_k} \left(x_j\sqrt{1-y} - \sqrt{x_j - y}\right) L_k\left(y; x(1-x)Z\right) dy \Bigg\} = \frac{\psi(x_j)}{x_j(1-x_j)}, \quad [3.13]$$

$$j = 1, 2, \ldots, N-1, \quad Z(x_N) = 0.$$

We study the accuracy of the mesh scheme [3.12] in the following proposition.

THEOREM 3.3.– *Let the solution $u(x)$ of problem [3.1] satisfy the condition $u \in C^2_{[0,1]}$ and let the parameter α be such that $1 - \alpha/\sqrt{\pi} > 0$. Then, the accuracy of the mesh scheme [3.12] is characterized by the weighted estimate taking into account the boundary effect:*

$$\left\| \frac{u - u^h}{x(1-x)} \right\|_{\infty, \bar{\omega}} \leq M h^2 \|u\|_{2,\infty},$$

where $M = \left(1 - \dfrac{\alpha}{\sqrt{\pi}}\right)^{-1} \dfrac{2\alpha}{\sqrt{\pi}}$ *is a positive constant independent of h and $u(x)$.*

PROOF.– Making use of the formula

$$u(x) - L_k(x; u) = \frac{u''(\bar{x}_k)}{2!}(x - x_{k-1})(x - x_k), \quad \bar{x}_k \in (x_{k-1}, x_k),$$

from [3.13], we have the inequality

$$|Z(x_j)| \leq \frac{2\alpha}{\sqrt{\pi x_j(1-x_j)}} \left\{ x_j \int_{x_j}^{1} \sqrt{1-y}\, dy + (1-x_j)\int_{0}^{x_j} \frac{1}{\sqrt{1-y}}\, dy \right\} \frac{1}{4} \|Z\|_{\infty, \bar{\omega}} +$$

$$+ \frac{2\alpha}{\sqrt{\pi x_j(1-x_j)}} \left\{ x_j \int_{x_j}^{1} \sqrt{1-y}\, dy + (1-x_j)\int_{0}^{x_j} \frac{1}{\sqrt{1-y}}\, dy \right\} \frac{h^2}{2} \|u\|_{2,\infty},$$

$$j = 1, 2, \ldots, N-1,$$

which combined with [3.11] gives the estimate

$$\|Z\|_{\infty, \bar{\omega}} \leq \frac{\alpha}{\sqrt{\pi}} \|Z\|_{\infty, \bar{\omega}} + \frac{2\alpha}{\sqrt{\pi}} h^2 \|u\|_{2,\infty}.$$

Since $1 - \frac{\alpha}{\sqrt{\pi}} > 0$, then

$$\|Z\|_{\infty, \bar{\omega}} = \left\| \frac{z(x)}{x(1-x)} \right\|_{\infty, \bar{\omega}} = \left\| \frac{u - u^h}{x(1-x)} \right\|_{\infty, \bar{\omega}} \leq h^2 \left(1 - \frac{\alpha}{\sqrt{\pi}}\right)^{-1} \frac{2\alpha}{\sqrt{\pi}} \|u\|_{2,\infty}.$$

The theorem is proved.

3.1.3. Conclusion

In this section, we will briefly summarize what has been done above.

In Theorem 3.1, we obtained the weighted estimate for the exact solution of the Dirichlet boundary value problem for the ODE with a fractional derivative of order $1/2$.

Next, we approximated this problem using two mesh schemes and studied their accuracy in the discrete norm $C(\omega)$ with a particular weight function. The weighted estimate obtained in Theorem 3.2 shows that the error of the mesh scheme [3.6] is $O(h^2)$ near the boundary points of the interval $(0,1)$, whereas it is $O(h)$ in its inner points. Similarly, the weighted estimate obtained in Theorem 3.3 indicates that the error of the mesh scheme [3.12] is $O(h^3)$ near the boundary points of the interval $(0,1)$, whereas it is $O(h^2)$ inside of the interval.

In the next section, some of these results are generalized for the case of the fractional derivative of order $\alpha \in (0,1)$.

3.2. BVP for a differential equation with constant coefficients and a fractional derivative of order $\alpha \in (0,1)$

3.2.1. *A scale of weighted estimates for the exact solution*

We address now the problem

$$u''(x) - (D_{0+}^{\alpha} u)(x) = -f(x), \quad x \in (0,1),$$
$$u(0) = 0, \quad u(1) = 0,$$
[3.14]

with the Riemann–Liouville derivative of order $\alpha > 0$ (Samko et al. 1993):

$$(D_{a+}^{\alpha} f)(x) = \frac{1}{\Gamma(n-\alpha)} \frac{d^n}{dx^n} \int_a^x \frac{f(t)}{(x-t)^{\alpha-n+1}} dt,$$

where $n = \lfloor \alpha \rfloor + 1$, $\lfloor \alpha \rfloor$ is the floor function and $\Gamma(\cdot)$ is the gamma function.

We will use the standard notation (Evans 2010) for norm in the spaces $C_{[0,1]}^k$ and $C_{[0,1]}^{k,\lambda}$, namely:

$$\|u\|_{\infty} \equiv \|u\|_{C_{[0,1]}} = \sup_{0<x<1} |u(x)|,$$

$$\|u\|_{k,\infty} \equiv \|u\|_{C_{[0,1]}^k} = \sum_{l=0}^{k} \|u^{(l)}\|_{C_{[0,1]}}, \quad k \in \mathbb{N},$$

$$\|u\|_{k,\lambda} \equiv \|u\|_{C_{[0,1]}^{k,\lambda}} = \|u\|_{C_{[0,1]}^k} + |u|_{C_{[0,1]}^{k,\lambda}}, \quad k \in \mathbb{N} \cup \{0\},$$

$$|u|_{C_{[0,1]}^{k,\lambda}} = \sup_{\substack{x,y \in [0,1] \\ x \neq y}} \frac{|u^{(k)}(x) - u^{(k)}(y)|}{|x-y|^{\lambda}},$$

where $|u|_{k,\lambda} \equiv |u|_{C_{[0,1]}^{k,\lambda}}$ is a seminorm in the Hölder space $C_{[0,1]}^{k,\lambda}$ with the exponent λ ($0 < \lambda \leq 1$).

We start with problem [3.14] for $\alpha = 0$:

$$u''(x) - u(x) = -f(x), \quad x \in (0,1),$$
$$u(0) = 0, \quad u(1) = 0. \qquad [3.15]$$

It is known (e.g. Bitsadze (1981)) that for $f(x) \in C_{[0,1]}$ problem [3.15] has a unique solution $u(x) \in C^2_{(0,1)} \cap C_{[0,1]}$, which is also a solution of the following integral equation:

$$u(x) + \int_0^1 G(x,\xi) u(\xi) d\xi = \int_0^1 G(x,\xi) f(\xi) d\xi, \qquad [3.16]$$

where

$$G(x,\xi) = \begin{cases} x(1-\xi), & x \le \xi, \\ \xi(1-x), & \xi \le x, \end{cases} \qquad [3.17]$$

is Green's function of the BVP

$$u''(x) = -f(x), \quad x \in (0,1), \ u(0) = 0, \ u(1) = 0.$$

Substituting $v(x) = \dfrac{u(x)}{x(1-x)}$ in [3.16], we obtain the equation

$$v(x) + \frac{1}{x(1-x)} \left\{ (1-x) \int_0^x \xi^2 (1-\xi) v(\xi) d\xi + x \int_x^1 \xi (1-\xi)^2 v(\xi) d\xi \right\} =$$
$$= \frac{1}{x(1-x)} \int_0^1 G(x,\xi) f(\xi) d\xi, \quad x \in (0,1). \qquad [3.18]$$

This leads to the inequality

$$\|v\|_\infty \le \sup_{0<x<1} \frac{1}{x(1-x)} \left\{ (1-x) \int_0^x \xi^2 (1-\xi) d\xi + x \int_x^1 \xi (1-\xi)^2 d\xi \right\} \|v\|_\infty +$$
$$+ \sup_{0<x<1} \frac{1}{x(1-x)} \int_0^1 G(x,\xi) d\xi \, \|f\|_\infty . \qquad [3.19]$$

Now we find

$$\sup_{0<x<1} \frac{1}{x(1-x)} \left\{ (1-x)\int_0^x \xi^2(1-\xi)d\xi + x\int_x^1 \xi(1-\xi)^2 d\xi \right\} =$$

$$= \sup_{0<x<1} \frac{1}{x(1-x)} \left\{ (1-x)\left(\frac{x^3}{3} - \frac{x^4}{4}\right) + x\left(\frac{(1-x)^3}{3} - \frac{(1-x)^4}{4}\right) \right\} =$$

$$= \sup_{0<x<1} \left(\frac{x^2}{3} - \frac{x^3}{4} + \frac{(1-x)^2}{3} - \frac{(1-x)^3}{4} \right) = \sup_{0<x<1} \frac{1}{12}(1+x-x^2) = \frac{5}{48}$$

and

$$\sup_{0<x<1} \frac{1}{x(1-x)} \int_0^1 G(x,\xi)d\xi = \sup_{0<x<1} \frac{1}{x(1-x)} \left\{ (1-x)\int_0^x \xi d\xi + x\int_x^1 (1-\xi)d\xi \right\} = \frac{1}{2}$$

and then rewrite [3.19] as follows:

$$\|v\|_\infty \le \frac{5}{48} \|v\|_\infty + \frac{1}{2} \|f\|_\infty .$$

Hence, for the solution $u(x)$ of problem [3.15] we obtain the weighted estimate, which takes into account the boundary effect:

$$\left\| \frac{u(x)}{x(1-x)} \right\|_\infty \le \frac{24}{43} \|f\|_\infty . \qquad [3.20]$$

Next, we consider problem [3.14] for $\alpha = 1$. Note that for $n = \lfloor \alpha \rfloor + 1 = 2$, the Riemann–Liouville derivative takes the form

$$(D_{0+}^1 u)(x) = \frac{1}{\Gamma(1)} \frac{d^2}{dx^2} \int_0^x u(t)dt = u'(x) .$$

Then, problem [3.14] turns into the following:

$$u''(x) - u'(x) = -f(x), \quad x \in (0,1),$$
$$u(0) = 0, \quad u(1) = 0. \qquad [3.21]$$

Similar to problem [3.15], problem [3.21] for $f(x) \in C_{[0,1]}$ is uniquely solvable in the class $C^2_{(0,1)} \cap C_{[0,1]}$ and its solution $u(x)$ is a solution of the integral equation

$$u(x) + \left\{ -(1-x)\int_0^x u(\xi)d\xi + x\int_x^1 u(\xi)d\xi \right\} = \int_0^1 G(x,\xi)f(\xi)d\xi, \qquad [3.22]$$

with Green's function $G(x,\xi)$ defined in [3.17].

Substituting $v(x) = \dfrac{u(x)}{x(1-x)}$ in [3.22], we obtain the equation

$$v(x) + \frac{1}{x(1-x)}\left\{ -(1-x)\int_0^x \xi(1-\xi)v(\xi)d\xi + x\int_x^1 \xi(1-\xi)v(\xi)d\xi \right\} =$$

$$= \frac{1}{x(1-x)}\int_0^1 G(x,\xi)f(\xi)d\xi, \quad x \in (0,1). \qquad [3.23]$$

This gives the inequality

$$\|v\|_\infty \le \sup_{0<x<1} \frac{1}{x(1-x)}\left\{ (1-x)\int_0^x \xi(1-\xi)d\xi + x\int_x^1 (1-\xi)\xi\,d\xi \right\} \|v\|_\infty +$$

$$+ \sup_{0<x<1} \frac{1}{x(1-x)} \int_0^1 G(x,\xi)d\xi \, \|f\|_\infty . \qquad [3.24]$$

Due to the calculations

$$\sup_{0<x<1} \frac{1}{x(1-x)}\left\{ (1-x)\int_0^x \xi(1-\xi)d\xi + x\int_x^1 \xi(1-\xi)d\xi \right\} =$$

$$= \sup_{0<x<1} \frac{1}{x(1-x)}\left\{ (1-x)\left(\frac{x^2}{2} - \frac{x^3}{3}\right) + x\left(\frac{(1-x)^2}{2} - \frac{(1-x)^3}{3}\right) \right\} =$$

$$= \sup_{0<x<1}\left(\frac{x}{2} - \frac{x^2}{3} + \frac{1-x}{2} - \frac{(1-x)^2}{3} \right) = \sup_{0<x<1} \frac{1}{6}(1 + 4x - 4x^2) = \frac{1}{3}$$

and $\sup_{0<x<1} \dfrac{1}{x(1-x)} \int_0^1 G(x,\xi)d\xi = \dfrac{1}{2}$, we can rewrite [3.24] as follows:

$$\|v\|_\infty \leq \frac{1}{3}\|v\|_\infty + \frac{1}{2}\|f\|_\infty. \qquad [3.25]$$

Hence, for the solution $u(x)$ of problem [3.21], the following estimate holds true:

$$\left\|\frac{u(x)}{x(1-x)}\right\|_\infty \leq \frac{3}{4}\|f\|_\infty. \qquad [3.26]$$

and it takes into account the boundary effect.

Finally, we will consider the main case – problem [3.14] for $0 < \alpha < 1$:

$$Lu \equiv u''(x) - \frac{1}{\Gamma(1-\alpha)}\frac{d}{dx}\int_0^x \frac{u(t)dt}{(x-t)^\alpha} = -f(x), \quad x \in (0,1), \qquad [3.27]$$
$$u(0) = 0, \quad u(1) = 0.$$

First, we prove the maximum principle.

LEMMA 3.1.– *If $u(x) \in C^2_{(0,1)} \cap C_{[0,1]}$ is the solution of the homogeneous equation $Lu = 0$ and different from a constant, then the function $u(x)$ can reach its positive maximum (negative minimum) only at the ends of the interval $[0,1]$.*

PROOF.– We will assume the opposite. Let, on the contrary,

$$\max_{0 \leq x \leq 1} u(x) = u(x_0) > 0, \quad 0 < x_0 < 1.$$

Then, $u'(x_0) = 0$, $u''(x_0) \leq 0$. Due to the relation (e.g. Samko et al. (1993))

$$(D_{0+}^\alpha u)(x) = \frac{1}{\Gamma(1-\alpha)}\left(\frac{u(x)}{x^\alpha} + \lim_{t\to x}\frac{u(t)-u(x)}{(x-t)^\alpha} + \alpha\int_0^x \frac{u(x)-u(t)}{(x-t)^{\alpha+1}}dt\right) =$$

$$= \frac{1}{\Gamma(1-\alpha)}\left(\frac{u(x)}{x^\alpha} + \alpha\int_0^x \frac{u(x)-u(t)}{(x-t)^{\alpha+1}}dt\right), \quad x \in (0,1),$$

we have $(D_{0+}^{\alpha}u)(x_0) > 0$. Since for $f(x) \equiv 0$, equation [3.27] turns into

$$u''(x) = (D_{0+}^{\alpha}u)(x), \quad x \in (0,1),$$

then $u''(x_0) > 0$, and we get the contradiction. Therefore, at the inner point of the interval [0,1], the function $u(x)$ cannot reach its positive maximum. Similarly, at the inner point of the interval [0,1], the function $u(x)$ cannot reach its negative minimum. The lemma is proved.

COROLLARY 3.1.– *The homogeneous (i.e. $f(x) \equiv 0$) problem [3.27] has only a trivial solution $u(x) \equiv 0$ in the class $C_{(0,1)}^2 \cap C_{[0,1]}$.*

THEOREM 3.4.– *For $f(x) \in C_{[0,1]}$, the Dirichlet BVP [3.27] is uniquely solvable in the class $C_{(0,1)}^2 \cap C_{[0,1]}$.*

PROOF.– Let $u(x)$ be a solution of problem [3.27]. Performing the transformations

$$\int_0^1 u''(\xi)G(x,\xi)d\xi = (1-x)\int_0^x \xi u''(\xi)d\xi + x\int_x^1 (1-\xi)u''(\xi)d\xi = -u(x)$$

and

$$\int_0^1 G(x,\xi)\frac{d}{d\xi}\int_0^{\xi}\frac{u(t)dt}{(\xi-t)^{\alpha}}d\xi = (1-x)\int_0^x \xi \frac{d}{d\xi}\int_0^{\xi}\frac{u(t)dt}{(\xi-t)^{\alpha}}d\xi + x\int_x^1 (1-\xi)\frac{d}{d\xi}\int_0^{\xi}\frac{u(t)dt}{(\xi-t)^{\alpha}}d\xi =$$

$$= (1-x)\left(\xi\int_0^{\xi}\frac{u(t)dt}{(\xi-t)^{\alpha}}\bigg|_{\xi=0}^{\xi=x} - \int_0^x d\xi \int_0^{\xi}\frac{u(t)dt}{(\xi-t)^{\alpha}}\right) + x\left((1-\xi)\int_0^{\xi}\frac{u(t)dt}{(\xi-t)^{\alpha}}\bigg|_{\xi=x}^{\xi=1} + \int_x^1 d\xi \int_0^{\xi}\frac{u(t)dt}{(\xi-t)^{\alpha}}\right) =$$

$$= (1-x)\left(x\int_0^x \frac{u(t)dt}{(x-t)^{\alpha}} - \int_0^x u(t)dt \int_t^x \frac{d\xi}{(\xi-t)^{\alpha}}\right) +$$

$$+ x\left(-(1-x)\int_0^x \frac{u(t)dt}{(x-t)^{\alpha}} + \int_0^x u(t)dt \int_x^1 \frac{d\xi}{(\xi-t)^{\alpha}} + \int_x^1 u(t)dt \int_t^1 \frac{d\xi}{(\xi-t)^{\alpha}}\right) =$$

$$= -\frac{1-x}{1-\alpha}\int_0^x u(t)(x-t)^{1-\alpha}dt$$

$$+\frac{x}{1-\alpha}\left(\int_0^x u(t)\left((1-t)^{1-\alpha} - (x-t)^{1-\alpha}\right)dt + \int_x^1 u(t)(1-t)^{1-\alpha}dt\right),$$

we get that $u(x)$ satisfies the Fredholm integral equation of the second kind

$$u(x) + \frac{1}{\Gamma(2-\alpha)}\left(-(1-x)\int_0^x (x-t)^{1-\alpha}u(t)dt + \right.$$

$$\left. + x\int_0^x \left((1-t)^{1-\alpha} - (x-t)^{1-\alpha}\right)u(t)dt + x\int_x^1 (1-t)^{1-\alpha}u(t)dt\right) = \quad [3.28]$$

$$= \int_0^1 G(x,\xi)f(\xi)d\xi, \quad x \in [0,1].$$

It is easy to show that the inverse statement is also true, i.e. a solution of the integral equation [3.28] is a solution of BVP [3.27]. Thus, the Dirichlet BVP [3.27] is equivalent to the integral equation [3.28].

Since the homogeneous problem [3.27] has only a trivial solution $u(x) \equiv 0$, then the homogeneous (i.e. for $f(x) \equiv 0$) Fredholm equation [3.28] also has only a trivial solution. Due to the Fredholm alternative, the inhomogeneous Fredholm equation [3.28] is uniquely solvable and therefore the Dirichlet problem [3.27] is also uniquely solvable. The theorem is proved.

REMARK 3.1.– *The method of Green's function is also used for the study of the BVP*

$$u''(x) - a(x)(D_{0+}^{\alpha}u)(x) = -f(x), \quad x \in (0,1),$$
$$u(0) = 0, \quad u(1) = 0,$$

with the variable coefficient $a(x) \in C_{[0,1]}$, $a(x) \geq 0$ *(Samko et al. 1993).*

Next, we will obtain a weighted estimate similar to [3.20] and [3.26] in the class of less smooth functions.

Let $f(x) = \varphi''(x)$ with $\varphi \in C_{[0,1]}$. We find

$$\int_0^1 G(x,\xi)f(\xi)d\xi = \int_0^1 G(x,\xi)\varphi''(\xi)d\xi = (1-x)\int_0^x \xi\varphi''(\xi)d\xi + x\int_x^1 (1-\xi)\varphi''(\xi)d\xi =$$

$$= (1-x)\left(\xi\varphi'(\xi)|_0^x - \int_0^x \varphi'(\xi)d\xi\right) + x\left((1-\xi)\varphi'(\xi)|_x^1 + \int_x^1 \varphi'(\xi)d\xi\right) =$$

$$= (1-x)(x\varphi'(x) - \varphi(x) + \varphi(0)) + x(-(1-x)\varphi'(x) + \varphi(1) - \varphi(x)) =$$
$$= -(1-x)(\varphi(x) - \varphi(0)) + x(\varphi(1) - \varphi(x)).$$

DEFINITION 3.1.– *A solution* $u(x) \in C_{[0,1]}$ *of the Fredholm integral equation*

$$u(x) + \frac{1}{\Gamma(2-\alpha)}\left(-(1-x)\int_0^x (x-t)^{1-\alpha}u(t)dt + \right.$$

$$\left. +x\int_0^x \left((1-t)^{1-\alpha} - (x-t)^{1-\alpha}\right)u(t)dt + x\int_x^1 (1-t)^{1-\alpha}u(t)dt\right) = \qquad [3.29]$$

$$= -(1-x)(\varphi(x) - \varphi(0)) + x(\varphi(1) - \varphi(x)), \quad x \in [0,1],$$

is called a weak solution of the Dirichlet BVP [3.27].

It follows from Fredholm's theorems that for an arbitrary function $\varphi(x) \in C_{[0,1]}$ equation [3.29] has a unique solution $u(x) \in C_{[0,1]}$. From [3.29], we derive the inequality

$$|u(x)| \leq \frac{1}{\Gamma(2-\alpha)} \sup_{0<x<1} \left((1-x)\int_0^x (x-t)^{1-\alpha}dt + \right.$$

$$\left. +x\int_0^x \left((1-t)^{1-\alpha} - (x-t)^{1-\alpha}\right)dt + x\int_x^1 (1-t)^{1-\alpha}dt\right) \|u\|_{C_{[0,1]}} +$$

$$+ \sup_{0<x<1} \left|-(1-x)(\varphi(x) - \varphi(0)) + x(\varphi(1) - \varphi(x))\right|, \quad x \in [0,1].$$

Since

$$\sup_{0<x<1}\left((1-x)\int_0^x (x-t)^{1-\alpha}dt + x\int_0^x\left((1-t)^{1-\alpha} - (x-t)^{1-\alpha}\right)dt + x\int_x^1 (1-t)^{1-\alpha}dt\right) =$$

$$= \sup_{0<x<1} \frac{(1-x)x^{2-\alpha} + x\left(1-(1-x)^{2-\alpha} - x^{2-\alpha}\right) + x(1-x)^{2-\alpha}}{(2-\alpha)} =$$

$$= \sup_{0<x<1} \frac{x^{2-\alpha} + x - 2x^{3-\alpha}}{(2-\alpha)} \leq \frac{1}{2(2-\alpha)},$$

then for $u(x)$, we have the inequality

$$\|u\|_{C_{[0,1]}} \leq \left(1 - \frac{1}{2\Gamma(3-\alpha)}\right)^{-1} 4\|\varphi\|_{C_{[0,1]}}.$$

In the following proposition, we will obtain a weighted estimate for $u(x)$.

THEOREM 3.5.– *Let $\varphi(x) \in C_{[0,1]}^{0,\beta}$ and let β satisfy the condition*

$$\frac{1}{\ln 4} \cdot \ln \frac{3-\alpha}{\Gamma(3-\alpha)} < \beta \leq 1 \quad (0 < \alpha < 1). \qquad [3.30]$$

Then, a weak solution $u(x)$ of problem [3.27] satisfies the condition $u(x) \in C_{[0,1]}^{0,\beta}$ and is characterized by the following weighted estimate, which takes into account the boundary effect:

$$\left\|\frac{u(x)}{[x(1-x)]^\beta}\right\|_\infty \leq 2^\beta \left(1 - \frac{3-\alpha}{4^\beta \Gamma(3-\alpha)}\right)^{-1} |\varphi|_{C_{[0,1]}^{0,\beta}}. \qquad [3.31]$$

PROOF.– Substituting $v(x) = \dfrac{u(x)}{[x(1-x)]^\beta}$ in [3.29], we obtain the equation

$$v(x) + \frac{1}{\Gamma(2-\alpha)[x(1-x)]^\beta}\left(-(1-x)\int_0^x (x-t)^{1-\alpha}[t(1-t)]^\beta v(t)dt + \right. \quad [3.32]$$

$$\left. +x\int_0^x \left((1-t)^{1-\alpha} - (x-t)^{1-\alpha}\right)[t(1-t)]^\beta v(t)dt + x\int_x^1 (1-t)^{1-\alpha}[t(1-t)]^\beta v(t)dt\right) =$$

$$= \frac{-(1-x)(\varphi(x)-\varphi(0)) + x(\varphi(1)-\varphi(x))}{[x(1-x)]^\beta}, \quad x \in (0,1).$$

This implies the inequality

$$\|v\|_\infty \leq \frac{1}{\Gamma(2-\alpha)} \sup_{0<x<1} \frac{1}{[x(1-x)]^\beta}\left((1-x)\int_0^x (x-t)^{1-\alpha}dt + \right. \quad [3.33]$$

$$+ \sup_{0<x<1} \frac{\left|-(1-x)(\varphi(x)-\varphi(0)) + x(\varphi(1)-\varphi(x))\right|}{[x(1-x)]^\beta}.$$

We find

$$\sup_{0<x<1} \frac{(1-x)\int_0^x (x-t)^{1-\alpha}dt + x\int_0^x \left((1-t)^{1-\alpha} - (x-t)^{1-\alpha}\right)dt + x\int_x^1 (1-t)^{1-\alpha}dt}{[x(1-x)]^\beta} =$$

$$= \sup_{0<x<1} \frac{(1-x)x^{2-\alpha} + x\left(1-(1-x)^{2-\alpha} - x^{2-\alpha}\right) + x(1-x)^{2-\alpha}}{(2-\alpha)[x(1-x)]^\beta} = \quad [3.34]$$

$$= \sup_{0<x<1} \frac{x^{2-\alpha} + x - 2x^{3-\alpha}}{(2-\alpha)[x(1-x)]^\beta} \leq \sup_{0<x<1} \frac{x^{2-\alpha} + x - 2x^{3-\alpha}}{(2-\alpha)x(1-x)} = \sup_{0<x<1} \frac{x^{1-\alpha} + 1 - 2x^{2-\alpha}}{(2-\alpha)(1-x)} =$$

$$= \lim_{x \to 1-0} \frac{x^{1-\alpha} + 1 - 2x^{2-\alpha}}{(2-\alpha)(1-x)} = \lim_{x \to 1-0} \frac{(1-\alpha)x^{1-\alpha} - (2-\alpha)2x^{1-\alpha}}{-(2-\alpha)} = \frac{3-\alpha}{2-\alpha}$$

(here we applied *L'Hôpital's rule*) and

$$\sup_{0<x<1} \frac{\left| -(1-x)(\varphi(x) - \varphi(0)) + x(\varphi(1) - \varphi(x)) \right|}{[x(1-x)]^{\beta}} =$$

$$= \sup_{0<x<1} \left| -(1-x)^{1-\beta} \frac{\varphi(x) - \varphi(0)}{x^{\beta}} + x^{1-\beta} \frac{\varphi(1) - \varphi(x)}{(1-x)^{\beta}} \right| \leq \qquad [3.35]$$

$$\leq \sup_{0<x<1} \left((1-x)^{1-\beta} + x^{1-\beta} \right) |\varphi|_{C^{0,\beta}_{[0,1]}} \leq 2^{\beta} |\varphi|_{C^{0,\beta}_{[0,1]}}.$$

Using estimates [3.34] and [3.35], from [3.33] we obtain

$$\|v\|_{\infty} \leq \frac{3-\alpha}{\Gamma(3-\alpha)} \frac{1}{4^{\beta}} \|v\|_{\infty} + 2^{\beta} |\varphi|_{C^{0,\beta}_{[0,1]}}. \qquad [3.36]$$

Due to condition [3.30], we have $\dfrac{3-\alpha}{\Gamma(3-\alpha)} \dfrac{1}{4^{\beta}} < 1$. Then, inequality [3.36] implies estimate [3.31].

Now we will show that $u(x) \in C^{0,\beta}_{[0,1]}$. First, we write the integral equation [3.29] as follows:

$$u(x) + \frac{1}{\Gamma(2-\alpha)} \left(x \int_0^1 (1-t)^{1-\alpha} u(t) dt - \int_0^x (x-t)^{1-\alpha} u(t) dt \right) =$$

$$= -\varphi(x) + \varphi(0) + x(\varphi(1) - \varphi(0)), \quad x \in [0,1].$$

Then, for $x_1, x_2 \in [0,1]$, $x_1 \neq x_2$ we get

$$\frac{u(x_2) - u(x_1)}{|x_2 - x_1|^{\beta}} + \frac{1}{\Gamma(2-\alpha)} \frac{x_2 - x_1}{|x_2 - x_1|^{\beta}} \int_0^1 (1-t)^{1-\alpha} u(t) dt -$$

$$-\frac{1}{\Gamma(2-\alpha)} \frac{\int\limits_0^{x_2}(x_2-t)^{1-\alpha}u(t)dt - \int\limits_0^{x_1}(x_1-t)^{1-\alpha}u(t)dt}{|x_2-x_1|^\beta} =$$

$$= \frac{\varphi(x_1) - \varphi(x_2)}{|x_2-x_1|^\beta} + (\varphi(1) - \varphi(0))\frac{x_2-x_1}{|x_2-x_1|^\beta}.$$

This gives the inequality

$$\frac{|u(x_2)-u(x_1)|}{|x_2-x_1|^\beta} \le \frac{1}{\Gamma(2-\alpha)}|x_2-x_1|^{1-\beta}\left|\int\limits_0^1(1-t)^{1-\alpha}u(t)dt\right| +$$

$$+ \frac{1}{\Gamma(2-\alpha)} \frac{\left|\int\limits_0^{x_2}(x_2-t)^{1-\alpha}u(t)dt - \int\limits_0^{x_1}(x_1-t)^{1-\alpha}u(t)dt\right|}{|x_2-x_1|^\beta} + \quad [3.37]$$

$$+ \frac{|\varphi(x_2)-\varphi(x_1)|}{|x_2-x_1|^\beta} + (\varphi(1)-\varphi(0))|x_2-x_1|^{1-\beta}.$$

Next, we consider the second term on the right-hand side of [3.37]. For the function

$$g(x) = \int\limits_0^x (x-t)^{1-\alpha}u(t)dt, \quad x \in [0,1],$$

we have $g(x) \in C^1_{[0,1]}$ i $g'(x) = (1-\alpha)\int\limits_0^x (x-t)^{-\alpha}u(t)dt,$

$$|g(x_2)-g(x_1)| = |g'(\xi)(x_2-x_1)| = (1-\alpha)\left|\int\limits_0^\xi (\xi-t)^{-\alpha}u(t)dt\right||x_2-x_1| \le$$

$$\le \|u\|_{C_{[0,1]}}|x_2-x_1| \quad (x_1 \overset{\le}{\ge} \xi \overset{\le}{\ge} x_2).$$

Then, [3.37] leads to the estimate

$$|u|_{C^{0,\beta}_{[0,1]}} \le \left(\frac{1}{\Gamma(3-\alpha)} + \frac{1}{\Gamma(2-\alpha)}\right) \|u\|_{C_{[0,1]}} + |\varphi|_{C^{0,\beta}_{[0,1]}} + |\varphi(1)-\varphi(0)|,$$

and therefore $u \in C^{0,\beta}_{[0,1]}$. The theorem is proved.

REMARK 3.2.– *The set of values of the exponent β that satisfies condition [3.30] is not empty, since for $0 \le \alpha \le 1$ it holds*

$$0.2924812504\ldots \le \frac{1}{\ln 4} \cdot \ln\frac{3-\alpha}{\Gamma(3-\alpha)} \le 0.5024543610\ldots.$$

3.2.2. *The scale of weighted estimates for approximate solutions*

To solve equation [3.29], we will use a mesh method. We introduce the mesh set

$$\omega = \{x_j = jh,\ j = 1,\ldots,N-1,\ h = 1/N\},\quad \bar\omega = \omega \cup \{0\} \cup \{1\},$$

where $N \ge 2$ is an integer number, and set $x = x_j$ in [3.29]:

$$u(x_j) + \frac{1}{\Gamma(2-\alpha)}\left\{-(1-x_j)\sum_{k=1}^{j}\int_{x_{k-1}}^{x_k}(x_j-t)^{1-\alpha}u(t)dt\right.$$

$$+ x_j \sum_{k=1}^{j}\int_{x_{k-1}}^{x_k}\left((1-t)^{1-\alpha} - (x_j-t)^{1-\alpha}\right)u(t)dt + x_j \sum_{k=j+1}^{N}\int_{x_{k-1}}^{x_k}(1-t)^{1-\alpha}u(t)dt\Bigg\} =$$

$$= -(1-x_j)\bigl(\varphi(x_j) - \varphi(0)\bigr) + x_j\bigl(\varphi(1) - \varphi(x_j)\bigr),\quad j = 1,2,\ldots,N-1.$$

This gives the mesh scheme

$$u^h(x_j) + \frac{1}{\Gamma(2-\alpha)}\left\{-(1-x_j)\sum_{k=1}^{j}u^h(x_k)\int_{x_{k-1}}^{x_k}(x_j-t)^{1-\alpha}dt + \right.$$ [3.38]

$$+x_j \sum_{k=1}^{j} u^h(x_k) \int_{x_{k-1}}^{x_k} \left((1-t)^{1-\alpha} - (x_j - t)^{1-\alpha}\right) dt + x_j \sum_{k=j+1}^{N} u^h(x_k) \int_{x_{k-1}}^{x_k} (1-t)^{1-\alpha} dt \Bigg\} =$$

$$= -(1-x_j)\left(\varphi(x_j) - \varphi(0)\right) + x_j \left(\varphi(1) - \varphi(x_j)\right), \quad j = 1, 2, ..., N-1, \quad u^h(x_N) = 0.$$

The error $z(x) = u^h(x) - u(x)$ is a solution of the mesh problem

$$z(x_j) + \frac{1}{\Gamma(2-\alpha)} \Bigg\{ -(1-x_j) \sum_{k=1}^{j} z(x_k) \int_{x_{k-1}}^{x_k} (x_j - t)^{1-\alpha} dt + \qquad [3.39]$$

$$+ x_j \sum_{k=1}^{j} z(x_k) \int_{x_{k-1}}^{x_k} \left((1-t)^{1-\alpha} - (x_j - t)^{1-\alpha}\right) dt + x_j \sum_{k=j+1}^{N} z(x_k) \int_{x_{k-1}}^{x_k} (1-t)^{1-\alpha} dt \Bigg\} =$$

$$= \psi(x_j), \quad j = 1, 2, ..., N-1, \quad z(x_N) = 0,$$

where $\psi(x_j)$ is the approximation error:

$$\psi(x_j) = \frac{1}{\Gamma(2-\alpha)} \Bigg\{ -(1-x_j) \sum_{k=1}^{j} \int_{x_{k-1}}^{x_k} (x_j - t)^{1-\alpha} \left(u(t) - u(x_k)\right) dt +$$

$$+ x_j \sum_{k=1}^{j} \int_{x_{k-1}}^{x_k} \left((1-t)^{1-\alpha} - (x_j - t)^{1-\alpha}\right)\left(u(t) - u(x_k)\right) dt + \qquad [3.40]$$

$$+ x_j \sum_{k=j+1}^{N} \int_{x_{k-1}}^{x_k} (1-t)^{1-\alpha} \left(u(t) - u(x_k)\right) dt \Bigg\}.$$

Using the norm $\| w \|_{\infty, \omega} = \max_{x \in \omega} |w(x)|$, we prove the following result.

THEOREM 3.6.– *Let the assumptions of Theorem 3.5 be fulfilled. Then, the accuracy of the mesh scheme [3.38] is characterized by the weighted estimate, which takes into account the boundary effect:*

$$\left\| \frac{u - u^h}{[x(1-x)]^\beta} \right\|_{\infty, \omega} \leq M h^\beta \, |u|_{0,\beta}, \qquad [3.41]$$

where $M = \left(1 - \dfrac{3-\alpha}{4^{\beta}\,\Gamma(3-\alpha)}\right)^{-1} \dfrac{3-\alpha}{\Gamma(3-\alpha)}$ is a positive constant independent of h and $u(x)$.

PROOF.– Substituting $Z(x) = \dfrac{z(x)}{[x(1-x)]^{\beta}}$ in [3.39], we obtain the mesh problem

$$Z(x_j) + \frac{1}{\Gamma(2-\alpha)\left[x_j(1-x_j)\right]^{\beta}} \Bigg\{ -(1-x_j) \sum_{k=1}^{j} Z(x_k)[x_k(1-x_k)]^{\beta} \int_{x_{k-1}}^{x_k} (x_j - t)^{1-\alpha} dt +$$

$$+ x_j \sum_{k=1}^{j} Z(x_k)[x_k(1-x_k)]^{\beta} \int_{x_{k-1}}^{x_k} \left((1-t)^{1-\alpha} - (x_j - t)^{1-\alpha}\right) dt + \qquad [3.42]$$

$$+ x_j \sum_{k=j+1}^{N} Z(x_k)[x_k(1-x_k)]^{\beta} \int_{x_{k-1}}^{x_k} (1-t)^{1-\alpha} dt \Bigg\} = \frac{\psi(x_j)}{\left[x_j(1-x_j)\right]^{\beta}},$$

$$j = 1, 2, \ldots, N-1, \; Z(x_N) = 0.$$

Next, we will find the estimate for the norm $\left\| \dfrac{\psi(x_j)}{\left[x_j(1-x_j)\right]^{\beta}} \right\|_{\infty,\omega}$ with $\psi(x_j)$ defined in [3.40].

If $u \in C_{[0,1]}^{0,\beta}\ (0 < \beta \le 1)$, then

$$|u(t) - u(x_k)| = \left| \frac{u(t) - u(x_k)}{(x_k - t)^{\beta}} (x_k - t)^{\beta} \right| \le h^{\beta}\, |u|_{0,\beta}. \qquad [3.43]$$

If $u(x) \in C_{[0,1]}^{1}$, then

$$|u(t) - u(x_k)| = |u'(\bar{x}_k)(t - x_k)| \le h\, \|u\|_{1,\infty} \quad (\bar{x}_k \in (x_{k-1}, x_k)). \qquad [3.44]$$

Bearing in mind inequalities [3.34], [3.43] and [3.44], we obtain from [3.42] the following estimates:

$$\| Z \|_{\infty,\omega} \leq \frac{3-\alpha}{\Gamma(3-\alpha)} \frac{1}{4^\beta} \| Z \|_{\infty,\omega} + \frac{3-\alpha}{\Gamma(3-\alpha)} h^\beta \, |u|_{0,\beta} \quad (0 < \beta \leq 1),$$

$$\| Z \|_{\infty,\omega} \leq \frac{3-\alpha}{\Gamma(3-\alpha)} \frac{1}{4^\beta} \| Z \|_{\infty,\omega} + \frac{3-\alpha}{\Gamma(3-\alpha)} h \| u \|_{1,\infty}.$$

This leads to estimate [3.41] and completes the proof.

REMARK 3.3.– If $u(x) \in C^1_{[0,1]}$, then the error $z(x) = u^h(x) - u(x)$ of the mesh scheme [3.38] is characterized by the weighted estimate

$$\left\| \frac{u - u^h}{x(1-x)} \right\|_{\infty,\omega} \leq Mh \| u \|_{1,\infty} \qquad [3.45]$$

with the positive constant $M = \left(1 - \frac{3-\alpha}{4\,\Gamma(3-\alpha)}\right)^{-1} \cdot \frac{3-\alpha}{\Gamma(3-\alpha)}$ independent of h and $u(x)$.

Now we will construct and study the mesh scheme of the second order of approximation:

$$u^h(x_j) + \frac{1}{\Gamma(2-\alpha)} \Bigg\{ -(1-x_j) \sum_{k=1}^{j} \int_{x_{k-1}}^{x_k} (x_j - t)^{1-\alpha} L_k(t; u^h) dt + \qquad [3.46]$$

$$+ x_j \sum_{k=1}^{j} \int_{x_{k-1}}^{x_k} \left((1-t)^{1-\alpha} - (x_j - t)^{1-\alpha}\right) L_k(t; u^h) dt + x_j \sum_{k=j+1}^{N} \int_{x_{k-1}}^{x_k} (1-t)^{1-\alpha} L_k(t; u^h) dt \Bigg\} =$$

$$= -(1-x_j)\big(\varphi(x_j) - \varphi(0)\big) + x_j \big(\varphi(1) - \varphi(x_j)\big), \quad j = 1, 2, \ldots, N-1, \ u^h(x_N) = 0,$$

where $L_k(t; w)$ is the Lagrange interpolating polynomial of the first degree for the function $w(x)$:

$$L_k(t; w) = \frac{t - x_{k-1}}{x_k - x_{k-1}} w(x_k) + \frac{x_k - t}{x_k - x_{k-1}} w(x_{k-1}), \quad t \in [x_{k-1}, x_k].$$

For the error $z(x) = u^h(x) - u(x)$, we have the difference scheme

$$u^h(x_j) + \frac{1}{\Gamma(2-\alpha)}\left\{-(1-x_j)\sum_{k=1}^{j}\int_{x_{k-1}}^{x_k}(x_j-t)^{1-\alpha}L_k(t;u^h)dt + \right.$$

$$+x_j\sum_{k=1}^{j}\int_{x_{k-1}}^{x_k}\left((1-t)^{1-\alpha} - (x_j-t)^{1-\alpha}\right)L_k(t;u^h)dt + \qquad [3.47]$$

$$\left. +x_j\sum_{k=j+1}^{N}\int_{x_{k-1}}^{x_k}(1-t)^{1-\alpha}L_k(t;u^h)dt\right\} = \psi(x_j), \quad j = 1,2,\ldots,N-1, \quad u^h(x_N) = 0,$$

where $\psi(x_j)$ is the approximation error

$$\psi(x_j) = \frac{1}{\Gamma(2-\alpha)}\left\{-(1-x_j)\sum_{k=1}^{j}\int_{x_{k-1}}^{x_k}(x_j-t)^{1-\alpha}\left(u(t) - L_k(t;u)\right)dt + \right.$$

$$+x_j\sum_{k=1}^{j}\int_{x_{k-1}}^{x_k}\left((1-t)^{1-\alpha} - (x_j-t)^{1-\alpha}\right)\left(u(t) - L_k(t;u)\right)dt + \qquad [3.48]$$

$$\left. +x_j\sum_{k=j+1}^{N}\int_{x_{k-1}}^{x_k}(1-t)^{1-\alpha}\left(u(t) - L_k(t;u)\right)dt\right\}.$$

THEOREM 3.7.– *Let a weak solution $u(x)$ of problem [3.27] satisfy the condition $u \in C_{[0,1]}^{k,\beta}$ ($k = 0,1$; $0 < \beta \le 1$) and let [3.30] be fulfilled. Then, the accuracy of the mesh scheme [3.46] is characterized by the weighted estimate, which takes into account the boundary effect:*

$$\left\|\frac{u - u^h}{[x(1-x)]^\beta}\right\|_{\infty,\omega} \le M_k h^{k+\beta} \, |u|_{k,\beta}, \qquad [3.49]$$

where M_0 and M_1 are positive constants independent of h and $u(x)$:

$$M_0 = \left(1 - \frac{3-\alpha}{4^\beta \Gamma(3-\alpha)}\right)^{-1} \frac{3-\alpha}{\Gamma(3-\alpha)}, \quad M_1 = \left(1 - \frac{3-\alpha}{4^\beta \Gamma(3-\alpha)}\right)^{-1} \frac{3-\alpha}{4\Gamma(3-\alpha)}.$$

PROOF.— It follows from [3.47] that the function $Z(x) = \dfrac{z(x)}{[x(1-x)]^\beta}$ is a solution of the discrete problem

$$Z(x_j) + \frac{1}{\Gamma(2-\alpha)[x_j(1-x_j)]^\beta} \left\{ -(1-x_j) \sum_{k=1}^{j} \int_{x_{k-1}}^{x_k} (x_j - t)^{1-\alpha} L_k\!\left(t; [x(1-x)]^\beta Z\right) dt \right.$$

$$+ x_j \sum_{k=1}^{j} \int_{x_{k-1}}^{x_k} \left((1-t)^{1-\alpha} - (x_j - t)^{1-\alpha}\right) L_k\!\left(t; [x(1-x)]^\beta Z\right) dt + \quad [3.50]$$

$$\left. + x_j \sum_{k=j+1}^{N} \int_{x_{k-1}}^{x_k} (1-t)^{1-\alpha} L_k\!\left(t; x(1-x)Z\right) dt \right\} = \frac{\psi(x_j)}{[x_j(1-x_j)]^\beta},$$

$$j = 1, 2, \ldots, N-1, \quad u^h(x_N) = 0.$$

We will now obtain an estimate for $\left\|\dfrac{\psi(x_j)}{[x_j(1-x_j)]^\beta}\right\|_{\infty,\omega}$ with $\psi(x_j)$ defined in [3.48]. If $u \in C^{0,\beta}_{[0,1]}$ ($0 < \beta \le 1$), then we have

$$|u(t) - L_k(t;u)| = \left|u(t) - \frac{t - x_{k-1}}{x_k - x_{k-1}} u(x_k) - \frac{x_k - t}{x_k - x_{k-1}} u(x_{k-1})\right| =$$

$$= \left|\frac{t - x_{k-1}}{x_k - x_{k-1}}(u(t) - u(x_k)) + \frac{x_k - t}{x_k - x_{k-1}}(u(t) - u(x_{k-1}))\right| = \quad [3.51]$$

$$= \left|\frac{t - x_{k-1}}{x_k - x_{k-1}} \frac{u(t) - u(x_k)}{(x_k - t)^\beta}(x_k - t)^\beta + \frac{x_k - t}{x_k - x_{k-1}} \frac{u(t) - u(x_{k-1})}{(t - x_{k-1})^\beta}(t - x_{k-1})^\beta\right| \le$$

$$h^\beta \, |u|_{0,\beta};$$

if $u \in C^1_{[0,1]}$, then we have

$$\left|u(t)-L_k(t;u)\right| = \left|\frac{t-x_{k-1}}{x_k-x_{k-1}}\big(u(t)-u(x_k)\big) + \frac{x_k-t}{x_k-x_{k-1}}\big(u(t)-u(x_{k-1})\big)\right| \le$$

$$= \left|\frac{t-x_{k-1}}{x_k-x_{k-1}}u'(\bar{x}_k)(t-x_k) + \frac{x_k-t}{x_k-x_{k-1}}u'(\bar{\bar{x}}_k)(t-x_{k-1})\right| \le \qquad [3.52]$$

$$= \frac{2(t-x_{k-1})(x_k-t)}{x_k-x_{k-1}}\|u\|_{1,\infty} \le \frac{h}{2}\|u\|_{1,\infty} \quad \left(\bar{x}_k, \bar{\bar{x}}_k \in (x_{k-1}, x_k)\right);$$

if $u \in C^{1,\beta}_{[0,1]}$ $(0 < \beta \le 1)$, then we have

$$\left|u(t)-L_k(t;u)\right| = \left|\frac{t-x_{k-1}}{x_k-x_{k-1}}\big(u(t)-u(x_k)\big) + \frac{x_k-t}{x_k-x_{k-1}}\big(u(t)-u(x_{k-1})\big)\right| =$$

$$= \left|\frac{t-x_{k-1}}{x_k-x_{k-1}}u'(\bar{x}_k)(t-x_k) + \frac{x_k-t}{x_k-x_{k-1}}u'(\bar{\bar{x}}_k)(t-x_{k-1})\right| = \qquad [3.53]$$

$$= \left|\frac{(t-x_{k-1})(x_k-t)}{x_k-x_{k-1}} \frac{u'(\bar{\bar{x}}_k)-u'(\bar{x}_k)}{(\bar{\bar{x}}_k-\bar{x}_k)^\beta}(\bar{\bar{x}}_k-\bar{x}_k)^\beta\right| \le \frac{h^{\beta+1}}{4}|u|_{1,\beta}$$

$$\left(\bar{x}_k, \bar{\bar{x}}_k \in (x_{k-1}, x_k)\right);$$

if $u \in C^2_{[0,1]}$, then due to the formula

$$u(t) - L_k(t;u) = \frac{u''(\bar{x}_k)}{2!}(t-x_{k-1})(t-x_k), \quad \bar{x}_k \in (x_{k-1}, x_k),$$

we have

$$\left|u(t) - L_k(t;u)\right| \le \frac{h^2}{8}\|u\|_{2,\infty}. \qquad [3.54]$$

Using inequalities [3.34] and [3.51]–[3.54], we derive from equation [3.50] the following estimates:

$$\|Z\|_{\infty,\omega} \le \frac{3-\alpha}{\Gamma(3-\alpha)}\frac{1}{4^\beta}\|Z\|_{\infty,\omega} + \frac{3-\alpha}{\Gamma(3-\alpha)}h^\beta\,|u|_{0,\beta} \quad (0 < \beta \le 1),$$

$$\| Z \|_{\infty,\omega} \le \frac{3-\alpha}{\Gamma(3-\alpha)}\frac{1}{4}\| Z \|_{\infty,\omega} + \frac{3-\alpha}{\Gamma(3-\alpha)}\frac{h}{2}\| u \|_{1,\infty},$$

$$\| Z \|_{\infty,\omega} \le \frac{3-\alpha}{\Gamma(3-\alpha)}\frac{1}{4^\beta}\| Z \|_{\infty,\omega} + \frac{3-\alpha}{\Gamma(3-\alpha)}\frac{h^{1+\beta}}{4}| u |_{1,\beta} \quad (0<\beta\le 1),$$

$$\| Z \|_{\infty,\omega} \le \frac{3-\alpha}{\Gamma(3-\alpha)}\frac{1}{4}\| Z \|_{\infty,\omega} + \frac{3-\alpha}{\Gamma(3-\alpha)}\frac{h^2}{8}\| u \|_{2,\infty},$$

which completes the proof of the theorem.

REMARK 3.4.– If $u(x) \in C^1_{[0,1]}$, then the error $z(x) = u^h(x) - u(x)$ of the mesh scheme [3.46] can be characterized by the weighted estimate

$$\left\| \frac{u-u^h}{x(1-x)} \right\|_{\infty,\omega} \le M_1 h \| u \|_{1,\infty};$$

if $u(x) \in C^2_{[0,1]}$, then the following estimate holds true

$$\left\| \frac{u-u^h}{x(1-x)} \right\|_{\infty,\omega} \le M_2 h^2 \| u \|_{2,\infty},$$

where M_1 and M_2 are positive constants independent of h and $u(x)$:

$$M_1 = \left(1 - \frac{3-\alpha}{4\Gamma(3-\alpha)}\right)^{-1} \cdot \frac{3-\alpha}{2\Gamma(3-\alpha)}, \quad M_2 = \left(1 - \frac{3-\alpha}{4\Gamma(3-\alpha)}\right)^{-1} \cdot \frac{3-\alpha}{8\Gamma(3-\alpha)}.$$

3.2.3. A numerical example and conclusion

We consider here problem [3.27] for $\alpha = 1/2$, i.e.

$$u''(x) - \frac{1}{\sqrt{\pi}}\frac{d}{dx}\int_0^x \frac{u(t)dt}{\sqrt{x-t}} = -f(x), \quad 0 < x < 1, \quad [3.55]$$
$$u(0) = 0, \quad u(1) = 0.$$

To solve the problem numerically, we will use the mesh scheme [3.38] of the first order of approximation:

$$\left\{1 + \frac{2}{\sqrt{\pi}} \int_{x_{j-1}}^{x_j} \left(x_j\sqrt{1-t} - \sqrt{x_j - t}\right) dt\right\} u^h(x_j) + \qquad [3.56]$$

$$+ \frac{2}{\sqrt{\pi}} \sum_{k=1}^{j-1} u^h(x_k) \int_{x_{k-1}}^{x_k} \left(x_j\sqrt{1-t} - \sqrt{x_j - t}\right) dt + \frac{2}{\sqrt{\pi}} x_j \sum_{k=j+1}^{N-1} u^h(x_k) \int_{x_{k-1}}^{x_k} \sqrt{1-t}\, dt =$$

$$= (1 - x_j) \int_0^{x_j} \xi f(\xi) d\xi + x_j \int_{x_j}^1 (1 - \xi) f(\xi) d\xi, \quad j = 1, 2, ..., N-1,$$

where $\omega_h = \{x_j = jh,\ j = 1, ..., N-1,\ h = 1/N\}$, $N \geq 2$ is an integer number.

It is easy to verify that for $f(x) = 6x + \frac{1}{\sqrt{\pi}}\left(2x^{1/2} - \frac{16}{5}x^{5/2}\right)$ problem [3.55] has the exact solution $u(x) = x(1 - x^2)$.

The right-hand side of system [3.56] takes the form

$$(1 - x_j) \int_0^{x_j} \xi \left(6\xi + \frac{2}{\sqrt{\pi}} \xi^{1/2} - \frac{16}{5\sqrt{\pi}} \xi^{5/2}\right) d\xi$$

$$+ x_j \int_{x_j}^1 (1 - \xi)\left(6\xi + \frac{2}{\sqrt{\pi}} \xi^{1/2} - \frac{16}{5\sqrt{\pi}} \xi^{5/2}\right) d\xi.$$

and estimate [3.45] is the following:

$$\left\|\frac{u - u^h}{x(1-x)}\right\|_{\infty, \omega} \leq Mh \|u\|_{1,\infty},$$

where $\|u\|_{1,\infty} = \|1 - 3x^2\|_\infty = 2$ and

$$M = \left(1 - \frac{3-\alpha}{4\Gamma(3-\alpha)}\right)^{-1} \frac{3-\alpha}{\Gamma(3-\alpha)} = \frac{20}{6\sqrt{\pi} - 5} = 3.549420196....$$

For numerical calculations we used the package Maple 18. The results are presented in Table 3.1.

N	$\text{err}_h = \left\| \dfrac{u^h(x) - u(x)}{x(1-x)} \right\|_{\infty, \omega_h}$	$M h \| u \|_{1,\infty}$	$p = \log_2 \dfrac{\text{err}_h}{\text{err}_{h/2}}$
4	0.108432317883416745723	1.774710098	—
8	0.054715270109418209186	0.887355049	0.986779384943758484362 5
16	0.027543141645125317190	0.443677524	0.990250394523665314608 8
32	0.013821907007419234579	0.221838762	0.994736448571623860157 0
64	0.006923504629151874448	0.110919381	0.997382268069024220886 2
128	0.003464796324459534748	0.055459690	0.998731957725250175509 7
256	0.001733131580104288423	0.027729845	0.999389358381004698470 8
512	0.000866742870244009382	0.013864922	0.999705219593735010043 0

Table 3.1. *Numerical example for problem [3.55]*

Thus, for the Dirichlet boundary value problem [3.27] in Theorem 3.4 a unique solvability in the class $C^2_{(0,1)} \cap C_{[0,1]}$ and in Theorem 3.5 a scale of weighted estimates for a weak solution $u(x)$ are proved. To solve the problem numerically, the mesh schemes of the first and second order of approximation are constructed and their accuracy is characterized by a priori estimates taking into account the boundary effect. These estimates are consistent with the smoothness of the exact solution in the sense of the book by Samarskii et al. (1987).

The weighted estimates obtained in Theorems 3.6 and 3.7 indicate that the accuracy of the approximate solution is higher near the boundary points of the interval (0,1) than it is in the inner points.

3.3. BVP for a differential equation with variable coefficients and a fractional derivative of order $\alpha \in (0,1)$

3.3.1. Differential properties of the exact solution

We will use the standard notation for norms and seminorms in Sobolev spaces W_p^k ($W_2^k \equiv H^k$ for $p = 2$). For example, in the case of the interval $\Omega = (0,1)$ we have

$$\|u\| \equiv \|u\|_{L_2(\Omega)} = \left(\int_0^1 u^2(x)dx\right)^{1/2}, \quad \|u\|_{L_\infty(\Omega)} = \operatorname*{vrai\,max}_{x\in\Omega} |u(x)|,$$

$$\|u\|_{H^k(\Omega)} = \left(\sum_{j=0}^k \|u^{(j)}\|^2_{L_2(\Omega)}\right)^{1/2}, \quad |u|_{H^k(\Omega)} = \|u^{(k)}\|_{L_2(\Omega)},$$

$$\|u\|_{W^1_\infty(\Omega)} = \|u\|_{L_\infty(\Omega)} + \|u'\|_{L_\infty(\Omega)}, \quad |u|_{W^1_\infty(\Omega)} = \|u'\|_{L_\infty(\Omega)}.$$

The spaces $\overset{\circ}{H}{}^k(\Omega)$ and $\overset{\circ}{W}{}^1_\infty(\Omega)$ consist of functions $u \in H^k(\Omega)$ and $u \in W^1_\infty(\Omega)$ respectively, satisfying the additional condition $u(0) = 0$, $u(1) = 0$.

We consider the problem

$$\frac{d^2u(x)}{dx^2} + k_1(x)D^\alpha u(x) + k_2(x)u(x) = f(x), \quad x \in (0,1), \qquad [3.57]$$

$$u(0) = 0, \quad u(1) = 0,$$

where $D^\alpha u(x)$ is the Riemann–Liouville derivative of order $\alpha \in (0,1)$:

$$D^\alpha u(x) = \frac{1}{\Gamma(1-\alpha)} \frac{d}{dx} \int_0^x \frac{u(t)dt}{(x-t)^\alpha}.$$

Changing x for ξ and integrating both sides of the equation from $\xi = \eta$ to $\xi = x$ and then from $\eta = 0$ to $\eta = 1$, we finally get

$$u'(x) + \frac{1}{\Gamma(1-\alpha)} \int_0^1 d\eta \int_\eta^x k_1(\xi) \frac{d}{d\xi} \int_0^\xi \frac{u(t)dt}{(\xi-t)^\alpha} d\xi + \int_0^1 d\eta \int_\eta^x k_2(\xi)u(\xi)d\xi =$$

$$= \int_0^1 d\eta \int_\eta^x f(\xi)d\xi. \qquad [3.58]$$

Integrating by parts and changing the order of integration, we transform the second term on the left-hand side of this equation in the following way:

$$\int_0^1 d\eta \int_\eta^x k_1(\xi) \frac{d}{d\xi} \int_0^\xi \frac{u(t)dt}{(\xi-t)^\alpha} d\xi = \int_0^1 d\eta \left[k_1(\xi) \int_0^\xi \frac{u(t)dt}{(\xi-t)^\alpha} \bigg|_\eta^x - \int_\eta^x k_1'(\xi) \int_0^\xi \frac{u(t)dt}{(\xi-t)^\alpha} d\xi \right] =$$

$$= k_1(x) \int_0^x \frac{u(t)dt}{(x-t)^\alpha} - \int_0^1 k_1(\eta) \int_0^\eta \frac{u(t)dt}{(\eta-t)^\alpha} d\eta - \int_0^1 d\eta \int_\eta^x k_1'(\xi) \int_0^\xi \frac{u(t)dt}{(\xi-t)^\alpha} d\xi =$$

$$= -\frac{1}{1-\alpha} k_1(x) \left[(x-t)^{1-\alpha} u(t) \bigg|_0^x - \int_0^x (x-t)^{1-\alpha} u'(t)dt \right] +$$

$$+ \frac{1}{1-\alpha} \int_0^1 k_1(\eta) d\eta \left[\underbrace{(\eta-t)^{1-\alpha} u(t) \bigg|_0^\eta}_{=0} - \int_0^\eta (\eta-t)^{1-\alpha} u'(t)dt \right] +$$

$$+ \frac{1}{1-\alpha} \int_0^1 d\eta \int_\eta^x k_1'(\xi) d\xi \left[(\xi-t)^{1-\alpha} u(t) \bigg|_0^\xi - \int_0^\xi (\xi-t)^{1-\alpha} u'(t)dt \right] =$$

$$= \frac{1}{1-\alpha} k_1(x) \int_0^x (x-t)^{1-\alpha} u'(t)dt - \frac{1}{1-\alpha} \int_0^1 k_1(\eta) d\eta \int_0^\eta (\eta-t)^{1-\alpha} u'(t)dt -$$

$$- \frac{1}{1-\alpha} \int_0^1 d\eta \int_\eta^x k_1'(\xi) d\xi \int_0^\xi (\xi-t)^{1-\alpha} u'(t)dt =$$

$$= \frac{1}{1-\alpha} k_1(x) \int_0^x (x-t)^{1-\alpha} u'(t)dt - \frac{1}{1-\alpha} \int_0^1 u'(t)dt \int_t^1 (\eta-t)^{1-\alpha} k_1(\eta) d\eta -$$

$$- \frac{1}{1-\alpha} \int_0^x k_1'(\xi) d\xi \int_0^\xi d\eta \int_0^\xi (\xi-t)^{1-\alpha} u'(t)dt + \frac{1}{1-\alpha} \int_x^1 k_1'(\xi) d\xi \int_\xi^1 d\eta \int_0^\xi (\xi-t)^{1-\alpha} u'(t)dt =$$

$$= \frac{1}{1-\alpha} k_1(x) \int_0^x (x-t)^{1-\alpha} u'(t)dt - \frac{1}{1-\alpha} \int_0^1 u'(t)dt \int_t^1 (\eta-t)^{1-\alpha} k_1(\eta) d\eta -$$

$$- \frac{1}{1-\alpha} \int_0^x \xi k_1'(\xi) d\xi \int_0^\xi (\xi-t)^{1-\alpha} u'(t)dt + \frac{1}{1-\alpha} \int_x^1 (1-\xi) k_1'(\xi) d\xi \int_0^\xi (\xi-t)^{1-\alpha} u'(t)dt =$$

$$= \frac{1}{1-\alpha} k_1(x) \int_0^x (x-t)^{1-\alpha} u'(t) dt - \frac{1}{1-\alpha} \int_0^1 u'(t) dt \int_t^1 (\eta - t)^{1-\alpha} k_1(\eta) d\eta -$$

$$- \frac{1}{1-\alpha} \int_0^x u'(t) dt \int_t^x (\xi - t)^{1-\alpha} \xi k_1'(\xi) d\xi + \frac{1}{1-\alpha} \int_0^x u'(t) dt \int_x^1 (\xi - t)^{1-\alpha} (1-\xi) k_1'(\xi) d\xi +$$

$$+ \frac{1}{1-\alpha} \int_x^1 u'(t) dt \int_t^1 (\xi - t)^{1-\alpha} (1-\xi) k_1'(\xi) d\xi =$$

$$= \frac{1}{1-\alpha} \int_0^x u'(t) dt \left[(x-t)^{1-\alpha} k_1(x) - \int_t^1 (\eta - t)^{1-\alpha} k_1(\eta) d\eta - \int_t^x (\xi - t)^{1-\alpha} \xi k_1'(\xi) d\xi + \right.$$

$$\left. + \left\{ \int_t^1 - \int_t^x \right\} (\xi - t)^{1-\alpha} (1-\xi) k_1'(\xi) d\xi \right] +$$

$$+ \frac{1}{1-\alpha} \int_x^1 u'(t) dt \left[-\int_t^1 (\eta - t)^{1-\alpha} k_1(\eta) d\eta + \int_t^1 (\xi - t)^{1-\alpha} (1-\xi) k_1'(\xi) d\xi \right] =$$

$$= \frac{1}{1-\alpha} \int_0^x u'(t) dt \left[(x-t)^{1-\alpha} k_1(x) + \int_t^1 (\eta - t)^{1-\alpha} \left(k_1(\eta)(1-\eta) \right)' d\eta - \right.$$

$$\left. - \int_t^x (\eta - t)^{1-\alpha} k_1'(\eta) d\eta \right] + \frac{1}{1-\alpha} \int_x^1 u'(t) dt \int_t^1 (\eta - t)^{1-\alpha} \left(k_1(\eta)(1-\eta) \right)' d\eta .$$

Changing the order of integration and performing some simple transformations in the third term on the left-hand side of [3.58], we get

$$\int_0^1 d\eta \int_\eta^x k_2(\xi) u(\xi) d\xi = \int_0^x k_2(\xi) u(\xi) d\xi \int_0^\xi d\eta - \int_x^1 k_2(\xi) u(\xi) d\xi \int_\xi^1 d\eta =$$

$$= \int_0^x \xi k_2(\xi) u(\xi) d\xi - \int_x^1 (1-\xi) k_2(\xi) u(\xi) d\xi =$$

$$= \int_0^x \xi k_2(\xi) d\xi \int_0^\xi u'(t) dt - \int_x^1 (1-\xi) k_2(\xi) d\xi \int_0^\xi u'(t) dt =$$

$$= \int_0^x u'(t)dt \int_t^x \xi k_2(\xi)d\xi - \int_0^x u'(t)dt \int_x^1 (1-\xi)k_2(\xi)d\xi - \int_x^1 u'(t)dt \int_t^1 (1-\xi)k_2(\xi)d\xi =$$

$$= \int_0^x u'(t)dt \left(\int_t^x k_2(\xi)d\xi - \int_t^1 (1-\xi)k_2(\xi)d\xi \right) - \int_x^1 u'(t)dt \int_t^1 (1-\xi)k_2(\xi)d\xi.$$

Then, we obtain the Fredholm integral equation of the second kind for the unknown function $u'(x)$:

$$u'(x) + \int_0^1 K(x,t)u'(t)dt = \int_0^1 d\eta \int_\eta^x f(\xi)d\xi, \quad x \in [0,1], \qquad [3.59]$$

with the kernel

$$K(x,t) = \frac{1}{\Gamma(2-\alpha)} \begin{cases} (x-t)^{1-\alpha} k_1(x) + \int_t^1 (\eta-t)^{1-\alpha} (k_1(\eta)(1-\eta))' d\eta - \\ -\int_t^x (\eta-t)^{1-\alpha} k_1'(\eta)d\eta + \int_t^x k_2(\eta)d\eta - \int_t^1 (1-\eta)k_2(\eta)d\eta, & t \leq x, \\ \int_t^1 (\eta-t)^{1-\alpha} (k_1(\eta)(1-\eta))' d\eta - \int_t^1 (1-\eta)k_2(\eta)d\eta, & t \geq x. \end{cases}$$

Next, we prove the following auxiliary result.

LEMMA 3.2.– *Let* $u(x) \in C^1_{[0,1]}$. *Then, the following weighted estimate holds true:*

$$\max_{x \in [0,1]} \left| \frac{u(x)}{\sqrt{\rho_1(x)}} \right| \leq |u|_{H^1(0,1)}, \quad \rho_1(x) = \min(x, 1-x). \qquad [3.60]$$

Note that for the function $u(x) = \rho_1(x) = \min(x, 1-x) = 0.5 - |x - 0.5|$ $\in \overset{\circ}{H}{}^1(0,1)$ this inequality turns into the equality, i.e. estimate [3.60] is exact.

PROOF.– Making use of the Cauchy–Bunyakovsky inequality for the integrals

$$u(x) = \int_0^x u'(\xi)d\xi, \quad u(x) = -\int_x^1 u'(\xi)d\xi,$$

we have the estimates

$$|u(x)| \leq \sqrt{x} \, |u|_{H^1(0,x)}, \quad x \in [0,1/2],$$

$$u(x) \leq \sqrt{1-x} \, |u|_{H^1(x,1)}, \quad x \in [1/2,1],$$

and that proves the lemma.

First, we will consider the case of the constant coefficients and prove the existence and uniqueness theorem.

THEOREM 3.8.– *Let* $f(x) = f_0(x) + f_1'(x)$ *with* $f_0(x), f_1(x) \in L_2(0,1)$ *and let the coefficients* $k_1 = \text{const}$ *and* $k_2 = \text{const}$ *satisfy the condition*

$$\frac{1}{6}k_1^2 + \frac{1}{30}k_2^2 + \frac{1}{10}k_1 k_2 \leq 1. \quad [3.61]$$

Then, for all $\alpha \in [0,1]$ *the integral equation [3.59] has a unique solution* $u'(x) \in L_2(0,1)$ *and therefore BVP [3.57] has a unique solution* $u(x) \in \overset{\circ}{H}{}^1(0,1)$, *and the following estimate holds true*:

$$|u|_{H^1(0,1)} \leq \frac{1}{1-q}\left(\sqrt{2}\,\|f_0\| + 2\,\|f_1\|\right), \quad [3.62]$$

where $q = \left(\int_0^1\int_0^1 K^2(x,t)dt dx\right)^{1/2}.$

PROOF.– If the coefficients $k_1(x)$ and $k_2(x)$ in equation [3.57] are constant, then the kernel $K(x,t)$ is the following:

Differential Equations with Fractional Derivatives   151

$$K(x,t) = \frac{k_1}{\Gamma(2-\alpha)}\left[(x-t)^{1-\alpha} - \int_t^1 (\eta-t)^{1-\alpha}d\eta\right] + k_2\left[(x-t) - \int_t^1 (1-\eta)d\eta\right] =$$

$$= \frac{k_1}{\Gamma(2-\alpha)}\left[(x-t)^{1-\alpha} - \frac{(1-t)^{2-\alpha}}{2-\alpha}\right] + k_2\left[(x-t) - \frac{(1-t)^2}{2}\right] =$$

$$= k_1\left[\frac{(x-t)^{1-\alpha}}{\Gamma(2-\alpha)} - \frac{(1-t)^{2-\alpha}}{\Gamma(3-\alpha)}\right] + k_2\left[(x-t) - \frac{(1-t)^2}{2}\right] \quad \text{for} \quad t \le x;$$

$$K(x,t) = \frac{-k_1}{\Gamma(2-\alpha)}\int_t^1 (\eta-t)^{1-\alpha}d\eta - k_2\int_t^1 (1-\eta)d\eta = \frac{-k_1}{\Gamma(2-\alpha)}\frac{(1-t)^{2-\alpha}}{2-\alpha} - k_2\frac{(1-t)^2}{2} =$$

$$= -k_1\frac{(1-t)^{2-\alpha}}{\Gamma(3-\alpha)} - k_2\frac{(1-t)^2}{2} \quad \text{for} \quad t \ge x.$$

Next, we find the L_2-norm of $K(x,t)$:

$$q^2 = \int_0^1\int_0^1 K^2(x,t)dtdx = \int_0^1 dx\int_0^x K^2(x,t)dt + \int_0^1 dx\int_x^1 K^2(x,t)dt =$$

$$= \int_0^1 dx\int_0^x \left\{k_1\left[\frac{(x-t)^{1-\alpha}}{\Gamma(2-\alpha)} - \frac{(1-t)^{2-\alpha}}{\Gamma(3-\alpha)}\right] + k_2\left[(x-t) - \frac{(1-t)^2}{2}\right]\right\}^2 dt +$$

$$+ \int_0^1 dx\int_x^1 \left[-k_1\frac{(1-t)^{2-\alpha}}{\Gamma(3-\alpha)} - k_2\frac{(1-t)^2}{2}\right]^2 dt =$$

$$= k_1^2\int_0^1 dx\int_0^x\left[\frac{(x-t)^{2-2\alpha}}{\Gamma^2(2-\alpha)} + \frac{(1-t)^{4-2\alpha}}{\Gamma^2(3-\alpha)} - 2\frac{(x-t)^{1-\alpha}}{\Gamma(2-\alpha)}\frac{(1-t)^{2-\alpha}}{\Gamma(3-\alpha)}\right]dt +$$

$$+ k_2^2\int_0^1 dx\int_0^x\left[(x-t)^2 + \frac{(1-t)^4}{4} - (x-t)(1-t)^2\right]dt +$$

$$+ 2k_1k_2\int_0^1 dx\int_0^x\left[\frac{(x-t)^{2-\alpha}}{\Gamma(2-\alpha)} + \frac{(1-t)^{4-\alpha}}{2\Gamma(3-\alpha)} - \frac{(x-t)^{1-\alpha}}{\Gamma(2-\alpha)}\frac{(1-t)^2}{2} - \frac{(1-t)^{2-\alpha}}{\Gamma(3-\alpha)}(x-t)\right]dt +$$

$$+k_1^2\int_0^1 dx\int_x^1\frac{(1-t)^{4-2\alpha}}{\Gamma^2(3-\alpha)}dt+k_2^2\int_0^1 dx\int_x^1\frac{(1-t)^4}{4}dt+2k_1k_2\int_0^1 dx\int_x^1\frac{(1-t)^{4-\alpha}}{2\Gamma(3-\alpha)}dt=$$

$$=k_1^2\left[\frac{2-\alpha}{2(3-2\alpha)\Gamma^2(3-\alpha)}+\frac{1}{(6-2\alpha)\Gamma^2(3-\alpha)}-\frac{2}{(5-2\alpha)\Gamma^2(3-\alpha)}\right]+\frac{k_2^2}{40}+$$

$$+2k_1k_2\left[\frac{1}{(3-\alpha)(4-\alpha)\Gamma(2-\alpha)}+\frac{1}{2(6-\alpha)\Gamma(3-\alpha)}-\frac{1}{(5-\alpha)\Gamma(3-\alpha)}\right]+$$

$$+\frac{k_1^2}{(5-2\alpha)(6-2\alpha)\Gamma^2(3-\alpha)}+\frac{k_2^2}{120}+\frac{2k_1k_2}{2(5-\alpha)(6-\alpha)\Gamma(3-\alpha)}=$$

$$=\frac{(3-\alpha)(-2\alpha^3+11\alpha^2-19\alpha+12)}{2(3-2\alpha)(5-2\alpha)\Gamma^2(4-\alpha)}k_1^2+\frac{k_2^2}{30}+\frac{-\alpha^3+13\alpha^2-50\alpha+48}{\Gamma(7-\alpha)}k_1k_2,$$

Since for $\alpha\in[0,1]$

$$\frac{1}{30}\leq\frac{(3-\alpha)(-2\alpha^3+11\alpha^2-19\alpha+12)}{2(3-2\alpha)(5-2\alpha)\Gamma^2(4-\alpha)}\leq\frac{1}{6},$$

$$\frac{1}{15}\leq\frac{-\alpha^3+13\alpha^2-50\alpha+48}{\Gamma(7-\alpha)}\leq 0.09428271480\ldots<\frac{1}{10},$$

then we obtain the estimate

$$q^2=\int_0^1\int_0^1 K^2(x,t)dtdx<\frac{1}{6}k_1^2+\frac{1}{30}k_2^2+\frac{1}{10}k_1k_2.$$

Due to condition [3.61], we have $q<1$. Then, by the fixed-point iteration method, we prove the statement of the theorem and obtain the inequality

$$|u|_{H^1(0,1)}\leq\frac{1}{1-q}\left\|\int_0^1 d\eta\int_\eta^x f(\xi)d\xi\right\|.$$ [3.63]

Using the representation from Lions and Magenes (1972)

$$f(x) = f_0(x) + f_1'(x), \text{ where } f_0(x), f_1(x) \in L_2(0,1),$$

we find

$$\left\| \int_0^1 d\eta \int_\eta^x f(\xi)d\xi \right\| = \left\| \int_0^1 d\eta \int_\eta^x [f_0(\xi) + f_1'(\xi)]d\xi \right\| =$$

$$= \left\| \int_0^1 d\eta \int_\eta^x f_0(\xi)d\xi + f_1(x) - \int_0^1 f_1(\eta)d\eta \right\| \le \sqrt{2} \| f_0 \| + 2 \| f_1 \|,$$

which together with [3.63] leads to estimate [3.62] and thus proves the theorem.

The next statement follows from Theorem 3.8 and Lemma 3.2.

THEOREM 3.9.– *Let the assumptions of Theorem 3.8 be fulfilled. Then, the solution $u(x)$ of problem [3.57] can be characterized by the weighted estimate*

$$\max_{x \in [0,1]} \left| \frac{u(x)}{\sqrt{\rho_1(x)}} \right| \le \frac{1}{1-q} \left(\sqrt{2} \| f_0 \| + 2 \| f_1 \| \right)$$

with $\rho_1(x) = \min(x, 1-x)$ *and* $q = \left(\int_0^1 \int_0^1 K^2(x,t) dt dx \right)^{1/2}$.

PROOF.– Using the Cauchy–Bunyakovsky inequality, we obtain the estimates

$$|u(x)| = \left| \int_0^x u'(t)dt \right| \le \sqrt{x} \sqrt{\int_0^x |u'(t)|^2 \, dt} \le \sqrt{x} \, |u|_{H^1(0,1)}, \quad x \in [0,1/2],$$

$$|u(x)| = \left| \int_x^1 u'(t)dt \right| \le \sqrt{1-x} \, |u|_{H^1(0,1)}, \quad x \in [1/2,1].$$

Combining them with [3.62], we prove the theorem.

Now we will address the case of the variable coefficients. First, we need the following auxiliary proposition.

LEMMA 3.3.– Let $u(x) \in \overset{\circ}{W}^1_\infty(0,1)$. Then, the weighted estimate holds true:

$$\max_{x \in [0,1]} \left| \frac{u(x)}{\rho_1(x)} \right| \leq |u|_{W^1_\infty(0,1)}, \quad \rho_1(x) = \min(x, 1-x),$$

which is unimprovable in the sense that there exists problem [3.57] with the solution $u(x) = 0.5 - |0.5 - x| \in \overset{\circ}{W}^1_\infty(0,1)$ such that the estimate turns into an equality.

PROOF.– The proof immediately follows from the inequalities

$$|u(x)| = \left| \int_0^x u'(t)dt \right| \leq x |u|_{W^1_\infty(0,1)}, \quad x \in [0, 1/2],$$

$$|u(x)| = \left| \int_x^1 u'(t)dt \right| \leq (1-x)|u|_{W^1_\infty(0,1)}, \quad x \in [1/2, 1].$$

THEOREM 3.10.– Let

$$k_1(x) \in W^1_\infty(0,1), \quad k_2(x) \in L_\infty(0,1), \quad f(x) = f'_0(x), \quad f_0(x) \in L_\infty(0,1)$$

and let the condition

$$\frac{13}{6} \|k_1\|_{W^1_\infty(\Omega)} + \frac{2}{3} \|k_2\|_{L_\infty(\Omega)} < 1 \qquad [3.64]$$

be fulfilled. Then, for all $\alpha \in [0,1]$, the integral equation [3.59] has a unique solution $u'(x) \in L_\infty(0,1)$ and therefore problem [3.57] has a unique solution $u(x) \in \overset{\circ}{W}^1_\infty(0,1)$, and the following estimate holds true:

$$|u|_{W^1_\infty(0,1)} \leq \frac{2}{1-q} \|f_0\|_{L_\infty(0,1)}, \qquad [3.65]$$

where $q = \max\limits_{x \in [0,1]} \int_0^1 |K(x,t)| \, dt$.

PROOF.– We find

$$|K(x,t)| \leq \frac{1}{\Gamma(2-\alpha)}\left[(x-t)^{1-\alpha} + \int_t^1 (\eta-t)^{1-\alpha}(2-\eta)d\eta + \int_t^x (\eta-t)^{1-\alpha}d\eta\right] \times$$

$$\times \|k_1\|_{W_\infty^1(0,1)} + \left[(x-t) + \int_t^1 (1-\eta)d\eta\right]\|k_2\|_{L_\infty(0,1)} =$$

$$= \left[\frac{(x-t)^{1-\alpha}}{\Gamma(2-\alpha)} + \frac{(1-t)^{3-\alpha}}{\Gamma(4-\alpha)} + \frac{(1-t)^{2-\alpha}}{\Gamma(3-\alpha)} + \frac{(x-t)^{2-\alpha}}{\Gamma(3-\alpha)}\right]\|k_1\|_{W_\infty^1(0,1)} +$$

$$+ \left[(x-t) + \frac{(1-t)^2}{2}\right]\|k_2\|_{L_\infty(0,1)} \quad \text{for } t \leq x,$$

$$|K(x,t)| \leq \frac{1}{\Gamma(2-\alpha)}\int_t^1 (\eta-t)^{1-\alpha}(2-\eta)d\eta \, \|k_1\|_{W_\infty^1(0,1)} + \int_t^1 (1-\eta)d\eta \, \|k_2\|_{L_\infty(0,1)} =$$

$$= \left[\frac{(1-t)^{3-\alpha}}{\Gamma(4-\alpha)} + \frac{(1-t)^{2-\alpha}}{\Gamma(3-\alpha)}\right]\|k_1\|_{W_\infty^1(0,1)} + \frac{(1-t)^2}{2}\|k_2\|_{L_\infty(0,1)} \quad \text{for } t \geq x.$$

Then

$$\int_0^1 |K(x,t)|\,dt = \int_0^x |K(x,t)|\,dt + \int_x^1 |K(x,t)|\,dt \leq \quad [3.66]$$

$$\leq \int_0^x \left[\frac{(x-t)^{1-\alpha}}{\Gamma(2-\alpha)} + \frac{(1-t)^{3-\alpha}}{\Gamma(4-\alpha)} + \frac{(1-t)^{2-\alpha}}{\Gamma(3-\alpha)} + \frac{(x-t)^{2-\alpha}}{\Gamma(3-\alpha)}\right]dt \, \|k_1\|_{W_\infty^1(0,1)} +$$

$$+ \int_0^x \left[(x-t) + \frac{(1-t)^2}{2}\right]dt \, \|k_2\|_{L_\infty(0,1)} +$$

$$+ \int_x^1 \left[\frac{(1-t)^{3-\alpha}}{\Gamma(4-\alpha)} + \frac{(1-t)^{2-\alpha}}{\Gamma(3-\alpha)}\right]dt \, \|k_1\|_{W_\infty^1(0,1)} + \int_x^1 \frac{(1-t)^2}{2}\,dt \, \|k_2\|_{L_\infty(0,1)} =$$

$$= \left[\frac{x^{2-\alpha}}{\Gamma(3-\alpha)} + \frac{1-(1-x)^{4-\alpha}}{\Gamma(5-\alpha)} + \frac{1-(1-x)^{3-\alpha}}{\Gamma(4-\alpha)} + \frac{x^{3-\alpha}}{\Gamma(4-\alpha)}\right] \| k_1 \|_{W^1_\infty(0,1)} +$$

$$+ \left[\frac{x^2}{2} + \frac{1-(1-x)^3}{6}\right] \| k_2 \|_{L_\infty(0,1)} +$$

$$+ \left[\frac{(1-x)^{4-\alpha}}{\Gamma(5-\alpha)} + \frac{(1-x)^{3-\alpha}}{\Gamma(4-\alpha)}\right] \| k_1 \|_{W^1_\infty(0,1)} + \frac{(1-x)^3}{6} \| k_2 \|_{L_\infty(0,1)} =$$

$$= \left[\frac{x^{2-\alpha}}{\Gamma(3-\alpha)} + \frac{1}{\Gamma(5-\alpha)} + \frac{1}{\Gamma(4-\alpha)} + \frac{x^{3-\alpha}}{\Gamma(4-\alpha)}\right] \| k_1 \|_{W^1_\infty(0,1)} +$$

$$\left[\frac{x^2}{2} + \frac{1}{6}\right] \| k_2 \|_{L_\infty(0,1)}.$$

From here we obtain the estimate

$$q = \max_{x \in [0,1]} \int_0^1 | K(x,t) | \, dt \le$$

$$\le \left[\frac{1}{\Gamma(3-\alpha)} + \frac{1}{\Gamma(5-\alpha)} + \frac{2}{\Gamma(4-\alpha)}\right] \| k_1 \|_{W^1_\infty(0,1)} + \frac{2}{3} \| k_2 \|_{L_\infty(0,1)} \le$$

$$\le \frac{13}{6} \| k_1 \|_{W^1_\infty(0,1)} + \frac{2}{3} \| k_2 \|_{L_\infty(0,1)} \text{ for all } \alpha \in [0,1].$$

Due to condition [3.64], we have $q < 1$. Applying the fixed-point iteration method, we come to the estimate

$$| u |_{W^1_\infty(0,1)} \le \frac{1}{1-q} \left\| \int_0^1 d\eta \int_\eta^x f(\xi) d\xi \right\|_{L_\infty(0,1)}.$$

Combining it with the estimate

$$\left\| \int_0^1 d\eta \int_\eta^x f(\xi) d\xi \right\|_{L_\infty(0,1)} = \left\| \int_0^1 d\eta \int_\eta^x f_0'(\xi) d\xi \right\|_{L_\infty(0,1)} =$$

$$= \left\| f_0(x) - \int_0^1 f_0(\eta) d\eta \right\|_{L_\infty(0,1)} \leq 2 \| f_0 \|_{L_\infty(0,1)},$$

we complete the proof of the theorem.

Lemma 3.3 and Theorem 3.10 imply the following result.

THEOREM 3.11.— *Let the assumptions of Theorem 3.10 be fulfilled. Then, for the solution $u(x)$ of problem [3.57], the weighted estimate holds true:*

$$\max_{x \in [0,1]} \left| \frac{u(x)}{\rho_1(x)} \right| \leq \frac{2}{1-q} \| f_0 \|_{L_\infty(0,1)},$$

where $\rho_1(x) = \min(x, 1-x)$ and $q = \max_{x \in [0,1]} \int_0^1 |K(x,t)| \, dt$.

Now we will find the sufficient conditions under which $u \in \overset{\circ}{H}{}^1(0,1)$.

THEOREM 3.12.— *Let $k_1(x) \in W_\infty^1(0,1)$, $k_2(x) \in L_\infty(0,1)$,*

$$f(x) = f_0(x) + f_1'(x) \in \overset{\circ}{H}{}^{-1}(0,1), \text{ where } f_0(x), f_1(x) \in L_2(0,1),$$

and let the condition

$$\frac{169}{160} \| k_1 \|_{W_\infty^1(0,1)}^2 + \frac{7}{30} \| k_2 \|_{L_\infty(0,1)}^2 + \frac{31}{40} \| k_1 \|_{W_\infty^1(0,1)} \| k_2 \|_{L_\infty(0,1)} < 1 \quad [3.67]$$

be fulfilled. Then, for all $\alpha \in [0,1]$, the integral equation [3.59] has a unique solution $u' \in L_2(0,1)$ and therefore problem [3.57] has a unique solution $u \in \overset{\circ}{H}{}^1(0,1)$ and the following estimate holds true:

$$|u|_{H^1(0,1)} \le \frac{1}{1-q}\left(\sqrt{2}\, \|f_0\| + 2\,\|f_1\|\right), \qquad [3.68]$$

where $q = \sqrt{\int_0^1\int_0^1 K^2(x,t)\,dt\,dx}$.

PROOF.– Using relation [3.66], we have

$$q^2 = \int_0^1\int_0^1 K^2(x,t)\,dt\,dx = \int_0^1 dx \int_0^x K^2(x,t)\,dt + \int_0^1 dx \int_x^1 K^2(x,t)\,dt \le$$

$$\le \int_0^1 dx \int_0^x \left\{\left[\frac{(x-t)^{1-\alpha}}{\Gamma(2-\alpha)} + \frac{(1-t)^{3-\alpha}}{\Gamma(4-\alpha)} + \frac{(1-t)^{2-\alpha}}{\Gamma(3-\alpha)} + \frac{(x-t)^{2-\alpha}}{\Gamma(3-\alpha)}\right]\|k_1\|_{W^1_\infty(0,1)} + \right.$$

$$\left. + \left[(x-t) + \frac{(1-t)^2}{2}\right]\|k_2\|_{L_\infty(0,1)}\right\}^2 dt +$$

$$+ \int_0^1 dx \int_x^1 \left\{\left[\frac{(1-t)^{3-\alpha}}{\Gamma(4-\alpha)} + \frac{(1-t)^{2-\alpha}}{\Gamma(3-\alpha)}\right]\|k_1\|_{W^1_\infty(0,1)} + \frac{(1-t)^2}{2}\|k_2\|_{L_\infty(0,1)}\right\}^2 dt =$$

$$= \|k_1\|^2_{W^1_\infty(0,1)} \int_0^1 dx \int_0^x \left[\frac{(x-t)^{2-2\alpha}}{\Gamma^2(2-\alpha)} + \frac{(1-t)^{6-2\alpha}}{\Gamma^2(4-\alpha)} + \frac{(1-t)^{4-2\alpha}}{\Gamma^2(3-\alpha)} + \frac{(x-t)^{4-2\alpha}}{\Gamma^2(3-\alpha)} + \right.$$

$$+ 2\frac{(x-t)^{1-\alpha}(1-t)^{3-\alpha}}{\Gamma(2-\alpha)\Gamma(4-\alpha)} + 2\frac{(x-t)^{1-\alpha}(1-t)^{2-\alpha}}{\Gamma(2-\alpha)\Gamma(3-\alpha)} + 2\frac{(x-t)^{3-2\alpha}}{\Gamma(2-\alpha)\Gamma(3-\alpha)} +$$

$$\left. + 2\frac{(1-t)^{5-2\alpha}}{\Gamma(3-\alpha)\Gamma(4-\alpha)} + 2\frac{(1-t)^{3-\alpha}(x-t)^{2-\alpha}}{\Gamma(3-\alpha)\Gamma(4-\alpha)} + 2\frac{(1-t)^{2-\alpha}(x-t)^{2-\alpha}}{\Gamma^2(3-\alpha)}\right]dt +$$

$$+ \|k_2\|^2_{L_\infty(0,1)} \int_0^1 dx \int_0^x \left[(x-t)^2 + \frac{(1-t)^4}{4} + (x-t)(1-t)^2\right]dt +$$

$$+2 \| k_1 \|_{W_\infty^1(0,1)} \| k_2 \|_{L_\infty(0,1)} \int_0^1 dx \int_0^x \left[\frac{(x-t)^{2-\alpha}}{\Gamma(2-\alpha)} + \frac{(x-t)^{1-\alpha}(1-t)^2}{2\Gamma(2-\alpha)} + \frac{(x-t)(1-t)^{3-\alpha}}{\Gamma(4-\alpha)} + \right.$$

$$\left. + \frac{(1-t)^{5-\alpha}}{2\Gamma(4-\alpha)} + \frac{(x-t)(1-t)^{2-\alpha}}{\Gamma(3-\alpha)} + \frac{(1-t)^{4-\alpha}}{2\Gamma(3-\alpha)} + \frac{(x-t)^{3-\alpha}}{\Gamma(3-\alpha)} + \frac{(x-t)^{2-\alpha}(1-t)^2}{2\Gamma(3-\alpha)} \right] dt +$$

$$+ \| k_1 \|_{W_\infty^1(0,1)}^2 \int_0^1 dx \int_x^1 \left[\frac{(1-t)^{6-2\alpha}}{\Gamma^2(4-\alpha)} + \frac{(1-t)^{4-2\alpha}}{\Gamma^2(3-\alpha)} + 2 \frac{(1-t)^{5-2\alpha}}{\Gamma(3-\alpha)\Gamma(4-\alpha)} \right] dt +$$

$$+ \| k_2 \|_{L_\infty(0,1)}^2 \int_0^1 dx \int_x^1 \frac{(1-t)^4}{4} dt +$$

$$+ 2 \| k_1 \|_{W_\infty^1(0,1)} \| k_2 \|_{L_\infty(0,1)} \int_0^1 dx \int_x^1 \left[\frac{(1-t)^{5-\alpha}}{2\Gamma(4-\alpha)} + \frac{(1-t)^{4-\alpha}}{2\Gamma(3-\alpha)} \right] dt =$$

$$= \| k_1 \|_{W_\infty^1(0,1)}^2 \times$$

$$\times \left[\frac{1}{(3-2\alpha)(4-2\alpha)\Gamma^2(2-\alpha)} + \frac{1}{(8-2\alpha)\Gamma^2(4-\alpha)} + \frac{1}{(6-2\alpha)\Gamma^2(3-\alpha)} + \right.$$

$$+ \frac{1}{(5-2\alpha)(6-2\alpha)\Gamma^2(3-\alpha)} + \frac{1}{\Gamma^2(4-\alpha)} +$$

$$+ \frac{3}{(5-2\alpha)\Gamma^2(3-\alpha)} + \frac{2}{(7-2\alpha)\Gamma(4-\alpha)\Gamma(3-\alpha)} +$$

$$+ \frac{2}{(7-2\alpha)\Gamma^2(4-\alpha)} + \frac{1}{\Gamma^2(4-\alpha)} + \frac{1}{(7-2\alpha)(8-2\alpha)\Gamma^2(4-\alpha)} +$$

$$\left. + \frac{1}{(5-2\alpha)(6-2\alpha)\Gamma^2(3-\alpha)} + \frac{1}{(7-2\alpha)\Gamma^2(4-\alpha)} \right] +$$

$$+ \| k_2 \|_{L_\infty(0,1)}^2 \left[\frac{1}{12} + \frac{1}{24} + \frac{1}{10} + \frac{1}{120} \right] +$$

$$+2\|k_1\|_{W^1_\infty(0,1)}\|k_2\|_{L_\infty(0,1)}\left[\frac{1}{(3-\alpha)(4-\alpha)\Gamma(2-\alpha)}+\frac{1}{2(5-\alpha)\Gamma(3-\alpha)}+\right.$$

$$+\frac{1}{2(6-\alpha)\Gamma(4-\alpha)}+\frac{1}{2(7-\alpha)\Gamma(4-\alpha)}+\frac{1}{2(5-\alpha)\Gamma(3-\alpha)}+\frac{1}{2(6-\alpha)\Gamma(3-\alpha)}+$$

$$+\frac{1}{(4-\alpha)(5-\alpha)\Gamma(3-\alpha)}+\frac{1}{2(6-\alpha)\Gamma(4-\alpha)}+$$

$$\left.+\frac{1}{2(6-\alpha)(7-\alpha)\Gamma(4-\alpha)}+\frac{1}{2(5-\alpha)(6-\alpha)\Gamma(3-\alpha)}\right]=$$

$$=\frac{-4\alpha^5+88\alpha^4-715\alpha^3+2688\alpha^2-4644\alpha+2925}{2(3-2\alpha)(5-2\alpha)(7-2\alpha)\Gamma^2(4-\alpha)}\|k_1\|^2_{W^1_\infty(0,1)}+\frac{7}{30}\|k_2\|^2_{L_\infty(0,1)}+$$

$$+\frac{-5\alpha^3+70\alpha^2-311\alpha+432}{\Gamma(7-\alpha)}\|k_1\|_{W^1_\infty(0,1)}\|k_2\|_{L_\infty(0,1)}.$$

Since for all $\alpha \in [0,1]$, it holds

$$\frac{65}{168}\le\frac{-4\alpha^5+88\alpha^4-715\alpha^3+2688\alpha^2-4644\alpha+2925}{2(3-2\alpha)(5-2\alpha)(7-2\alpha)\Gamma^2(4-\alpha)}\le\frac{169}{160},$$

$$\frac{3}{10}\le\frac{-5\alpha^3+70\alpha^2-311\alpha+432}{\Gamma(7-\alpha)}\le\frac{31}{40},$$

then we obtain the estimate

$$q^2=\int_0^1\int_0^1 K^2(x,t)dtdx\le$$

$$\le\frac{169}{160}\|k_1\|^2_{W^1_\infty(0,1)}+\frac{7}{30}\|k_2\|^2_{L_\infty(0,1)}+\frac{31}{40}\|k_1\|_{W^1_\infty(0,1)}\|k_2\|_{L_\infty(0,1)}.$$

Due to condition [3.67], we have $q < 1$. Applying the fixed-point iteration method, we obtain the inequality

$$|u|_{H^1(0,1)} \le \frac{1}{1-q}\left\|\int_0^1 d\eta \int_\eta^x f(\xi)d\xi\right\|.$$

Next, we use the representation from Lions and Magenes (1972):

$$f(x) = f_0(x) + f_1'(x), \text{ where } f_0(x), f_1(x) \in L_2(0,1),$$

and find

$$\left\|\int_0^1 d\eta \int_\eta^x f(\xi)d\xi\right\| = \left\|\int_0^1 d\eta \int_\eta^x [f_0(\xi) + f_1'(\xi)]d\xi\right\| =$$

$$= \left\|\int_0^1 d\eta \int_\eta^x f_0(\xi)d\xi + f_1(x) - \int_0^1 f_1(\eta)d\eta\right\| \le \sqrt{2}\,\|f_0\| + 2\,\|f_1\|.$$

The theorem is proved.

THEOREM 3.13.– *Let the assumptions of Theorem 3.12 be fulfilled. Then, the weighted estimate holds true:*

$$\max_{x\in[0,1]} \left|\frac{u(x)}{\sqrt{\rho_1(x)}}\right| \le \frac{1}{1-q}\left(\sqrt{2}\,\|f_0\| + 2\,\|f_1\|\right),$$

where $q = \sqrt{\int_0^1\int_0^1 K^2(x,t)dtdx}$ and $\rho_1(x) = \min(x, 1-x)$.

The proof easily follows from Lemma 3.2 and estimate [3.68].

THEOREM 3.14.– *Let the assumptions of Theorem 3.12 be fulfilled and let $f(x) \in L_2(0,1)$. Then, the solution $u(x)$ of problem [3.57] satisfies the condition $u(x) \in H^2(0,1) \cap \overset{\circ}{H}{}^1(0,1)$ and the following estimate holds true:*

$$\max_{x\in[0,1]} \left|\frac{u(x)}{\rho_1(x)}\right| \le |u|_{H^2(0,1)},$$

where $\rho_1(x) = \min(x, 1-x)$.

The proof follows directly from the inequality

$$|u(x)| = \left|(1-x)\int_0^x \xi u''(\xi)d\xi + x\int_x^1 (1-\xi)u''(\xi)d\xi\right| \leq x(1-x)|u|_{H^2(0,1)}.$$

3.3.2. *The accuracy of the mesh scheme*

Here, we use the standard notation of the theory of finite-difference schemes from Samarskii (2001).

We introduce a uniform mesh on the interval [0,1]:

$$\omega_h = \{x_i = ih, \ i = 0,1,\ldots,N, \ h = 1/N\}, \text{ where } N \geq 1 \text{ is an integer number,}$$

and denote a left-side difference derivative $y_{\bar{x}}(x_i) = \dfrac{y(x_i) - y(x_{i-1})}{h}$ and the mesh

norm $\|y_{\bar{x}}\| = \left(\sum_{i=1}^{N} h y_{\bar{x}}^2(x_i)\right)^{1/2}$.

Next, we write on the mesh ω_h the corollary of the integral equation [3.59]:

$$u_{\bar{x}}(x_i) + \frac{1}{h}\sum_{p=1}^{N} \int_{x_{p-1}}^{x_p} u'(t)dt \int_{x_{i-1}}^{x_i} K(x,t)dx = \varphi(x_i), \quad i = 1,\ldots,N,$$

where $\varphi(x_i) = \dfrac{1}{h}\int_0^1 d\eta \int_{x_{i-1}}^{x_i} dx \int_\eta^x f(\xi)d\xi$.

Now we take a mesh scheme in the form

$$y_{\bar{x}}(x_i) + \sum_{p=1}^{N} a_{i,p} y_{\bar{x}}(x_p) = \varphi(x_i), \quad i = 1,\ldots,N, \qquad [3.69]$$

where $a_{i,p} = \dfrac{1}{h}\int_{x_{p-1}}^{x_p} dt \int_{x_{i-1}}^{x_i} K(x,t)dx$ and $\varphi(x_i) = \dfrac{1}{h}\int_0^1 d\eta \int_{x_{i-1}}^{x_i} dx \int_\eta^x f(\xi)d\xi$.

For the error $z(x_i) = y(x_i) - u(x_i)$, we have the mesh scheme

$$z_{\bar{x}}(x_i) + \sum_{p=1}^{N} a_{i,p} z_{\bar{x}}(x_p) = \psi(x_i), \quad i = 1,...,N, \qquad [3.70]$$

where $\psi(x_i)$ is the approximation error:

$$\psi(x_i) = \frac{1}{h} \sum_{p=1}^{N} \int_{x_{p-1}}^{x_p} \left[u'(t) - y_{\bar{x}}(x_p) \right] dt \int_{x_{i-1}}^{x_i} K(x,t) dx =$$

$$= \frac{1}{h^2} \sum_{p=1}^{N} \int_{x_{p-1}}^{x_p} dt \int_{x_{p-1}}^{x_p} [u'(t) - y'(s)] ds \int_{x_{i-1}}^{x_i} K(x,t) dx =$$

$$= \frac{1}{h^2} \sum_{p=1}^{N} \int_{x_{p-1}}^{x_p} dt \int_{x_{p-1}}^{x_p} ds \int_{s}^{t} u''(\xi) d\xi \int_{x_{i-1}}^{x_i} K(x,t) dx.$$

Applying the technique from Samarskii et al. (1987), we prove the following proposition.

THEOREM 3.15.– *Let the assumptions of Theorem 3.14 be fulfilled. Then, the accuracy of the mesh scheme [3.69] is characterized by the estimate*

$$\| z_{\bar{x}} \| \leq h \frac{q}{\sqrt{3(1-q)}} |u|_{H^2(0,1)}, \qquad [3.71]$$

and by the estimate that takes into account the boundary effect:

$$\max_{i=0,...,N} \left| \frac{z(x_i)}{\sqrt{\rho_1(x_i)}} \right| \leq h \frac{q}{\sqrt{3(1-q)}} |u|_{H^2(0,1)}, \qquad [3.72]$$

where $q = \sqrt{\int_0^1 \int_0^1 K^2(x,t) dt dx}$ *and* $\rho_1(x) = \min(x, 1-x)$.

PROOF.– Equation [3.70] yields the inequality

$$\|z_{\bar{x}}\| \leq \left\{\sum_{i=1}^{N} h \left[\sum_{p=1}^{N} a_{i,p} z_{\bar{x}}(x_p)\right]^2\right\}^{1/2} + \|\psi\| \leq \left\{\sum_{i=1}^{N} h \left[\sum_{p=1}^{N} a_{i,p}^2\right]\left[\sum_{p=1}^{N} z_{\bar{x}}^2(x_p)\right]\right\}^{1/2} =$$

$$= \left\{\sum_{i=1}^{N} \sum_{p=1}^{N} a_{i,p}^2\right\}^{1/2} \|z_{\bar{x}}\| + \|\psi\| \leq q \|z_{\bar{x}}\| + \|\psi\|$$

since

$$\left\{\sum_{i=1}^{N} \sum_{p=1}^{N} a_{i,p}^2\right\}^{1/2} = \left\{\sum_{i=1}^{N} \sum_{p=1}^{N} \frac{1}{h^2} \left(\int_{x_{p-1}}^{x_p} \int_{x_{i-1}}^{x_i} K(x,t)dxdt\right)^2\right\}^{1/2} \leq$$

$$\leq \left\{\sum_{i=1}^{N} \sum_{p=1}^{N} \int_{x_{p-1}}^{x_p} \int_{x_{i-1}}^{x_i} K^2(x,t)dxdt\right\}^{1/2} = \left\{\int_{0}^{1}\int_{0}^{1} K^2(x,t)dxdt\right\}^{1/2} = q.$$

This gives the estimate

$$\|z_{\bar{x}}\| \leq \frac{\|\psi\|}{1-q}. \qquad [3.73]$$

Now we find

$$\|\psi\|^2 = \sum_{i=1}^{N} h \left\{\frac{1}{h^2} \sum_{p=1}^{N} \int_{x_{p-1}}^{x_p} \int_{x_{i-1}}^{x_i} K(x,t)dx \int_{x_{p-1}}^{x_p} \int_{s}^{t} u''(\xi)d\xi ds dt\right\}^2 \leq$$

$$\leq \frac{1}{h^3} \sum_{i=1}^{N} \left\{\sum_{p=1}^{N} \left[\int_{x_{p-1}}^{x_p} \left(\int_{x_{i-1}}^{x_i} K(x,t)dx\right)^2 dt\right]^{1/2} \left[\int_{x_{p-1}}^{x_p} \left(\int_{x_{p-1}}^{x_p} \int_{s}^{t} u''(\xi)d\xi ds\right)^2 dt\right]^{1/2}\right\}^2 \leq$$

$$\leq \frac{1}{h^3} \sum_{i=1}^{N} \left(\sum_{p=1}^{N} \int_{x_{p-1}}^{x_p} \left(\int_{x_{i-1}}^{x_i} K(x,t)dx\right)^2 dt\right) \left(\sum_{p=1}^{N} \int_{x_{p-1}}^{x_p} \left(\int_{x_{p-1}}^{x_p} \int_{s}^{t} u''(\xi)d\xi ds\right)^2 dt\right) \leq$$

$$\le \frac{1}{h^3} \sum_{i=1}^{N} \left(\sum_{p=1}^{N} \int_{x_{p-1}}^{x_p} \left(h \int_{x_{i-1}}^{x_i} K^2(x,t) dx \right) dt \right) \left(\sum_{p=1}^{N} \int_{x_{p-1}}^{x_p} \left(h \int_{x_{p-1}}^{x_p} \left(\int_s^t u''(\xi) d\xi \right)^2 ds \right) dt \right) \le$$

$$\le \frac{1}{h} \left(\sum_{i=1}^{N} \sum_{p=1}^{N} \int_{x_{p-1}}^{x_p} \int_{x_{i-1}}^{x_i} K^2(x,t) dx dt \right) \left(\sum_{p=1}^{N} \int_{x_{p-1}}^{x_p} \left(\int_{x_{p-1}}^{x_p} \left(\int_s^t u''(\xi) d\xi \right)^2 ds \right) dt \right) =$$

$$= \frac{q^2}{h} \sum_{p=1}^{N} \int_{x_{p-1}}^{x_p} \left(\int_{x_{p-1}}^{x_p} \left(\int_s^t u''(\xi) d\xi \right)^2 ds \right) dt \le \frac{q^2}{h} \sum_{p=1}^{N} \int_{x_{p-1}}^{x_p} \left(\int_{x_{p-1}}^{x_p} |t-s| \left| \int_s^t u''^2(\xi) d\xi \right| ds \right) dt \le$$

$$\le \frac{q^2}{h} \sum_{p=1}^{N} \int_{x_{p-1}}^{x_p} u''^2(\xi) d\xi \int_{x_{p-1}}^{x_p} \int_{x_{p-1}}^{x_p} |t-s| ds dt = \frac{q^2}{h} \sum_{p=1}^{N} \int_{x_{p-1}}^{x_p} u''^2(\xi) d\xi \frac{h^3}{3} = \frac{q^2 h^2}{3} |u|^2_{H^2(0,1)},$$

which in combination with [3.73] leads to estimate [3.71].

To prove the weighted estimate [3.72], we use the inequalities

$$|z(x_i)| = \left| \sum_{k=1}^{i} h z_{\bar{x}}(x_k) \right| \le \left(\sum_{k=1}^{i} h \right)^{1/2} \left(\sum_{k=1}^{i} h z_{\bar{x}}^2(x_k) \right)^{1/2} \le \sqrt{x_i} \, \| z_{\bar{x}}]|,$$

$$|z(x_i)| = \left| -\sum_{k=i+1}^{N} h z_{\bar{x}}(x_k) \right| \le \left(\sum_{k=i+1}^{N} h \right)^{1/2} \left(\sum_{k=i+1}^{N} h z_{\bar{x}}^2(x_k) \right)^{1/2} \le \sqrt{1-x_i} \, \| z_{\bar{x}}]|.$$

Then, bearing [3.71] in mind, we obtain

$$\max_{i=0,\dots,N} \left| \frac{z(x_i)}{\sqrt{\rho(x_i)}} \right| \le \| z_{\bar{x}}]| \le h \frac{q}{\sqrt{3}(1-q)} |u|_{H^2(0,1)}.$$

The theorem is proved.

REMARK 3.5.– *Estimate [3.71] is consistent in the sense of the book by Samarskii et al. (1987).*

REMARK 3.6.– *For solving the discrete problem [3.69], we can use the fixed-point iteration method that converges with the rate of a geometric progression.*

3.3.3. Conclusion

Here, we make some concluding comments about the results of this section.

In Theorem 3.8, we found the sufficient condition for the solution $u(x)$ of BVP [3.57] to belong to the space $\overset{\circ}{H}{}^1(0,1)$ in the case of the constant coefficients of the equation. In Theorems 3.10, 3.12 and 3.14, we found the sufficient condition for the solution to belong to the spaces $\overset{\circ}{W}{}^1_\infty(0,1)$, $\overset{\circ}{H}{}^1(0,1)$ and $H^2(0,1)$ respectively if the coefficients are variable.

In Theorems 3.11 and 3.14, for the solution $u(x)$ of the differential problem [3.57], we obtained the weighted estimates in the uniform norm $C[0,1]$ with the weight function $\rho_1^{-1}(x)$, $\rho_1(x) = \min(x, 1-x)$. The weighted estimates with the weight function $\rho_1^{-1/2}(x)$ are obtained in Theorems 3.9 and 3.13.

The weighted estimate obtained in Theorem 3.15 shows that in the discrete norm $C(\omega)$ the error of the mesh scheme [3.69] is $O(h\sqrt{h})$ near the boundary points of the interval $[0,1]$ and $O(h)$ in the inner points.

3.4. Two-dimensional differential equation with a fractional derivative

3.4.1. A weighted estimate for the exact solution

Here, we consider the Dirichlet boundary value problem for the two-dimensional Poisson equation with the derivative of order $1/2$ with respect to the variable x:

$$\frac{\partial^2 u(x,y)}{\partial x^2} - \frac{\alpha}{\sqrt{\pi}} \frac{\partial}{\partial x} \int_0^x \frac{u(t,y)}{\sqrt{x-t}} dt + \frac{\partial^2 u(x,y)}{\partial y^2} = -f(x,y), \quad x \in \Omega = (0,1)^2, \quad [3.74]$$
$$u(x,y) = 0, \quad (x,y) \in \Gamma = \partial\Omega.$$

To obtain the weighted estimate for the solution $u(x,y)$, we use Green's function of the first boundary problem for the Laplace operator:

$$G(x,\xi;y,\eta) = \frac{4}{\pi^2} \sum_{m,n=1}^{\infty} \frac{\sin(n\pi x)\sin(n\pi\xi)\sin(m\pi y)\sin(m\pi\eta)}{n^2 + m^2} = \quad [3.75]$$

$$= \frac{2}{\pi} \sum_{n=1}^{\infty} \frac{\sin(n\pi x)\sin(n\pi \xi)\sinh\left(n\pi\left(1 - \frac{y+\eta+|y-\eta|}{2}\right)\right)\sinh\left(n\pi \frac{y+\eta-|y-\eta|}{2}\right)}{n\sinh(n\pi)}$$

(e.g. Gradshteyn and Ryzhik (2014)).

Changing in [3.74] x for ξ and y for η, multiplying both sides of the equation by Green's function [3.75], integrating over Ω and performing some easy transformations, we come to the Fredholm integral equation of the second kind

$$u(x,y) - \frac{\alpha}{\sqrt{\pi}} \int_0^1\int_0^1 u(t,\eta) \int_t^1 \frac{\partial G(x,\xi;y,\eta)}{\partial \xi} \frac{d\xi}{\sqrt{\xi-t}} d\eta dt = \int_0^1\int_0^1 G(x,\xi;y,\eta) f(\xi,\eta) d\eta d\xi,$$

i.e.

$$u(x,y) - \int_0^1\int_0^1 K(x,t;y,\eta)u(t,\eta) d\eta dt = \int_0^1\int_0^1 G(x,\xi;y,\eta) f(\xi,\eta) d\eta d\xi \qquad [3.76]$$

with the kernel

$$K(x,t;y,\eta) = \frac{\alpha}{\sqrt{\pi}} \int_t^1 \frac{\partial G(x,\xi;y,\eta)}{\partial \xi} \frac{d\xi}{\sqrt{\xi-t}} = \qquad [3.77]$$

$$= \frac{2\alpha}{\pi\sqrt{\pi}} \sum_{n=1}^{\infty} \frac{\sin(n\pi x)v_n(t)\sinh\left(n\pi\left(1 - \frac{y+\eta+|y-\eta|}{2}\right)\right)\sinh\left(n\pi \frac{y+\eta-|y-\eta|}{2}\right)}{n\sinh(n\pi)},$$

where

$$v_n(t) = \int_t^1 n\pi \cos(n\pi\xi) \frac{d\xi}{\sqrt{\xi-t}} =$$

$$= \pi\sqrt{2n}\left[\cos(n\pi t)C(\sqrt{2n(1-t)}) - \sin(n\pi t)S(\sqrt{2n(1-t)})\right]$$

and $C(x) = \int_0^x \cos\frac{\pi\varphi^2}{2} d\varphi$, $S(x) = \int_0^x \sin\frac{\pi\varphi^2}{2} d\varphi$ are the Fresnel integrals.

We will use the following norms:

$$\|w\|_\infty = \max_{(x,y)\in \bar{\Omega}} |w(x,y)|,$$

$$\|w\|_{1,\infty} = \left\|\frac{\partial w}{\partial x}\right\|_\infty + \left\|\frac{\partial w}{\partial y}\right\|_\infty, \quad \|w\|_{2,\infty} = \left\|\frac{\partial^2 w}{\partial x^2}\right\|_\infty + \left\|\frac{\partial^2 w}{\partial y^2}\right\|_\infty.$$

Taking into account the asymptotic inequalities

$$C(x) = \int_0^x \cos\frac{\pi\varphi^2}{2}\,d\varphi = \frac{1}{2} + O\!\left(\frac{1}{x}\right), \quad S(x) = \int_0^x \sin\frac{\pi\varphi^2}{2}\,d\varphi = \frac{1}{2} + O\!\left(\frac{1}{x}\right)$$

as $x \to +\infty$, we have the estimate

$$\|v_n\|_\infty \le \pi\sqrt{n}.\qquad [3.78]$$

THEOREM 3.16.– *Let a parameter α in [3.74] satisfy the conditions*

$$1 - \frac{2\alpha\zeta(3/2)}{\pi^{3/2}} = 1 - \alpha \cdot 0.9382979416\ldots > 0,\qquad [3.79]$$

$$1 - \frac{2\alpha}{(\sigma+1)\pi^{3/2-\sigma}}\zeta(3/2-\sigma) > 0,\qquad [3.80]$$

where σ is a number arbitrarily close to $1/2$ from below. Then, for the solution $u(x,y)$ of problem [3.74], the following weighted estimate holds true:

$$\left\|\frac{u(x,y)}{\rho(x,y)}\right\|_\infty \le M\,\|f\|_\infty,$$

where $\rho(x,y) = \min\{x^\sigma, (1-x)^\sigma, y, 1-y\}$, $\zeta(\cdot)$ is the Riemann zeta function and M is a positive constant independent of $u(x,y)$:

$$M = \max\left\{\frac{1}{2}\left(1 - \frac{2\alpha\zeta(3/2)}{\pi^{3/2}}\right)^{-1},\; \frac{4}{\pi^{3-\sigma}}\zeta(3-\sigma)\left(1 - \frac{2\alpha}{(\sigma+1)\pi^{3/2-\sigma}}\zeta(3/2-\sigma)\right)^{-1}\right\}.$$

PROOF.– Substituting $U(x,y) = \dfrac{u(x,y)}{y(1-y)}$ in [3.76], we obtain the equation

$$U(x,y) - \int_0^1\int_0^1 \frac{\eta(1-\eta)K(x,t;y,\eta)}{y(1-y)} U(t,\eta)dtd\eta = \int_0^1\int_0^1 \frac{G(x,t;y,\eta)}{y(1-y)} f(t,\eta)dtd\eta ,$$

which leads to the estimate

$$\|U\|_\infty - \max_{(x,y)\in\bar\Omega} \int_0^1\int_0^1 \frac{\eta(1-\eta)|K(x,t;y,\eta)|}{y(1-y)} dtd\eta \, \|U\|_\infty \le \qquad [3.81]$$

$$\le \max_{(x,y)\in\bar\Omega} \int_0^1\int_0^1 \frac{G(x,t;y,\eta)}{y(1-y)} dtd\eta \, \|f\|_\infty .$$

Next, we will need the following two inequalities

$$\int_0^1 \frac{\eta(1-\eta)}{y(1-y)} \cdot \frac{\sinh\left(n\pi\left(1 - \dfrac{y+\eta+|y-\eta|}{2}\right)\right) \sinh\left(n\pi \dfrac{y+\eta-|y-\eta|}{2}\right)}{n\sinh(n\pi)} d\eta =$$

$$= \frac{\sinh(n\pi(1-y))}{y(1-y)n\sinh(n\pi)} \int_0^y \eta(1-\eta)\sinh(n\pi\eta)d\eta +$$

$$+ \frac{\sinh(n\pi y)}{y(1-y)n\sinh(n\pi)} \int_y^1 \eta(1-\eta)\sinh(n\pi(1-\eta))d\eta =$$

$$= \frac{1}{y(1-y)n^2\pi\sinh(n\pi)} \left\{ y(1-y)\sinh(n\pi) + \frac{4\cosh\dfrac{n\pi(1-2y)}{2}\sinh\dfrac{n\pi}{2} - 2\sinh(n\pi)}{(n\pi)^2} \right\} \le$$

$$\le \frac{1}{n^2\pi}\left(1 - \frac{8\left(\cosh\dfrac{n\pi}{2} - 1\right)}{(n\pi)^2 \cosh\dfrac{n\pi}{2}}\right) < \frac{1}{n^2\pi} ,$$

and

$$\int_0^1 \frac{1}{y(1-y)} \frac{\sinh\left(n\pi\left(1-\frac{y+\eta+|y-\eta|}{2}\right)\right)\sinh\left(n\pi\frac{y+\eta-|y-\eta|}{2}\right)}{n\sinh(n\pi)} d\eta =$$

$$= \frac{\sinh(n\pi(1-y))}{y(1-y)n\sinh(n\pi)} \int_0^y \sinh(n\pi\eta) d\eta + \frac{\sinh(n\pi y)}{y(1-y)n\sinh(n\pi)} \int_y^1 \sinh(n\pi(1-\eta)) d\eta =$$

$$= \frac{2\sinh\frac{n\pi(1-y)}{2}\sinh\frac{n\pi y}{2}}{y(1-y)n^2\pi\cosh\frac{n\pi}{2}} < \frac{\tanh\frac{n\pi}{2}}{n} < \frac{1}{n}.$$

Taking into account inequality [3.78], we derive the estimates

$$\max_{(x,y)\in\bar{\Omega}} \int_0^1\int_0^1 \frac{\eta(1-\eta)|K(x,t;y,\eta)|}{y(1-y)} dt d\eta \leq \frac{2\alpha}{\pi\sqrt{\pi}} \sum_{n=1}^{\infty} \frac{\pi\sqrt{n}}{n^2\pi} = \frac{2\alpha}{\pi^{3/2}} \zeta(3/2)$$

and

$$\max_{(x,y)\in\bar{\Omega}} \int_0^1\int_0^1 \frac{G(x,t;y,\eta)}{y(1-y)} dt d\eta \leq \frac{2}{\pi} \sum_{n=1}^{\infty} \frac{1}{n} \int_0^1 \sin(n\pi t) dt =$$

$$= \frac{2}{\pi} \sum_{n=1}^{\infty} \frac{1-(-1)^n}{n^2\pi} = \frac{4}{\pi^2} \sum_{k=1}^{\infty} \frac{1}{(2k-1)^2} = \frac{4}{\pi^2} \frac{\pi^2}{8} = \frac{1}{2}.$$

Due to condition [3.79], inequality [3.81] implies the weighted estimate

$$\|U\|_\infty = \left\|\frac{u}{y(1-y)}\right\|_\infty \leq \frac{1}{2}\left(1 - \frac{2\alpha\zeta(3/2)}{\pi^{3/2}}\right)^{-1} \|f\|_\infty.$$

Substituting in [3.76]

$$V(x,y) = \frac{u(x,y)}{x^\sigma}, \quad \sigma < \frac{1}{2},$$

we have the equation

$$V(x,y) - \int_0^1\int_0^1 \frac{t^\sigma K(x,t;y,\eta)}{x^\sigma} V(t,\eta) dt d\eta = \int_0^1\int_0^1 \frac{G(x,t;y,\eta)}{x^\sigma} f(t,\eta) dt d\eta,$$

which gives

$$\|V\|_\infty - \max_{(x,y)\in\bar\Omega} \int_0^1\int_0^1 \frac{t^\sigma |K(x,t;y,\eta)|}{x^\sigma} dt d\eta \, \|V\|_\infty \leq \qquad [3.82]$$

$$\leq \max_{(x,y)\in\bar\Omega} \int_0^1\int_0^1 \frac{G(x,t;y,\eta)}{x^\sigma} dt d\eta \, \|f\|_\infty.$$

Using the inequality

$$\int_0^1 \frac{\sinh\left(n\pi\left(1 - \frac{y+\eta+|y-\eta|}{2}\right)\right)\sinh\left(n\pi \frac{y+\eta-|y-\eta|}{2}\right)}{n\sinh(n\pi)} d\eta =$$

$$= \frac{\sinh(n\pi(1-y))}{n\sinh(n\pi)} \int_0^y \sinh(n\pi\eta) d\eta + \frac{\sinh(n\pi y)}{n\sinh(n\pi)} \int_y^1 \sinh(n\pi(1-\eta)) d\eta = \quad [3.83]$$

$$= \frac{\sinh(n\pi) - 2\sinh\frac{n\pi}{2}\cosh\frac{n\pi(1-2y)}{2}}{n^2\pi\sinh(n\pi)} \leq \frac{1}{n^2\pi}$$

and taking [3.78] into account, we obtain the estimates

$$\max_{(x,y)\in\bar\Omega} \int_0^1\int_0^1 \frac{G(x,t;y,\eta)}{x^\sigma} dt d\eta \leq \frac{2}{\pi} \max_{0\leq x\leq 1} \sum_{n=1}^\infty \frac{1}{n^2\pi} \frac{|\sin(n\pi x)|}{x^\sigma} \int_0^1 \sin(n\pi t) dt =$$

$$= \frac{2}{\pi} \max_{0\leq x\leq 1} \sum_{n=1}^\infty \frac{1}{n^{2-\sigma}\pi^{1-\sigma}} \frac{|\sin(n\pi x)|}{(n\pi x)^\sigma} \frac{1-(-1)^n}{n\pi} \leq \frac{4}{\pi^{3-\sigma}} \sum_{n=1}^\infty \frac{1}{n^{3-\sigma}} = \frac{4}{\pi^{3-\sigma}} \zeta(3-\sigma)$$

and

$$\max_{(x,y)\in\bar\Omega} \int_0^1\int_0^1 \frac{t^\sigma |K(x,t;y,\eta)|}{x^\sigma} dt d\eta \leq \frac{2\alpha}{\pi\sqrt\pi} \max_{0\leq x\leq 1} \sum_{n=1}^\infty \frac{\pi\sqrt n}{n^2\pi} \frac{|\sin(n\pi x)|}{x^\sigma} \int_0^1 t^\sigma dt =$$

$$= \frac{2\alpha}{\pi\sqrt{\pi}} \max_{0 \le x \le 1} \sum_{n=1}^{\infty} \frac{\pi\sqrt{n}}{n^{2-\sigma}\pi^{1-\sigma}} \frac{|\sin(n\pi x)|}{(n\pi x)^{\sigma}} \frac{1}{\sigma+1} \le \frac{2\alpha}{\pi\sqrt{\pi}} \sum_{n=1}^{\infty} \frac{\pi\sqrt{n}}{n^{2-\sigma}\pi^{1-\sigma}(\sigma+1)} =$$

$$= \frac{2\alpha}{(\sigma+1)\pi^{3/2-\sigma}} \sum_{n=1}^{\infty} \frac{1}{n^{3/2-\sigma}} = \frac{2\alpha}{(\sigma+1)\pi^{3/2-\sigma}} \zeta(3/2-\sigma).$$

Let σ be arbitrarily close to $1/2$ from below and let α satisfy condition [3.80]. Then, [3.82] yields the estimate

$$\|V\|_{\infty} = \left\|\frac{u}{x^{\sigma}}\right\|_{\infty} \le \frac{4}{\pi^{3-\sigma}} \zeta(3-\sigma) \left(1 - \frac{2\alpha}{(\sigma+1)\pi^{3/2-\sigma}} \zeta(3/2-\sigma)\right)^{-1} \|f\|_{\infty}. \quad [3.84]$$

A similar estimate can be obtained if we change x^{σ} for $(1-x)^{\sigma}$.

The theorem is proved.

3.4.2. *A mesh scheme of the first order of accuracy*

To solve equation [3.76] numerically, we apply the mesh method. We introduce a set of nodes

$$\bar{\omega} = \{(x_i, y_j): x_i = ih, \ y_j = jh, \ i,j = 0,1,\ldots,N, \ h = 1/N\}$$

and put $x = x_i$, $y = y_p$ in the integral equation [3.76]:

$$u(x_i, y_p) - \int_0^1\int_0^1 K(x_i,t;y_p,\eta)u(t,\eta)d\eta dt = \qquad [3.85]$$

$$= \int_0^1\int_0^1 G(x_i,\xi;y_p,\eta) f(\xi,\eta) d\eta d\xi .$$

Then, we have the mesh scheme

$$u^h(x_i,y_p) - \sum_{k,j=1}^{N-1} u^h(x_k,y_j) \iint_{\Omega_{kj}} K(x_i,t;y_p,\eta)dtd\eta =$$

$$= \int_0^1\int_0^1 G(x_i,t;y_p,\eta)f(t,\eta)dtd\eta, \quad i,p = 1,2,...,N-1, \qquad [3.86]$$

where the rectangles $\Omega_{kj} = [x_{k-1},x_k] \times [y_{j-1},y_j]$, $k,j = 1,...,N$, form a partition of the unit square $\Omega = [0,1]^2$.

The error $z(x,y) = u^h(x,y) - u(x,y)$ is a solution of the discrete problem

$$z(x_i,y_p) - \sum_{k,j=1}^{N-1} z(x_k,y_j) \iint_{\Omega_{kj}} K(x_i,t;y_p,\eta)dtd\eta =$$

$$= \sum_{k,j=1}^{N-1} \iint_{\Omega_{kj}} \left[u(x_k,y_j) - u(t,\eta)\right] K(x_i,t;y_p,\eta)dtd\eta, \qquad [3.87]$$

$$i,p = 1,2,...,N-1.$$

Furthermore, we will need the mesh norm $\|w\|_{\infty,\bar{\omega}} = \max_{(x,y)\in\bar{\omega}} |w(x,y)|$.

THEOREM 3.17.— *Let the solution* $u(x,y)$ *of equation [3.76] satisfy the condition* $u \in C^1(\bar{\Omega})$ *and let the parameter* α *satisfy the inequalities*

$$1 - \frac{\alpha \cdot 0.72}{\sqrt{\pi}} \zeta(3/2 - \sigma) > 0, \quad \sigma \in [0,1/2), \qquad [3.88]$$

$$1 - \frac{2\alpha}{\pi^{3/2-\sigma}} \zeta(3/2 - \sigma) > 0, \quad \sigma \in [0,1/2). \qquad [3.89]$$

Then, the accuracy of the mesh scheme [3.86] is characterized by the weighted estimate, which takes into account the boundary effect:

$$\left\|\frac{u - u^h}{\rho(x,y)}\right\|_{\infty} \leq Mh \|u\|_{1,\infty},$$

where $\rho(x,y) = \min\{x^\sigma, (1-x)^\sigma, y^\sigma, (1-y)^\sigma\}$, $\zeta(\cdot)$ is the Riemann zeta function and M is a positive constant independent of h and $u(x,y)$:

$$M = \max\left\{\frac{\dfrac{\alpha \cdot 0.72}{\sqrt{\pi}}\zeta(3/2-\sigma)}{1-\dfrac{\alpha \cdot 0.72}{\sqrt{\pi}4^\sigma}\zeta(3/2-\sigma)}, \frac{\dfrac{2\alpha}{\pi^{3/2-\sigma}}\zeta(3/2-\sigma)}{1-\dfrac{2\alpha}{\pi^{3/2-\sigma}}\zeta(3/2-\sigma)}\right\}.$$

PROOF.– Substituting in [3.87]

$$Z(x,y) = \frac{z(x,y)}{[y(1-y)]^\sigma}, \quad 0 \leq \sigma < 1/2,$$

we come to the mesh scheme

$$Z(x_i, y_p) - \frac{1}{[y_p(1-y_p)]^\sigma} \sum_{k,j=1}^{N-1} [y_j(1-y_j)]^\sigma Z(x_k, y_j) \iint_{\Omega_{kj}} K(x_i, t; y_p, \eta)dtd\eta =$$

$$= \frac{1}{[y_p(1-y_p)]^\sigma} \sum_{k,j=1}^{N-1} \iint_{\Omega_{kj}} [u(x_k, y_j) - u(t,\eta)]K(x_i, t; y_p, \eta)dtd\eta, \quad [3.90]$$

$$i, p = 1, 2, \ldots, N-1.$$

This yields the inequality

$$\|Z\|_\infty - \max_{(x,y)\in\bar{\Omega}} \int_0^1\int_0^1 \frac{|K(x,t;y,\eta)|}{[y(1-y)]^\sigma}dtd\eta \frac{1}{4^\sigma}\|Z\|_\infty \leq$$

$$\leq \max_{(x,y)\in\bar{\Omega}} \int_0^1\int_0^1 \frac{|K(x,t;y,\eta)|}{[y(1-y)]^\sigma}dtd\eta\, h\, \|u\|_{1,\infty}. \quad [3.91]$$

Making use of the estimate

$$\int_0^1 \frac{\sinh\left(n\pi\left(1-\dfrac{y+\eta+|y-\eta|}{2}\right)\right)\sinh\left(n\pi\dfrac{y+\eta-|y-\eta|}{2}\right)}{[y(1-y)]^\sigma n\sinh(n\pi)}d\eta =$$

$$= \frac{\sinh(n\pi(1-y))}{[y(1-y)]^\sigma n \sinh(n\pi)} \int_0^y \sinh(n\pi\eta) d\eta + \frac{\sinh(n\pi y)}{[y(1-y)]^\sigma n \sinh(n\pi)}$$

$$\int_y^1 \sinh(n\pi(1-\eta)) d\eta =$$

$$= \frac{2 \sinh \frac{n\pi(1-y)}{2} \sinh \frac{n\pi y}{2}}{[y(1-y)]^\sigma n^2 \pi \cosh \frac{n\pi}{2}} \leq \frac{0.36}{n^{2-\sigma}}, \quad \sigma \in [0, 1/2),$$

and taking [3.77] and [3.78] into account, we obtain the estimate

$$\max_{(x,y) \in \bar{\Omega}} \int_0^1 \int_0^1 \frac{|K(x,t;y,\eta)|}{[y(1-y)]^\sigma} dt d\eta \leq \frac{2\alpha}{\pi\sqrt{\pi}} \sum_{n=1}^\infty \frac{0.36\pi\sqrt{n}}{n^{2-\sigma}} = \frac{0.72\alpha}{\sqrt{\pi}} \zeta(3/2 - \sigma). \quad [3.92]$$

Then, [3.91] leads to the estimate

$$\| Z \|_\infty \leq \frac{\alpha \cdot 0.72}{\sqrt{\pi}} \zeta(3/2 - \sigma) \frac{1}{4^\sigma} \| Z \|_\infty + \frac{\alpha \cdot 0.72}{\sqrt{\pi}} \zeta(3/2 - \sigma) h \| u \|_{1,\infty}. \quad [3.93]$$

Due to condition [3.88], we obtain from [3.93] the estimate

$$\| Z \|_\infty = \left\| \frac{z(x,y)}{[y(1-y)]^\sigma} \right\|_\infty = \left\| \frac{u - u^h}{[y(1-y)]^\sigma} \right\|_\infty \leq$$

$$\leq h \left(1 - \frac{\alpha \cdot 0.72}{\sqrt{\pi} 4^\sigma} \zeta(3/2 - \sigma)\right)^{-1} \frac{\alpha \cdot 0.72}{\sqrt{\pi}} \zeta(3/2 - \sigma) \| u \|_{1,\infty}.$$

[3.94]

Substituting in [3.87]

$$Z(x,y) = \frac{z(x,y)}{x^\sigma}, \quad 0 \leq \sigma < 1/2,$$

we obtain the mesh scheme

$$Z(x_i, y_p) - \frac{1}{x_i^\sigma} \sum_{k,j=1}^{N-1} x_k^\sigma Z(x_k, y_j) \iint_{\Omega_{kj}} K(x_i, t; y_p, \eta) dt d\eta =$$

$$= \frac{1}{x_i^\sigma} \sum_{k,j=1}^{N-1} \iint_{\Omega_{kj}} [u(x_k, y_j) - u(t,\eta)] K(x_i, t; y_p, \eta) dt d\eta, \quad i, p = 1, 2, ..., N-1.$$

From here it follows

$$\| Z \|_\infty - \max_{(x,y) \in \bar\Omega} \int_0^1 \int_0^1 \frac{|K(x,t;y,\eta)|}{x^\sigma} dt d\eta \, \| Z \|_\infty \le \quad [3.95]$$

$$\le \max_{(x,y) \in \bar\Omega} \int_0^1 \int_0^1 \frac{|K(x,t;y,\eta)|}{x^\sigma} dt d\eta \, h \, \| u \|_{1,\infty} .$$

Making use of inequality [3.83] and taking [3.77] and [3.78] into account, we arrive at the estimate

$$\max_{(x,y) \in \bar\Omega} \int_0^1 \int_0^1 \frac{|K(x,t;y,\eta)|}{x^\sigma} dt d\eta \le \frac{2\alpha}{\pi\sqrt\pi} \max_{0 \le x \le 1} \sum_{n=1}^\infty \frac{\pi\sqrt n}{n^2 \pi} \frac{|\sin(n\pi x)|}{x^\sigma} =$$

$$= \frac{2\alpha}{\pi\sqrt\pi} \max_{0 \le x \le 1} \sum_{n=1}^\infty \frac{\pi\sqrt n}{n^{2-\sigma}\pi^{1-\sigma}} \frac{|\sin(n\pi x)|}{(n\pi x)^\sigma} \le \frac{2\alpha}{\pi\sqrt\pi} \sum_{n=1}^\infty \frac{\pi\sqrt n}{n^{2-\sigma}\pi^{1-\sigma}} = \quad [3.96]$$

$$= \frac{2\alpha}{\pi^{3/2-\sigma}} \sum_{n=1}^\infty \frac{1}{\pi^{3/2-\sigma}} = \frac{2\alpha}{\pi^{3/2-\sigma}} \zeta(3/2 - \sigma) .$$

Then, [3.95] gives the inequality

$$\| Z \|_\infty \le \frac{2\alpha}{\pi^{3/2-\sigma}} \zeta(3/2 - \sigma) \| Z \|_\infty + \frac{2\alpha}{\pi^{3/2-\sigma}} \zeta(3/2 - \sigma) h \, \| u \|_{1,\infty} ,$$

which due to [3.89] leads to the following estimate:

$$\| Z \|_\infty = \left\| \frac{z(x,y)}{x^\sigma} \right\|_\infty = \left\| \frac{u - u^h}{x^\sigma} \right\|_\infty \le \quad [3.97]$$

$$\le h \left(1 - \frac{2\alpha}{\pi^{3/2-\sigma}} \zeta(3/2 - \sigma) \right)^{-1} \frac{2\alpha}{\pi^{3/2-\sigma}} \zeta(3/2 - \sigma) \| u \|_{1,\infty} .$$

The same estimate as [3.97] can be obtained if we replace here x^σ for $(1-x)^\sigma$.

The statement of the theorem now follows from estimates [3.94] and [3.97].

3.4.3. *A mesh scheme of the second order of accuracy*

From relation [3.85], we obtain the mesh scheme

$$u^h(x_i, y_p) - \sum_{k,j=1}^{N-1} \iint_{\Omega_{kj}} K(x_i, t; y_p, \eta) L_{kj}(t, \eta, u^h) dt d\eta =$$

$$= \int_0^1\int_0^1 G(x_i, t; y_p, \eta) f(t, \eta) dt d\eta, \quad i, p = 1, 2, ..., N-1,$$

[3.98]

where

$$L_{kj}(x, y; w) = \frac{x - x_{k-1}}{x_k - x_{k-1}} \left[\frac{y - y_{j-1}}{y_j - y_{j-1}} w(x_k, y_j) + \frac{y_j - y}{y_j - y_{j-1}} w(x_k, y_{j-1}) \right] +$$

$$+ \frac{x_k - x}{x_k - x_{k-1}} \left[\frac{y - y_{j-1}}{y_j - y_{j-1}} w(x_{k-1}, y_j) + \frac{y_j - y}{y_j - y_{j-1}} w(x_{k-1}, y_{j-1}) \right]$$

is the Lagrange interpolating polynomial (linear in each variable x and y) of the first degree for the function $w(x, y)$.

It is easy to verify that the error $z(x, y) = u^h(x, y) - u(x, y)$ is a solution of the discrete problem

$$z(x_i, y_p) - \sum_{k,j=1}^{N-1} \iint_{\Omega_{kj}} K(x_i, t; y_p, \eta) L_{kj}(t, \eta; z) dt d\eta =$$

$$= \sum_{k,j=1}^{N-1} \iint_{\Omega_{kj}} K(x_i, t; y_p, \eta) [L_{kj}(t, \eta; u) - u(t, \eta)] dt d\eta,$$

[3.99]

$$i, p = 1, 2, ..., N-1.$$

THEOREM 3.18.– *Let the solution $u(x,y)$ of equation [3.76] satisfy the condition $u \in C^1(\overline{\Omega})$ and let the parameter α satisfy the inequalities*

$$1 - \frac{\alpha \cdot 0.72}{\sqrt{\pi}} \zeta(3/2 - \sigma) > 0, \quad \sigma \in [0, 1/2), \qquad [3.100]$$

$$1 - \frac{2\alpha}{\pi^{3/2-\sigma}} \zeta(3/2 - \sigma) > 0, \quad \sigma \in [0, 1/2). \qquad [3.101]$$

Then, the accuracy of the mesh scheme [3.98] is characterized by the weighted estimate, which takes into account the boundary effect:

$$\left\| \frac{u - u^h}{\rho(x,y)} \right\|_\infty \leq M h^2 \|u\|_{2,\infty},$$

where $\rho(x,y) = \min\{x^\sigma, (1-x)^\sigma, y^\sigma, (1-y)^\sigma\}$, $\zeta(\cdot)$ is the Riemann zeta function and M is a positive constant independent of h and $u(x,y)$:

$$M = \max\left\{ \frac{\dfrac{\alpha \cdot 0.36}{\sqrt{\pi}} \zeta(3/2 - \sigma)}{1 - \dfrac{\alpha \cdot 0.72}{\sqrt{\pi}4^\sigma} \zeta(3/2 - \sigma)}, \frac{\dfrac{\alpha}{\pi^{3/2-\sigma}} \zeta(3/2 - \sigma)}{1 - \dfrac{2\alpha}{\pi^{3/2-\sigma}} \zeta(3/2 - \sigma)} \right\}.$$

PROOF.– Substituting in [3.99]

$$Z(x,y) = \frac{z(x,y)}{[y(1-y)]^\sigma}, \quad 0 \leq \sigma < 1/2,$$

we have the mesh scheme

$$Z(x_i, y_p) - \frac{1}{[y_p(1-y_p)]^\sigma} \sum_{k,j=1}^{N-1} \iint_{\Omega_{kj}} K(x_i, t; y_p, \eta) L_{kj}(t, \eta; [y(1-y)]^\sigma Z) dt d\eta =$$

$$= \frac{1}{[y_p(1-y_p)]^\sigma} \sum_{k,j=1}^{N-1} \iint_{\Omega_{kj}} K(x_i, t; y_p, \eta)[L_{kj}(t, \eta; u) - u(t, \eta)] dt d\eta,$$

$$i, p = 1, 2, \ldots, N-1.$$

This yields the inequality

$$\|Z\|_\infty - \max_{(x,y)\in\bar\Omega}\int_0^1\int_0^1\frac{|K(x,t;y,\eta)|}{[y(1-y)]^\sigma}dtd\eta\,\frac{1}{4^\sigma}\|Z\|_\infty \le$$

$$\le \max_{(x,y)\in\bar\Omega}\int_0^1\int_0^1\frac{|K(x,t;y,\eta)|}{[y(1-y)]^\sigma}dtd\eta\,\frac{h^2}{2}\|u\|_{2,\infty}, \qquad [3.102]$$

since

$$u(x,y) - L_{kj}(x,y;u) = u(x,y) - \frac{x-x_{k-1}}{x_k-x_{k-1}}u(x_k,y) - \frac{x_k-x}{x_k-x_{k-1}}u(x_{k-1},y) +$$

$$+\frac{x-x_{k-1}}{x_k-x_{k-1}}\left[u(x_k,y) - \frac{y-y_{j-1}}{y_j-y_{j-1}}u(x_k,y_j) - \frac{y_j-y}{y_j-y_{j-1}}u(x_k,y_{j-1})\right] +$$

$$+\frac{x_k-x}{x_k-x_{k-1}}\left[u(x_{k-1},y) - \frac{y-y_{j-1}}{y_j-y_{j-1}}u(x_{k-1},y_j) - \frac{y_j-y}{y_j-y_{j-1}}u(x_{k-1},y_{j-1})\right] =$$

$$= \frac{u''_{xx}(\bar x_k,y)}{2!}(x-x_{k-1})(x-x_k)$$

$$+\frac{x-x_{k-1}}{x_k-x_{k-1}}\frac{u''_{yy}(x_k,\bar y_j)}{2!}(y-y_{j-1})(y-y_j) +$$

$$+\frac{x_k-x}{x_k-x_{k-1}}\frac{u''_{yy}(x_{k-1},\bar{\bar y}_j)}{2!}(y-y_{j-1})(y-y_j),$$

where $\bar x_k \in (x_{k-1},x_k)$, $\bar y_j, \bar{\bar y}_j \in (y_{j-1},y_j)$, and the estimate

$$|u(x,y)-L_{kj}(x,y;u)| \le \frac{h^2}{2!}\|u''_{yy}\|_\infty + \frac{h^2}{2!}\|u''_{xx}\|_\infty = \frac{h^2}{2!}\|u\|_{2,\infty}. \qquad [3.103]$$

Using inequality [3.92], from [3.102], we come to the estimate

$$\|Z\|_\infty \le \frac{\alpha\cdot 0.72}{\sqrt\pi}\zeta(3/2-\sigma)\frac{1}{4^\sigma}\|Z\|_\infty + \frac{\alpha\cdot 0.72}{\sqrt\pi}\zeta(3/2-\sigma)\frac{h^2}{2}\|u\|_{2,\infty}.$$

Due to condition [3.100], we have

$$\|Z\|_\infty = \left\| \frac{z(x,y)}{[y(1-y)]^\sigma} \right\|_\infty = \left\| \frac{u - u^h}{[y(1-y)]^\sigma} \right\|_\infty \leq \qquad [3.104]$$

$$\leq \frac{h^2}{2}\left(1 - \frac{\alpha \cdot 0.72}{\sqrt{\pi}\, 4^\sigma} \zeta(3/2 - \sigma)\right)^{-1} \frac{\alpha \cdot 0.72}{\sqrt{\pi}} \zeta(3/2 - \sigma) \|u\|_{2,\infty}.$$

Next, substituting in [3.99]

$$Z(x,y) = \frac{z(x,y)}{x^\sigma}, \quad 0 \leq \sigma < 1/2,$$

we have the mesh scheme

$$Z(x_i, y_p) - \frac{1}{x_i^\sigma} \sum_{k,j=1}^{N-1} \iint_{\Omega_{kj}} K(x_i, t; y_p, \eta) L_{kj}(t, \eta; x^\sigma Z) dt d\eta =$$

$$= \frac{1}{x_i^\sigma} \sum_{k,j=1}^{N-1} \iint_{\Omega_{kj}} K(x_i, t; y_p, \eta)[L_{kj}(t, \eta; u) - u(t, \eta)] \, dt d\eta,$$

$$i, p = 1, 2, \ldots, N-1.$$

From here and due to [3.103], we obtain the estimate

$$\|Z\|_\infty - \max_{(x,y)\in\overline{\Omega}} \int_0^1\int_0^1 \frac{|K(x,t;y,\eta)|}{x^\sigma} dt d\eta \, \|Z\|_\infty \leq$$

$$\leq \max_{(x,y)\in\overline{\Omega}} \int_0^1\int_0^1 \frac{|K(x,t;y,\eta)|}{x^\sigma} dt d\eta \, \frac{h^2}{2} \|u\|_{2,\infty}.$$

Applying here inequality [3.96], we come to the following:

$$\|Z\|_\infty \leq \frac{2\alpha}{\pi^{3/2-\sigma}} \zeta(3/2 - \sigma) \|Z\|_\infty + \frac{2\alpha}{\pi^{3/2-\sigma}} \zeta(3/2 - \sigma) \frac{h^2}{2} \|u\|_{2,\infty}.$$

Due to assumption [3.101], we finally arrive at the estimate

$$\|Z\|_\infty = \left\|\frac{z(x,y)}{x^\sigma}\right\|_\infty = \left\|\frac{u-u^h}{x^\sigma}\right\|_\infty \le \qquad [3.105]$$

$$\le \frac{h^2}{2}\left(1 - \frac{2\alpha}{\pi^{3/2-\sigma}}\zeta(3/2-\sigma)\right)^{-1}\frac{2\alpha}{\pi^{3/2-\sigma}}\zeta(3/2-\sigma)\,\|u\|_{2,\infty}\,.$$

A similar estimate can be obtained if we change x^σ for $(1-x)^\sigma$.

Combining inequalities [3.104] and [3.105], we complete the proof of the theorem.

REMARK 3.7.– *Similar results can be proved for the fractional derivative with respect to the variable y.*

REMARK 3.8.– *The proposed approach and technique can also be applied in the general case of the fractional derivative. We have addressed the case of the fractional derivative of order $\alpha = 1/2$ only for brevity and simplicity.*

3.4.4. Conclusion

In Theorem 3.16, the solution $u(x,y)$ of the BVP [3.74] is characterized by the weighted estimate in the uniform norm $C(\bar{\Omega})$ with the weight function $\rho^{-1}(x,y)$, $\rho(x,y) = \min\{x^\sigma, (1-x)^\sigma, y, 1-y\}$.

The weighted estimate obtained in Theorem 3.17 indicates that in the discrete uniform norm $C(\omega)$ the error of the mesh schemes [3.86] is $O(h^{1+\sigma})$ near the boundary nodes, whereas it is $O(h)$ in the inner nodes away from the boundary of the domain. The weighted estimate obtained in Theorem 3.18 shows that in the discrete uniform norm $C(\omega)$ the error of the mesh schemes [3.98] is $O(h^{2+\sigma})$ near the boundary nodes, whereas it is $O(h^2)$ in the inner nodes away from the boundary of the domain.

3.5. The Goursat problem with fractional derivatives

3.5.1. *Properties of the exact solution*

We consider the problem (e.g. Cheung (1977))

$$\frac{\partial^2 u(x,y)}{\partial x \partial y} + k_1(x,y) D_x^\alpha (u(\cdot,y)) +$$

$$+ k_2(x,y) D_y^\alpha (u(x,\cdot)) + k_3(x,y) u(x,y) = f(x,y), \quad (x,y) \in \Omega = [0,T]^2, \quad [3.106]$$

$$u(x,0) = 0, \; x \in [0,T], \quad u(0,y) = 0, \; y \in [0,T],$$

where $0 < T < 1$, $\alpha \in (0,1)$ and

$$D_x^\alpha (u(\cdot,y)) = \frac{1}{\Gamma(1-\alpha)} \frac{\partial}{\partial x} \int_0^x \frac{u(t,y) dt}{(x-t)^\alpha}, \quad D_y^\alpha (u(x,\cdot)) = \frac{1}{\Gamma(1-\alpha)} \frac{\partial}{\partial y} \int_0^y \frac{u(x,t) dt}{(y-t)^\alpha}.$$

Assuming that the coefficients of the equation are smooth enough, we write the integral corollary of problem [3.106]:

$$u(x,y) + \frac{1}{\Gamma(1-\alpha)} \int_0^y \int_0^x u(t,\eta) \left[\frac{k_1(x,\eta)}{(x-t)^\alpha} - \int_t^x \frac{\partial k_1(\xi,\eta)}{\partial \xi} \frac{d\xi}{(\xi-t)^\alpha} \right] dt\, d\eta +$$

$$+ \int_0^x \int_0^y k_3(\xi,\eta) u(\xi,\eta) d\eta\, d\xi = \int_0^x \int_0^y f(\xi,\eta) d\eta\, d\xi.$$

This leads to the Volterra integral equation of the second kind

$$u(x,y) + \int_0^x \int_0^y K(x,y,\xi,\eta) u(\xi,\eta) d\eta\, d\xi = \varphi(x,y) \quad [3.107]$$

with the kernel

$$K(x,y,\xi,\eta) = \frac{1}{\Gamma(1-\alpha)} \left[\frac{k_1(x,\eta)}{(x-\xi)^\alpha} - \int_\xi^x \frac{\partial k_1(s,\eta)}{\partial s} \frac{ds}{(s-\xi)^\alpha} + \right.$$

$$\left. + \frac{k_2(\xi,y)}{(y-\eta)^\alpha} - \int_\eta^y \frac{\partial k_2(\xi,t)}{\partial t} \frac{dt}{(t-\eta)^\alpha} \right] + k_3(\xi,\eta)$$

and the right-hand side $\varphi(x,y) = \int_0^x \int_0^y f(\xi,\eta) d\eta\, d\xi$.

For the kernel $K(x,y,\xi,\eta)$, we have the estimate

$$|K(x,y,\xi,\eta)| \leq \|k_3\|_{L_\infty(\Omega)} + \qquad [3.108]$$

$$+ \frac{\|k_1\|_{W_\infty^1(\Omega)}}{\Gamma(1-\alpha)}\left[\frac{1}{(x-\xi)^\alpha}+\frac{(x-\xi)^{1-\alpha}}{1-\alpha}\right]+\frac{\|k_2\|_{W_\infty^1(\Omega)}}{\Gamma(1-\alpha)}\left[\frac{1}{(y-\eta)^\alpha}+\frac{(y-\eta)^{1-\alpha}}{1-\alpha}\right] =$$

$$= \frac{\|k_1\|_{W_\infty^1(\Omega)}}{\Gamma(1-\alpha)}\frac{1}{(x-\xi)^\alpha}\left[1+\frac{x-\xi}{1-\alpha}\right]+\frac{\|k_2\|_{W_\infty^1(\Omega)}}{\Gamma(1-\alpha)}\frac{1}{(y-\eta)^\alpha}\left[1+\frac{y-\eta}{1-\alpha}\right]+\|k_3\|_{L_\infty(\Omega)} \leq$$

$$\leq \frac{\|k_1\|_{W_\infty^1(\Omega)}}{\Gamma(1-\alpha)}\frac{1}{(x-\xi)^\alpha}\frac{2-\alpha}{1-\alpha}+\frac{\|k_2\|_{W_\infty^1(\Omega)}}{\Gamma(1-\alpha)}\frac{1}{(y-\eta)^\alpha}\frac{2-\alpha}{1-\alpha}+\|k_3\|_{L_\infty(\Omega)} \leq$$

$$\leq \frac{2-\alpha}{\Gamma(2-\alpha)}\left[\frac{\|k_1\|_{W_\infty^1(\Omega)}}{(x-\xi)^\alpha}+\frac{\|k_2\|_{W_\infty^1(\Omega)}}{(y-\eta)^\alpha}+\|k_3\|_{L_\infty(\Omega)}\right].$$

To solve equation [3.107] numerically, we apply the fixed-point iteration method

$$u_{n+1}(x,y) = -\int_0^x\int_0^y K(x,y,\xi,\eta)u_n(\xi,\eta)d\eta d\xi + \varphi(x,y), \quad n=0,1,\ldots, \qquad [3.109]$$

$$u_0(x,y) = 0.$$

Denoting $z_{n+1}(x,y) = u_{n+1}(x,y) - u_n(x,y)$, we obtain for $z_n(x,y)$ the recurrent sequence

$$z_{n+1}(x,y) = -\int_0^x\int_0^y K(x,y,\xi,\eta)z_n(\xi,\eta)d\eta d\xi, \quad n=1,2,\ldots, \quad z_1(x,y) = \varphi(x,y).$$

For brevity, we introduce the notation

$$\|v\|_{x,y} = \max_{t\in[0,x],\ s\in[0,y]} |v(t,s)|.$$

LEMMA 3.4.– Let $k_1(x,y), k_2(x,y) \in W_\infty^1(\Omega)$, $k_3(x,y), \varphi(x,y) \in L_\infty(\Omega)$. Then, the following estimate holds true:

$$\|z_{n+1}\|_{x,y} \leq R_{n+1}(xy)^{n(1-\alpha)}\|\varphi\|_{L_\infty(\Omega)}, \quad n=1,2,\ldots, \qquad [3.110]$$

where

$$R_{n+1} = \left[\frac{2-\alpha}{1-\alpha}\mu\right]^n \frac{1}{\Gamma(n+1-\alpha n)\prod_{p=1}^{n}[p-\alpha(p-1)]},$$

$$\mu = \|k_1\|_{W_\infty^1(\Omega)} + \|k_2\|_{W_\infty^1(\Omega)} + \|k_3\|_{L_\infty(\Omega)}.$$

PROOF.– We will prove the theorem by induction. For $n = 1$, inequality [3.110] is true since due to [3.108] we have

$$|z_2(x,y)| = \left|-\int_0^x\int_0^y K(x,y,\xi,\eta)z_1(\xi,\eta)d\eta d\xi\right| = \left|-\int_0^x\int_0^y K(x,y,\xi,\eta)\varphi(\xi,\eta)d\eta d\xi\right| \le$$

$$\le \int_0^x\int_0^y |K(x,y,\xi,\eta)||\varphi(\xi,\eta)|d\eta d\xi \le \int_0^x\int_0^y |K(x,y,\xi,\eta)|d\eta d\xi \|\varphi\|_{L_\infty(\Omega)} \le$$

$$\le \int_0^x\int_0^y \frac{2-\alpha}{\Gamma(2-\alpha)}\left[\frac{\|k_1\|_{W_\infty^1(\Omega)}}{(x-\xi)^\alpha} + \frac{\|k_2\|_{W_\infty^1(\Omega)}}{(y-\eta)^\alpha} + \|k_3\|_{L_\infty(\Omega)}\right]d\eta d\xi \|\varphi\|_{L_\infty(\Omega)} =$$

$$= \frac{2-\alpha}{\Gamma(2-\alpha)}\left[\|k_1\|_{W_\infty^1(\Omega)}\frac{x^{1-\alpha}y}{1-\alpha} + \|k_2\|_{W_\infty^1(\Omega)}\frac{xy^{1-\alpha}}{1-\alpha} + \|k_3\|_{L_\infty(\Omega)} xy\right]\|\varphi\|_{L_\infty(\Omega)} =$$

$$= \frac{(2-\alpha)(xy)^{1-\alpha}}{(1-\alpha)\Gamma(2-\alpha)}\left[\|k_1\|_{W_\infty^1(\Omega)} y^\alpha + \|k_2\|_{W_\infty^1(\Omega)} x^\alpha + (1-\alpha)\|k_3\|_{L_\infty(\Omega)} (xy)^\alpha\right] \times$$

$$\times \|\varphi\|_{L_\infty(\Omega)} \le \frac{(2-\alpha)\mu}{(1-\alpha)\Gamma(2-\alpha)}(xy)^{1-\alpha}\|\varphi\|_{L_\infty(\Omega)}.$$

This gives the inequality

$$\|z_2\|_{x,y} \le \frac{(2-\alpha)\mu}{(1-\alpha)\Gamma(2-\alpha)}(xy)^{1-\alpha}\|\varphi\|_{L_\infty(\Omega)}.$$

Next, we assume that inequality [3.110] is true for some $n \in \mathbb{N}$ and verify it for the next value $n + 1$:

$$|z_{n+2}(x,y)| = \left| -\int_0^x \int_0^y K(x,y,\xi,\eta) z_{n+1}(\xi,\eta) d\eta d\xi \right| \leq$$

$$\leq \int_0^x \int_0^y |K(x,y,\xi,\eta)| |z_{n+1}(\xi,\eta)| d\eta d\xi \leq$$

$$\leq \int_0^x \int_0^y \frac{2-\alpha}{\Gamma(2-\alpha)} \left[\frac{\|k_1\|_{W^1_\infty(\Omega)}}{(x-\xi)^\alpha} + \frac{\|k_2\|_{W^1_\infty(\Omega)}}{(y-\eta)^\alpha} + \|k_3\|_{L_\infty(\Omega)} \right] (\xi\eta)^{n(1-\alpha)} d\xi d\eta \times$$

$$\times R_{n+1} \|\varphi\|_{L_\infty(\Omega)} =$$

$$= \frac{2-\alpha}{\Gamma(2-\alpha)} R_{n+1} \|\varphi\|_{L_\infty(\Omega)} \left[\|k_1\|_{W^1_\infty(\Omega)} \int_0^x \xi^{n(1-\alpha)} (x-\xi)^{-\alpha} d\xi \int_0^y \eta^{n(1-\alpha)} d\eta + \right.$$

$$+ \|k_2\|_{W^1_\infty(\Omega)} \int_0^x \xi^{n(1-\alpha)} d\xi \int_0^y \eta^{n(1-\alpha)} (y-\eta)^{-\alpha} d\eta +$$

$$\left. + \|k_3\|_{L_\infty(\Omega)} \int_0^x \xi^{n(1-\alpha)} d\xi \int_0^y \eta^{n(1-\alpha)} d\eta \right]$$

Making here the substitution $\xi = xs$, $\eta = xs$, we then have

$$|z_{n+2}(x,y)| \leq \frac{2-\alpha}{\Gamma(2-\alpha)} R_{n+1} \|\varphi\|_{L_\infty(\Omega)} \times$$

$$\times \left[\frac{x^{(n+1)(1-\alpha)} \Gamma(n(1-\alpha)+1) \Gamma(1-\alpha)}{\Gamma((n+1)(1-\alpha)+1)} \frac{y^{n(1-\alpha)+1}}{n(1-\alpha)+1} \|k_1\|_{W^1_\infty(\Omega)} + \right.$$

$$+ \frac{x^{n(1-\alpha)+1}}{n(1-\alpha)+1} \frac{y^{(n+1)(1-\alpha)} \Gamma(n(1-\alpha)+1) \Gamma(1-\alpha)}{\Gamma((n+1)(1-\alpha)+1)} \|k_2\|_{W^1_\infty(\Omega)} +$$

$$\left. + \frac{(xy)^{n(1-\alpha)+1}}{(n(1-\alpha)+1)^2} \|k_3\|_{L_\infty(\Omega)} \right] =$$

$$= \frac{2-\alpha}{\Gamma(2-\alpha)} R_{n+1} \| \varphi \|_{L_\infty(\Omega)} (xy)^{(n+1)(1-\alpha)} \frac{\Gamma(n(1-\alpha)+1)\Gamma(1-\alpha)}{(n(1-\alpha)+1)\Gamma((n+1)(1-\alpha)+1)} \times$$

$$\times \left[\| k_1 \|_{W_\infty^1(\Omega)} + \| k_2 \|_{W_\infty^1(\Omega)} + \frac{\Gamma((n+1)(1-\alpha)+1)}{\Gamma((n+1)(1-\alpha)+2)\Gamma(1-\alpha)} \| k_3 \|_{L_\infty(\Omega)} \right] \le$$

$$\le \frac{2-\alpha}{\Gamma(2-\alpha)} R_{n+1} \| \varphi \|_{L_\infty(\Omega)} (xy)^{(n+1)(1-\alpha)} \frac{\Gamma(n+1-n\alpha)\Gamma(1-\alpha)}{(n+1-n\alpha)\Gamma(n+2-(n+1)\alpha)} \mu =$$

$$= \frac{2-\alpha}{\Gamma(2-\alpha)} \left[\frac{2-\alpha}{1-\alpha}\mu\right]^n \frac{\| \varphi \|_{L_\infty(\Omega)} (xy)^{(n+1)(1-\alpha)}}{\Gamma(n+1-\alpha n) \prod_{p=1}^{n}[p-\alpha(p-1)]} \times$$

$$\times \frac{\Gamma(n+1-n\alpha)\Gamma(1-\alpha)\mu}{(n+1-n\alpha)\Gamma(n+2-(n+1)\alpha)} =$$

$$= \left[\frac{2-\alpha}{1-\alpha}\mu\right]^{n+1} \frac{\| \varphi \|_{L_\infty(\Omega)} (xy)^{(n+1)(1-\alpha)}}{\Gamma(n+2-(n+1)\alpha) \prod_{p=1}^{n+1}[p-\alpha(p-1)]}$$

$$= R_{n+2} \| \varphi \|_{L_\infty(\Omega)} (xy)^{(n+1)(1-\alpha)},$$

and therefore

$$\| z_{n+2} \|_{x,y} \le R_{n+2} (xy)^{(n+1)(1-\alpha)} \| \varphi \|_{L_\infty(\Omega)}.$$

The lemma is proved.

Next, we prove the convergence of the recurrent sequence [3.109].

THEOREM 3.19.– *Let $k_1(x,y)$, $k_2(x,y) \in W_\infty^1(\Omega)$, $k_3(x,y)$, $\varphi(x,y) \in L_\infty(\Omega)$. Then, the fixed-point iteration method [3.109] uniformly converges to the unique solution $u(x,y) \in C(\overline{\Omega})$ of the integral equation [3.107] (a weak solution of the Goursat problem [3.106]) and is characterized by the estimate*

$$\| u \|_{x,y} \le (M(1-\alpha)+1) e^M \| \varphi \|_{L_\infty(\Omega)} \qquad [3.111]$$

with $M = \dfrac{(2-\alpha)\mu T^{2(1-\alpha)}}{(1-\alpha)^2}$ and $\mu = \|k_1\|_{W_\infty^1(\Omega)} + \|k_2\|_{W_\infty^1(\Omega)} + \|k_3\|_{L_\infty(\Omega)}$.

The accuracy of method [3.109] is characterized by the estimate

$$\|u - u_n\|_{x,y} \le \frac{M^{n-1}(M(1-\alpha)+1)e^M}{(n-1)!\,\Gamma(n(1-\alpha)+2)}\|\varphi\|_{L_\infty(\Omega)} \quad \forall n \in \mathbb{N}.$$

PROOF.– First, we will show that the sequence $(u_n(x,y))$ is fundamental in $C(\bar{\Omega})$. Applying Lemma 3.4 and using the inequality

$$R_{n+1} = \left[\frac{2-\alpha}{1-\alpha}\mu\right]^n \frac{1}{\Gamma(n+1-\alpha n)\prod_{p=1}^{n}[p-\alpha(p-1)]} \le$$

$$\le \left[\frac{(2-\alpha)\mu}{1-\alpha}\right]^n \frac{1}{\Gamma(n+1-\alpha n)\prod_{p=1}^{n}[p(1-\alpha)]} = \left[\frac{(2-\alpha)\mu}{(1-\alpha)^2}\right]^n \frac{1}{\Gamma(n+1-\alpha n)n!},$$

we have

$$\|u_{n+p} - u_n\|_{x,y} = \|(u_{n+p} - u_{n+p-1}) + (u_{n+p-1} - u_{n+p-2}) + \ldots + (u_{n+1} - u_n)\|_{x,y} \le$$

$$\le \sum_{k=n}^{n+p-1} \|z_{k+1}\|_{T,T} \le \sum_{k=n}^{n+p-1} R_{k+1} T^{2(1-\alpha)k} \|\varphi\|_{L_\infty(\Omega)} \le$$

$$\le \sum_{k=n}^{n+p-1} \left[\frac{(2-\alpha)\mu T^{2(1-\alpha)}}{(1-\alpha)^2}\right]^k \frac{1}{\Gamma(k+1-\alpha k)k!}\|\varphi\|_{L_\infty(\Omega)} = \qquad [3.112]$$

$$= \sum_{k=n}^{n+p-1} \frac{M^k(k(1-\alpha)+1)}{k!\,\Gamma(k(1-\alpha)+2)}\|\varphi\|_{L_\infty(\Omega)} \le \frac{\|\varphi\|_{L_\infty(\Omega)}}{\Gamma(n(1-\alpha)+2)} \sum_{k=n}^{n+p-1} \frac{M^k(k(1-\alpha)+1)}{k!} =$$

$$= \frac{\|\varphi\|_{L_\infty(\Omega)}}{\Gamma(n(1-\alpha)+2)}\left(M(1-\alpha)\sum_{k=n-1}^{n+p-2}\frac{M^k}{k!} + \sum_{k=n}^{n+p-1}\frac{M^k}{k!}\right) \le$$

$$\leq \frac{(M(1-\alpha)+1)\|\varphi\|_{L_\infty(\Omega)}}{\Gamma(n(1-\alpha)+2)} \sum_{k=n-1}^{n+p-1} \frac{M^k}{k!} \underset{n\to\infty}{\to} 0 \quad \forall p \in \mathbb{N}.$$

Then, due to the fullness of $C(\overline{\Omega})$, the sequence $(u_n(x,y))$ converges to an element $u^*(x,y) \in C(\overline{\Omega})$. It is easy to show that the function $u^*(x,y)$ is unique and satisfies the integral equation [3.107].

Taking the limit in [3.112] as $p \to +\infty$, we obtain the estimate

$$\|u - u_n\|_{x,y} \leq \frac{(M(1-\alpha)+1)\|\varphi\|_{L_\infty(\Omega)}}{\Gamma(n(1-\alpha)+2)} \sum_{k=n-1}^{\infty} \frac{M^k}{k!} \leq$$

$$\leq \frac{(M(1-\alpha)+1)\|\varphi\|_{L_\infty(\Omega)}}{\Gamma(n(1-\alpha)+2)} \frac{M^{n-1}}{(n-1)!} \sum_{k=0}^{\infty} \frac{M^k}{k!} = \frac{(M(1-\alpha)+1)\|\varphi\|_{L_\infty(\Omega)}}{\Gamma(n(1-\alpha)+2)} \frac{M^{n-1}}{(n-1)!} e^M.$$

Next, we derive estimate [3.111]. Since $u_{n+1} = \sum_{k=0}^{n} z_{k+1}$ then

$$\|u_{n+1}\|_{x,y} \leq \sum_{k=0}^{n} \|z_{k+1}\|_{x,y} \leq \sum_{k=0}^{n} \frac{M^k}{\Gamma(k+1-\alpha k)k!} \|\varphi\|_{L_\infty(\Omega)} =$$

$$= \sum_{k=0}^{n} \frac{M^k(k(1-\alpha)+1)}{\Gamma(k(1-\alpha)+2)k!} \|\varphi\|_{L_\infty(\Omega)} \leq \sum_{k=0}^{n} \frac{M^k(k(1-\alpha)+1)}{k!} \|\varphi\|_{L_\infty(\Omega)}.$$

Taking here the limit as $n \to +\infty$, we have

$$\|u\|_{x,y} \leq \sum_{k=0}^{\infty} \frac{M^k(k(1-\alpha)+1)}{k!} \|\varphi\|_{L_\infty(\Omega)} =$$

$$= \left[M(1-\alpha) \sum_{k=1}^{\infty} \frac{M^{k-1}}{(k-1)!} + \sum_{k=0}^{\infty} \frac{M^k}{k!} \right] \|\varphi\|_{L_\infty(\Omega)} = \left[M(1-\alpha)e^M + e^M \right] \|\varphi\|_{L_\infty(\Omega)}.$$

The theorem is proved.

Now we will write equation [3.106] as follows:

$$\frac{\partial^2 u(x,y)}{\partial x \partial y} + \frac{k_1(x,y)}{\Gamma(1-\alpha)} \int_0^x \frac{dt}{(x-t)^\alpha} \int_0^y \frac{\partial^2 u(t,s)}{\partial t \partial s} ds +$$

$$+ \frac{k_2(x,y)}{\Gamma(1-\alpha)} \int_0^y \frac{dt}{(y-t)^\alpha} \int_0^x \frac{\partial^2 u(s,t)}{\partial t \partial s} ds + k_3(x,y) \int_0^y \int_0^x \frac{\partial^2 u(t,s)}{\partial t \partial s} ds dt = f(x,y).$$

This leads to the Volterra integral equation of the second kind

$$\frac{\partial^2 u(x,y)}{\partial x \partial y} + \int_0^x \int_0^y Q(x,y,t,s) \frac{\partial^2 u(t,s)}{\partial t \partial s} ds dt = f(x,y) \qquad [3.113]$$

with the kernel

$$Q(x,y,t,s) = \frac{k_1(x,y)}{\Gamma(1-\alpha)} \frac{1}{(x-t)^\alpha} + \frac{k_2(x,y)}{\Gamma(1-\alpha)} \frac{1}{(y-s)^\alpha} + k_3(x,y).$$

To solve equation [3.113], we apply the fixed-point iteration method

$$v_{n+1}(x,y) = -\int_0^x \int_0^y Q(x,y,\xi,\eta) v_n(\xi,\eta) d\eta d\xi + f(x,y), \quad n = 0,1,\ldots, \qquad [3.114]$$

$$v_0(x,y) = 0.$$

Then, for $z_{n+1}(x,y) = v_{n+1}(x,y) - v_n(x,y)$, we have the recurrent sequence of the integral equations

$$z_{n+1}(x,y) = -\int_0^x \int_0^y Q(x,y,\xi,\eta) z_n(\xi,\eta) d\eta d\xi, \quad n = 1,2,\ldots, \quad z_1(x,y) = f(x,y).$$

LEMMA 3.5.– *Let* $k_1(x,y), k_2(x,y)\, k_3(x,y), f(x,y) \in L_\infty(\Omega)$. *Then, the estimate holds true:*

$$\| z_{n+1} \|_{x,y} \le R_{n+1}(xy)^{n(1-\alpha)} \| f \|_{L_\infty(\Omega)}, \quad n = 1,2,\ldots, \qquad [3.115]$$

where

$$R_{n+1} = \frac{\mu^n}{\Gamma(n+1-\alpha n)\prod_{p=1}^{n}[p-\alpha(p-1)]},$$

$$\mu = \|k_1\|_{L_\infty(\Omega)} + \|k_2\|_{L_\infty(\Omega)} + \|k_3\|_{L_\infty(\Omega)}.$$

PROOF.– Here, again, we apply the method of mathematical induction. For $n=1$ inequality [3.115] is true, indeed:

$$|z_2(x,y)| = \left|-\int_0^x\int_0^y Q(x,y,\xi,\eta)z_1(\xi,\eta)d\eta d\xi\right| = \left|-\int_0^x\int_0^y Q(x,y,\xi,\eta)f(\xi,\eta)d\eta d\xi\right| \le$$

$$\le \int_0^x\int_0^y |Q(x,y,\xi,\eta)||f(\xi,\eta)|d\eta d\xi \le \int_0^x\int_0^y |Q(x,y,\xi,\eta)|d\eta d\xi\, \|f\|_{L_\infty(\Omega)} \le$$

$$\le \int_0^x\int_0^y\left[\frac{\|k_1\|_{L_\infty(\Omega)}}{\Gamma(1-\alpha)(x-\xi)^\alpha} + \frac{\|k_2\|_{L_\infty(\Omega)}}{\Gamma(1-\alpha)(y-\eta)^\alpha} + \|k_3\|_{L_\infty(\Omega)}\right]d\eta d\xi\, \|f\|_{L_\infty(\Omega)} =$$

$$=\left[\frac{x^{1-\alpha}y}{\Gamma(2-\alpha)}\|k_1\|_{L_\infty(\Omega)} + \frac{xy^{1-\alpha}}{\Gamma(2-\alpha)}\|k_2\|_{L_\infty(\Omega)} + xy\|k_3\|_{L_\infty(\Omega)}\right]\|f\|_{L_\infty(\Omega)} =$$

$$= \frac{(xy)^{1-\alpha}}{\Gamma(2-\alpha)}\Big[\|k_1\|_{L_\infty(\Omega)} y^\alpha + \|k_2\|_{L_\infty(\Omega)} x^\alpha +$$

$$+\Gamma(2-\alpha)\|k_3\|_{L_\infty(\Omega)}(xy)^\alpha\Big]\|f\|_{L_\infty(\Omega)} \le \frac{(xy)^{1-\alpha}}{\Gamma(2-\alpha)}\mu\|f\|_{L_\infty(\Omega)},$$

which gives

$$\|z_2\|_{x,y} \le \frac{\mu}{\Gamma(2-\alpha)}(xy)^{1-\alpha}\|f\|_{L_\infty(\Omega)}.$$

Next, we assume that inequality [3.115] is true for some $n \in \mathbb{N}$ and verify that it is true for the next value $n+1$:

Differential Equations with Fractional Derivatives   191

$$|z_{n+2}(x,y)| = \left| -\int_0^x \int_0^y Q(x,y,\xi,\eta) z_{n+1}(\xi,\eta) d\eta d\xi \right| \le$$

$$\le \int_0^x \int_0^y |Q(x,y,\xi,\eta)| \, |z_{n+1}(\xi,\eta)| \, d\eta d\xi \le$$

$$\le \int_0^x \int_0^y \left[\frac{\|k_1\|_{L_\infty(\Omega)}}{\Gamma(1-\alpha)(x-\xi)^\alpha} + \frac{\|k_2\|_{L_\infty(\Omega)}}{\Gamma(1-\alpha)(y-\eta)^\alpha} + \|k_3\|_{L_\infty(\Omega)} \right] \times$$

$$\times (\xi\eta)^{n(1-\alpha)} d\xi d\eta \, R_{n+1} \|f\|_{L_\infty(\Omega)} =$$

$$= R_{n+1} \|f\|_{L_\infty(\Omega)} \left[\frac{\|k_1\|_{L_\infty(\Omega)}}{\Gamma(1-\alpha)} \int_0^x \xi^{n(1-\alpha)} (x-\xi)^{-\alpha} d\xi \int_0^y \eta^{n(1-\alpha)} d\eta + \right.$$

$$+ \frac{\|k_2\|_{L_\infty(\Omega)}}{\Gamma(1-\alpha)} \int_0^x \xi^{n(1-\alpha)} d\xi \int_0^y \eta^{n(1-\alpha)} (y-\eta)^{-\alpha} d\eta +$$

$$\left. + \|k_3\|_{L_\infty(\Omega)} \int_0^x \xi^{n(1-\alpha)} d\xi \int_0^y \eta^{n(1-\alpha)} d\eta \right].$$

Making the substitution $\xi = xs$, $\eta = xs$, we obtain

$$|z_{n+2}(x,y)| = R_{n+1} \|f\|_{L_\infty(\Omega)} \left[\frac{x^{(n+1)(1-\alpha)} \Gamma(n(1-\alpha)+1)}{\Gamma((n+1)(1-\alpha)+1)} \frac{y^{n(1-\alpha)+1}}{n(1-\alpha)+1} \|k_1\|_{L_\infty(\Omega)} + \right.$$

$$+ \frac{x^{n(1-\alpha)+1}}{n(1-\alpha)+1} \frac{y^{(n+1)(1-\alpha)} \Gamma(n(1-\alpha)+1)}{\Gamma((n+1)(1-\alpha)+1)} \|k_2\|_{L_\infty(\Omega)} + \frac{(xy)^{n(1-\alpha)+1}}{(n(1-\alpha)+1)^2} \|k_3\|_{L_\infty(\Omega)} \right] =$$

$$= R_{n+1} \|f\|_{L_\infty(\Omega)} (xy)^{(n+1)(1-\alpha)} \frac{\Gamma(n(1-\alpha)+1)}{(n(1-\alpha)+1)\Gamma((n+1)(1-\alpha)+1)} \times$$

$$\times \left[\|k_1\|_{L_\infty(\Omega)} + \|k_2\|_{L_\infty(\Omega)} + \frac{\Gamma((n+1)(1-\alpha)+1)}{\Gamma(n(1-\alpha)+2)} \|k_3\|_{L_\infty(\Omega)} \right] \le$$

$$\leq R_{n+1} \| f \|_{L_\infty(\Omega)} (xy)^{(n+1)(1-\alpha)} \frac{\Gamma(n+1-n\alpha)}{(n+1-n\alpha)\Gamma(n+2-(n+1)\alpha)} \mu =$$

$$= \mu^n \frac{\| f \|_{L_\infty(\Omega)} (xy)^{(n+1)(1-\alpha)}}{\Gamma(n+1-\alpha n)\prod_{p=1}^{n}[p-\alpha(p-1)]} \frac{\Gamma(n+1-n\alpha)}{(n+1-n\alpha)\Gamma(n+2-(n+1)\alpha)} \mu =$$

$$= \mu^{n+1} \frac{\| f \|_{L_\infty(\Omega)} (xy)^{(n+1)(1-\alpha)}}{\Gamma(n+2-(n+1)\alpha)\prod_{p=1}^{n+1}[p-\alpha(p-1)]} = R_{n+2} \| f \|_{L_\infty(\Omega)} (xy)^{(n+1)(1-\alpha)},$$

which gives

$$\| z_{n+2} \|_{x,y} \leq R_{n+2}(xy)^{(n+1)(1-\alpha)} \| f \|_{L_\infty(\Omega)}$$

and proves the lemma.

In the next proposition, we study the smoothness of the mixed derivative $\dfrac{\partial^2 u(x,y)}{\partial x \partial y}$.

THEOREM 3.20.– Let $k_1(x,y), k_2(x,y), k_3(x,y), f(x,y) \in L_\infty(\Omega)$. Then, $\dfrac{\partial^2 u}{\partial x \partial y} \in C(\bar{\Omega})$ and the following estimate holds true:

$$\left\| \frac{\partial^2 u}{\partial x \partial y} \right\|_{T,T} \leq (N(1-\alpha)+1)e^N \| f \|_{L_\infty(\Omega)}, \quad [3.116]$$

where $N = \dfrac{\mu T^{2(1-\alpha)}}{1-\alpha}$ and $\mu = \|k_1\|_{L_\infty(\Omega)} + \|k_2\|_{L_\infty(\Omega)} + \|k_3\|_{L_\infty(\Omega)}$.

PROOF.– First, we will show that the sequence $(v_n(x,y))$ is fundamental in $C(\bar{\Omega})$. Applying Lemma 3.5 and using the inequality

$$R_{n+1} = \frac{\mu^n}{\Gamma(n+1-\alpha n) \prod_{p=1}^{n} [p - \alpha(p-1)]}$$

$$\leq \frac{\mu^n}{\Gamma(n+1-\alpha n) \prod_{p=1}^{n} [p(1-\alpha)]} = \frac{\left[\frac{\mu}{1-\alpha}\right]^n}{\Gamma(n+1-\alpha n) n!},$$

we have

$$\| v_{n+p} - v_n \|_{x,y} = \| (v_{n+p} - v_{n+p-1}) + (v_{n+p-1} - v_{n+p-2}) + \ldots + (v_{n+1} - v_n) \|_{x,y} \leq$$

$$\leq \sum_{k=n}^{n+p-1} \| z_{k+1} \|_{T,T} \leq \sum_{k=n}^{n+p-1} R_{k+1} T^{2(1-\alpha)k} \| f \|_{L_\infty(\Omega)} \leq$$

$$\leq \sum_{k=n}^{n+p-1} \left[\frac{\mu T^{2(1-\alpha)}}{1-\alpha}\right]^k \frac{1}{\Gamma(k+1-\alpha k) k!} \| f \|_{L_\infty(\Omega)} = \qquad [3.117]$$

$$= \sum_{k=n}^{n+p-1} \frac{N^k (k(1-\alpha)+1)}{k! \Gamma(k(1-\alpha)+2)} \| f \|_{L_\infty(\Omega)} \leq \frac{\| f \|_{L_\infty(\Omega)}}{\Gamma(n(1-\alpha)+2)} \sum_{k=n}^{n+p-1} \frac{N^k (k(1-\alpha)+1)}{k!} =$$

$$= \frac{\| f \|_{L_\infty(\Omega)}}{\Gamma(n(1-\alpha)+2)} \left(N(1-\alpha) \sum_{k=n-1}^{n+p-2} \frac{N^k}{k!} + \sum_{k=n}^{n+p-1} \frac{N^k}{k!} \right) \leq$$

$$\leq \frac{(N(1-\alpha)+1) \| f \|_{L_\infty(\Omega)}}{\Gamma(n(1-\alpha)+2)} \sum_{k=n-1}^{n+p-1} \frac{N^k}{k!} \xrightarrow[n \to \infty]{} 0 \quad \forall p \in \mathbb{N}.$$

Due to the fullness of the space $C(\bar{\Omega})$, the sequence $(u_n(x,y))$ converges to an element $v^*(x,y) \in C(\bar{\Omega})$. It is easy to show that the function $v^*(x,y)$ is unique and satisfies the integral equation [3.113].

Taking in [3.117] the limit as $p \to +\infty$, we obtain the estimate

$$\| v - v_n \|_{x,y} \leq \frac{(N(1-\alpha)+1) \| f \|_{L_\infty(\Omega)}}{\Gamma(n(1-\alpha)+2)} \sum_{k=n-1}^{\infty} \frac{N^k}{k!} \leq$$

$$\leq \frac{(N(1-\alpha)+1) \| f \|_{L_\infty(\Omega)}}{\Gamma(n(1-\alpha)+2)} \frac{N^{n-1}}{(n-1)!} e^N.$$

Since $v_{n+1} = \sum_{k=0}^{n} z_{k+1}$, then

$$\| v_{n+1} \|_{x,y} \leq \sum_{k=0}^{n} \| z_{k+1} \|_{x,y} \leq \sum_{k=0}^{n} \frac{N^k}{\Gamma(k+1-\alpha k)k!} \| f \|_{L_\infty(\Omega)} =$$

$$= \sum_{k=0}^{n} \frac{N^k (k(1-\alpha)+1)}{\Gamma(k(1-\alpha)+2)k!} \| \varphi \|_{L_\infty(\Omega)} \leq \sum_{k=0}^{n} \frac{N^k (k(1-\alpha)+1)}{k!} \| \varphi \|_{L_\infty(\Omega)}.$$

Taking here the limit as $n \to +\infty$ and making some easy transformations, we come to the inequality

$$\| v \|_{x,y} \leq \sum_{k=0}^{\infty} \frac{N^k (k(1-\alpha)+1)}{k!} \| f \|_{L_\infty(\Omega)} =$$

$$= \left[N(1-\alpha) \sum_{k=1}^{\infty} \frac{N^{k-1}}{(k-1)!} + \sum_{k=0}^{\infty} \frac{N^k}{k!} \right] \| f \|_{L_\infty(\Omega)} = \left[N(1-\alpha)e^N + e^N \right] \| f \|_{L_\infty(\Omega)},$$

which leads to estimate [3.116] and completes the proof of the theorem.

THEOREM 3.21.– *Let the conditions of Theorem 3.20 be satisfied. Then, the following estimates hold true:*

$$\max_{x,y \in [0,T]} \left| \frac{u(x,y)}{xy} \right| \leq \left\| \frac{\partial^2 u}{\partial x \partial y} \right\|_{T,T}, \qquad [3.118]$$

$$\max_{x,y \in [0,T]} \left| \frac{u(x,y)}{xy} \right| \leq (N(1-\alpha)+1)e^N \| f \|_{L_\infty(\Omega)}. \qquad [3.119]$$

Estimate [3.118] is unimprovable in the sense that there exists problem [3.106] with the solution $u(x,y) = xy$ on which estimate [3.118] turns into an equality.

PROOF.– Estimate [3.118] follows from the inequality

$$|u(x,y)| = \left|\int_0^x\int_0^y \frac{\partial^2 u(\xi,\eta)}{\partial\xi\partial\eta} d\eta d\xi\right| \leq xy \left\|\frac{\partial^2 u}{\partial x \partial y}\right\|_{T,T}, \quad x,y \in [0,T].$$

Estimate [3.118] combined with estimate [3.116] gives [3.119]. The theorem is proved.

We denote by $C^{0,\lambda}(\overline{\Omega})$ a class of functions continuous on $\overline{\Omega}$ and satisfying Hölder's condition with the exponent λ. A seminorm in $C^{0,\lambda}(\overline{\Omega})$ is defined as follows:

$$|u|_{C^{0,\lambda}(\overline{\Omega})} = \sup_{(x,y),(s,t)\in\overline{\Omega}} \frac{|u(x,y) - u(s,t)|}{\sqrt{(x-s)^2 + (y-t)^2}}.$$

We will now prove a stronger result about the smoothness of the solution of problem [3.106].

THEOREM 3.22.– Let the coefficients and the right-hand side of equation [3.106] satisfy the condition $k_1(x,y), k_2(x,y)\ k_3(x,y), f(x,y) \in C^{0,1-\alpha}(\overline{\Omega})$. Then

$$\frac{\partial^2 u(x,y)}{\partial x \partial y} \in C^{0,1-\alpha}(\overline{\Omega}).$$

PROOF.– For $0 \leq x_1 < x_2 \leq T$ and $y \in [0,T]$, equation [3.113] yields

$$\frac{\partial^2 u(x_2,y)}{\partial x \partial y} - \frac{\partial^2 u(x_1,y)}{\partial x \partial y} + A + B + C = f(x_2,y) - f(x_1,y), \qquad [3.120]$$

where

$$A = \frac{k_1(x_2,y)}{\Gamma(1-\alpha)}\int_0^{x_2}\frac{d\xi}{(x_2-\xi)^\alpha}\int_0^y \frac{\partial^2 u(\xi,\eta)}{\partial\xi\partial\eta}d\eta - \frac{k_1(x_1,y)}{\Gamma(1-\alpha)}\int_0^{x_1}\frac{d\xi}{(x_1-\xi)^\alpha}\int_0^y \frac{\partial^2 u(\xi,\eta)}{\partial\xi\partial\eta}d\eta,$$

$$B = \frac{k_2(x_2,y)}{\Gamma(1-\alpha)} \int_0^y \frac{d\eta}{(y-\eta)^\alpha} \int_0^{x_2} \frac{\partial^2 u(\xi,\eta)}{\partial \xi \partial \eta} d\xi - \frac{k_2(x_1,y)}{\Gamma(1-\alpha)} \int_0^y \frac{d\eta}{(y-\eta)^\alpha} \int_0^{x_1} \frac{\partial^2 u(\xi,\eta)}{\partial \xi \partial \eta} d\xi,$$

$$C = k_3(x_2,y)u(x_2,y) - k_3(x_1,y)u(x_1,y).$$

We find

$$|A| = \left| \frac{k_1(x_2,y)}{\Gamma(1-\alpha)} \int_0^{x_2} \frac{d\xi}{(x_2-\xi)^\alpha} \int_0^y \frac{\partial^2 u(\xi,\eta)}{\partial \xi \partial \eta} d\eta - \frac{k_1(x_1,y)}{\Gamma(1-\alpha)} \int_0^{x_1} \frac{d\xi}{(x_1-\xi)^\alpha} \int_0^y \frac{\partial^2 u(\xi,\eta)}{\partial \xi \partial \eta} d\eta \right| =$$

$$= \left| \frac{k_1(x_2,y) - k_1(x_1,y)}{\Gamma(1-\alpha)} \int_0^{x_2} \frac{d\xi}{(x_2-\xi)^\alpha} \int_0^y \frac{\partial^2 u(\xi,\eta)}{\partial \xi \partial \eta} d\eta + \right.$$

$$\left. + \frac{k_1(x_1,y)}{\Gamma(1-\alpha)} \left[\int_{x_1}^{x_2} \frac{d\xi}{(x_2-\xi)^\alpha} \int_0^y \frac{\partial^2 u(\xi,\eta)}{\partial \xi \partial \eta} d\eta - \int_0^{x_1} \left(\frac{d\xi}{(x_1-\xi)^\alpha} - \frac{d\xi}{(x_2-\xi)^\alpha} \right) \int_0^y \frac{\partial^2 u(\xi,\eta)}{\partial \xi \partial \eta} d\eta \right] \right| \leq$$

$$\leq \frac{|k_1|_{C^{0,1-\alpha}(\bar\Omega)} (x_2-x_1)^{1-\alpha}}{\Gamma(1-\alpha)} \frac{x_2^{1-\alpha}}{1-\alpha} y \left\| \frac{\partial^2 u(x,y)}{\partial x \partial y} \right\|_{C(\bar\Omega)} +$$

$$+ \frac{\|k_1\|_{C(\bar\Omega)}}{\Gamma(1-\alpha)} \left[\frac{(x_2-x_1)^{1-\alpha}}{1-\alpha} y + \frac{x_1^{1-\alpha} - x_2^{1-\alpha} + (x_2-x_1)^{1-\alpha}}{1-\alpha} y \right] \left\| \frac{\partial^2 u(x,y)}{\partial x \partial y} \right\|_{C(\bar\Omega)} \leq$$

$$\leq \frac{(x_2-x_1)^{1-\alpha}}{\Gamma(2-\alpha)} \left[T^{2-\alpha} |k_1|_{C^{0,1-\alpha}(\bar\Omega)} + 2T \|k_1\|_{C(\bar\Omega)} \right] \left\| \frac{\partial^2 u(x,y)}{\partial x \partial y} \right\|_{C(\bar\Omega)}.$$

Similarly, we have

$$|B| = \left| \frac{k_2(x_2,y)}{\Gamma(1-\alpha)} \int_0^y \frac{d\eta}{(y-\eta)^\alpha} \int_0^{x_2} \frac{\partial^2 u(\xi,\eta)}{\partial \xi \partial \eta} d\xi - \frac{k_2(x_1,y)}{\Gamma(1-\alpha)} \int_0^y \frac{d\eta}{(y-\eta)^\alpha} \int_0^{x_1} \frac{\partial^2 u(\xi,\eta)}{\partial \xi \partial \eta} d\xi \right| =$$

$$= \left| \frac{k_2(x_2,y) - k_2(x_1,y)}{\Gamma(1-\alpha)} \int_0^y \frac{d\eta}{(y-\eta)^\alpha} \int_0^{x_2} \frac{\partial^2 u(\xi,\eta)}{\partial \xi \partial \eta} d\xi + \right.$$

$$+\frac{k_2(x_1,y)}{\Gamma(1-\alpha)}\int_0^y \frac{d\eta}{(y-\eta)^\alpha}\int_{x_1}^{x_2}\frac{\partial^2 u(\xi,\eta)}{\partial\xi\partial\eta}d\xi\Bigg| \le$$

$$\le \frac{|k_2|_{C^{0,1-\alpha}(\bar\Omega)}(x_2-x_1)^{1-\alpha}}{\Gamma(1-\alpha)}\frac{y^{1-\alpha}}{1-\alpha}x_2\left\|\frac{\partial^2 u(x,y)}{\partial x\partial y}\right\|_{C(\bar\Omega)}+$$

$$+\frac{\|k_2\|_{C^{0,1-\alpha}(\bar\Omega)}}{\Gamma(1-\alpha)}\frac{y^{1-\alpha}}{1-\alpha}(x_2-x_1)\left\|\frac{\partial^2 u(x,y)}{\partial x\partial y}\right\|_{C(\bar\Omega)} \le$$

$$\le \frac{(x_2-x_1)^{1-\alpha}}{\Gamma(2-\alpha)}\left[T^{2-\alpha}|k_2|_{C^{0,1-\alpha}(\bar\Omega)}+T\|k_2\|_{C(\bar\Omega)}\right]\left\|\frac{\partial^2 u(x,y)}{\partial x\partial y}\right\|_{C(\bar\Omega)}.$$

Next, we obtain

$$|C| = |k_3(x_2,y)u(x_2,y) - k_3(x_1,y)u(x_1,y)| =$$

$$=\left|[k_3(x_2,y)-k_3(x_1,y)]u(x_2,y)+k_3(x_1,y)[u(x_2,y)-u(x_1,y)]\right| \le$$

$$\le |k_3|_{C^{0,1-\alpha}(\bar\Omega)}(x_2-x_1)^{1-\alpha}\|u\|_{C(\bar\Omega)}+\|k_3\|_{C(\bar\Omega)}(x_2-x_1)\left\|\frac{\partial u}{\partial x}\right\|_{C(\bar\Omega)} \le$$

$$\le (x_2-x_1)^{1-\alpha}\left[|k_3|_{C^{0,1-\alpha}(\bar\Omega)}\|u\|_{C(\bar\Omega)}+T^\alpha\|k_3\|_{C(\bar\Omega)}\left\|\frac{\partial u}{\partial x}\right\|_{C(\bar\Omega)}\right].$$

We also have

$$|f(x_2,y)-f(x_1,y)| \le (x_2-x_1)^{1-\alpha}|f|_{C^{0,1-\alpha}(\bar\Omega)}.$$

The obtained inequalities mean that $\dfrac{\partial^2 u(x,y)}{\partial x\partial y}\in C^{0,1-\alpha}(\bar\Omega)$ as for the variable x. Similarly, we can show that $\dfrac{\partial^2 u(x,y)}{\partial x\partial y}\in C^{0,1-\alpha}(\bar\Omega)$ as for the variable y. Then, we have

$$\left| \frac{\partial^2 u(x_1,y_1)}{\partial x \partial y} - \frac{\partial^2 u(x_2,y_2)}{\partial x \partial y} \right| = $$

$$= \left| \frac{\frac{\partial^2 u(x_1,y_1)}{\partial x \partial y} - \frac{\partial^2 u(x_2,y_1)}{\partial x \partial y}}{\left[(x_1-x_2)^2+(y_1-y_2)^2\right]^{(1-\alpha)/2}} + \frac{\frac{\partial^2 u(x_2,y_1)}{\partial x \partial y} - \frac{\partial^2 u(x_2,y_2)}{\partial x \partial y}}{\left[(x_1-x_2)^2+(y_1-y_2)^2\right]^{(1-\alpha)/2}} \right| \leq$$

$$\leq \frac{\left| \frac{\partial^2 u(x_1,y_1)}{\partial x \partial y} - \frac{\partial^2 u(x_2,y_1)}{\partial x \partial y} \right|}{|x_2-x_1|^{1-\alpha}} + \frac{\left| \frac{\partial^2 u(x_2,y_1)}{\partial x \partial y} - \frac{\partial^2 u(x_2,y_2)}{\partial x \partial y} \right|}{|x_2-x_1|^{1-\alpha}},$$

which means $\dfrac{\partial^2 u(x,y)}{\partial x \partial y} \in C^{0,1-\alpha}(\bar{\Omega})$. The theorem is proved.

3.5.2. *The accuracy of the mesh scheme*

We introduce a set of nodes

$$x_i = ih, \quad y_j = jh, \quad i,j = 0,1,\ldots,N, \quad h = T/N,$$

where $N \geq 1$ is an integer number, and write the integral corollary of problem [3.106]:

$$u_{\bar{x}\bar{y}}(x_i,y_j) + \frac{1}{h^2} \int\limits_{x_{i-1}}^{x_i} \int\limits_{y_{j-1}}^{y_j} \frac{k_1(x,y)}{\Gamma(1-\alpha)} \int\limits_0^x \frac{1}{(x-t)^\alpha} \int\limits_0^y \frac{\partial^2 u(t,s)}{\partial t \partial s} ds dt dy dx +$$

$$+ \int\limits_{x_{i-1}}^{x_i} \int\limits_{y_{j-1}}^{y_j} \frac{k_2(x,y)}{\Gamma(1-\alpha)} \int\limits_0^y \frac{1}{(y-s)^\alpha} \int\limits_0^x \frac{\partial^2 u(t,s)}{\partial t \partial s} dt ds dy dx +$$

$$+\frac{1}{h^2}\int_{x_{i-1}}^{x_i}\int_{y_{j-1}}^{y_j} k_3(x,y)\int_0^y\int_0^x \frac{\partial^2 u(t,s)}{\partial t \partial s}dt ds dy dx = \frac{1}{h^2}\int_{x_{i-1}}^{x_i}\int_{y_{j-1}}^{y_j} f(x,y)dy dx,$$

$$i,j = 1,2,\ldots,N.$$

Changing the order of integration in each term of the left-hand side of the equality, we obtain

$$u_{\bar{x}\bar{y}}(x_i,y_j) + \sum_{r=1}^{3}\left\{\frac{1}{h^2}\int_0^{y_{j-1}}\int_0^{x_{i-1}} m_r(x_{i-1},y_{j-1},\tau_r)\frac{\partial^2 u(t,s)}{\partial t \partial s}dt ds +\right.$$

$$+\frac{1}{h^2}\int_0^{y_{j-1}}\int_{x_{i-1}}^{x_i} m_r(t,y_{j-1},\tau_r)\frac{\partial^2 u(t,s)}{\partial t \partial s}dt ds +$$

$$\left.+\frac{1}{h^2}\int_{y_{j-1}}^{y_j}\int_0^{x_{i-1}} m_r(x_{i-1},s,\tau_r)\frac{\partial^2 u(t,s)}{\partial t \partial s}dt ds + \frac{1}{h^2}\int_{y_{j-1}}^{y_j}\int_{x_{i-1}}^{x_i} m_r(t,s,\tau_r)\frac{\partial^2 u(t,s)}{\partial t \partial s}dt ds\right\} =$$

$$= \frac{1}{h^2}\int_{x_{i-1}}^{x_i}\int_{y_{j-1}}^{y_j} f(x,y)dy dx, \quad i,j = 1,2,\ldots,N, \qquad [3.121]$$

where

$$m_1(t,s,\xi) = \int_s^{y_j}\int_t^{x_i} \frac{k_1(x,y)}{\Gamma(1-\alpha)}\frac{dx}{(x-\xi)^\alpha}dy, \quad m_2(t,s,\eta) = \int_t^{x_i}\int_s^{y_j} \frac{k_2(x,y)}{\Gamma(1-\alpha)}\frac{dy}{(y-\eta)^\alpha}dx,$$

$$m_3(t,s,0) = \int_t^{x_i}\int_s^{y_j} k_3(x,y)dy dx, \quad \tau_1 = t,\ \tau_2 = s,\ \tau_3 = 0.$$

We approximate [3.121] with the following mesh scheme:

$$Lv_{i,j} \equiv v_{\bar{x}\bar{y},i,j} + \qquad [3.122]$$

$$+\sum_{r=1}^{3}\left\{\sum_{l=1}^{j-1}\sum_{n=1}^{i-1} {}_r a_{n,l}^{i,j} v_{\bar{x}\bar{y},n,l} + \sum_{l=1}^{j-1} {}_r b_{i,l}^{i,j} v_{\bar{x}\bar{y},i,l} + \sum_{n=1}^{i-1} {}_r c_{n,j}^{i,j} v_{\bar{x}\bar{y},n,j} + {}_r d_{i,j}^{i,j} v_{\bar{x}\bar{y},i,j}\right\} = \varphi_{i,j},$$

$$i,j = 1,2,\ldots,N,$$

where

$$\varphi_{i,j} = \frac{1}{h^2} \int_{x_{i-1}}^{x_i} \int_{y_{j-1}}^{y_j} f(x,y)\,dy\,dx;$$

$$_1a_{n,l}^{i,j} = \frac{1}{h^2} \int_{y_{l-1}}^{y_l} \int_{x_{n-1}}^{x_n} m_1(x_{i-1}, y_{j-1}, t)\,dt\,ds = \frac{1}{h^2} \int_{y_{l-1}}^{y_l} ds \int_{x_{n-1}}^{x_n} dt \int_{y_{j-1}}^{y_j} dy \int_{x_{i-1}}^{x_i} \frac{k_1(x,y)}{\Gamma(1-\alpha)} \frac{dx}{(x-t)^\alpha} =$$

$$= \frac{1}{h} \int_{x_{n-1}}^{x_n} dt \int_{y_{j-1}}^{y_j} dy \int_{x_{i-1}}^{x_i} \frac{k_1(x,y)}{\Gamma(1-\alpha)} \frac{dx}{(x-t)^\alpha},$$

$$_1b_{i,l}^{i,j} = \frac{1}{h^2} \int_{y_{l-1}}^{y_l} \int_{x_{i-1}}^{x_i} m_1(t, y_{j-1}, t)\,dt\,ds = \frac{1}{h^2} \int_{y_{l-1}}^{y_l} ds \int_{x_{i-1}}^{x_i} dt \int_{y_{j-1}}^{y_j} dy \int_{t}^{x_i} \frac{k_1(x,y)}{\Gamma(1-\alpha)} \frac{dx}{(x-t)^\alpha} =$$

$$= \frac{1}{h} \int_{x_{i-1}}^{x_i} dt \int_{y_{j-1}}^{y_j} dy \int_{t}^{x_i} \frac{k_1(x,y)}{\Gamma(1-\alpha)} \frac{dx}{(x-t)^\alpha},$$

$$_1c_{n,j}^{i,j} = \frac{1}{h^2} \int_{y_{j-1}}^{y_j} \int_{x_{n-1}}^{x_n} m_1(x_{i-1}, s, t)\,dt\,ds = \frac{1}{h^2} \int_{y_{j-1}}^{y_j} ds \int_{x_{n-1}}^{x_n} dt \int_{s}^{y_j} dy \int_{x_{i-1}}^{x_i} \frac{k_1(x,y)}{\Gamma(1-\alpha)} \frac{dx}{(x-t)^\alpha},$$

$$_1d_{i,j}^{i,j} = \frac{1}{h^2} \int_{y_{j-1}}^{y_j} \int_{x_{i-1}}^{x_i} m_1(t, s, t)\,dt\,ds = \frac{1}{h^2} \int_{x_{i-1}}^{x_i} dt \int_{y_{j-1}}^{y_j} ds \int_{s}^{y_j} dy \int_{t}^{x_i} \frac{k_1(x,y)}{\Gamma(1-\alpha)} \frac{dx}{(x-t)^\alpha};$$

$$_2a_{n,l}^{i,j} = \frac{1}{h^2} \int_{y_{l-1}}^{y_l} \int_{x_{n-1}}^{x_n} m_2(x_{i-1}, y_{j-1}, s)\,dt\,ds =$$

$$\frac{1}{h^2} \int_{y_{l-1}}^{y_l} ds \int_{x_{n-1}}^{x_n} dt \int_{x_{i-1}}^{x_i} dx \int_{y_{j-1}}^{y_j} \frac{k_2(x,y)}{\Gamma(1-\alpha)} \frac{dy}{(y-s)^\alpha} = \frac{1}{h} \int_{y_{l-1}}^{y_l} ds \int_{x_{i-1}}^{x_i} dx \int_{y_{j-1}}^{y_j} \frac{k_2(x,y)}{\Gamma(1-\alpha)} \frac{dy}{(y-s)^\alpha},$$

$$_2b_{i,l}^{i,j} = \frac{1}{h^2} \int_{y_{l-1}}^{y_l} \int_{x_{i-1}}^{x_i} m_2(t, y_{j-1}, s)\,dt\,ds = \frac{1}{h^2} \int_{y_{l-1}}^{y_l} ds \int_{x_{i-1}}^{x_i} dt \int_{t}^{x_i} dx \int_{y_{j-1}}^{y_j} \frac{k_2(x,y)}{\Gamma(1-\alpha)} \frac{dy}{(y-s)^\alpha},$$

$$_2c_{n,j}^{i,j} = \frac{1}{h^2}\int\limits_{y_{j-1}}^{y_j}\int\limits_{x_{n-1}}^{x_n} m_2(x_{i-1},s,s)dtds = \frac{1}{h^2}\int\limits_{y_{j-1}}^{y_j} ds\int\limits_{x_{n-1}}^{x_n} dt\int\limits_{x_{i-1}}^{x_i} dx\int\limits_{s}^{y_j}\frac{k_2(x,y)}{\Gamma(1-\alpha)}\frac{dy}{(y-s)^\alpha} =$$

$$= \frac{1}{h}\int\limits_{y_{j-1}}^{y_j} ds\int\limits_{x_{i-1}}^{x_i} dx\int\limits_{s}^{y_j}\frac{k_2(x,y)}{\Gamma(1-\alpha)}\frac{dy}{(y-s)^\alpha},$$

$$_2d_{i,j}^{i,j} = \frac{1}{h^2}\int\limits_{x_{i-1}}^{x_i}\int\limits_{y_{j-1}}^{y_j} m_2(t,s,s)dsdt = \frac{1}{h^2}\int\limits_{x_{i-1}}^{x_i} dt\int\limits_{y_{j-1}}^{y_j} ds\int\limits_{t}^{x_i} dx\int\limits_{s}^{y_j}\frac{k_2(x,y)}{\Gamma(1-\alpha)}\frac{dy}{(y-s)^\alpha};$$

$$_3a_{n,l}^{i,j} = \frac{1}{h^2}\int\limits_{y_{l-1}}^{y_l}\int\limits_{x_{n-1}}^{x_n} m_3(x_{i-1},y_{j-1},0)dtds = \frac{1}{h^2}\int\limits_{y_{l-1}}^{y_l} ds\int\limits_{x_{n-1}}^{x_n} dt\int\limits_{x_{i-1}}^{x_i} dx\int\limits_{y_{j-1}}^{y_j} k_3(x,y)dy =$$

$$= \int\limits_{x_{i-1}}^{x_i} dx\int\limits_{y_{j-1}}^{y_j} k_3(x,y)dy,$$

$$_3b_{i,l}^{i,j} = \frac{1}{h^2}\int\limits_{y_{l-1}}^{y_l}\int\limits_{x_{i-1}}^{x_i} m_3(t,y_{j-1},0)dtds = \frac{1}{h^2}\int\limits_{y_{l-1}}^{y_l} ds\int\limits_{x_{i-1}}^{x_i} dt\int\limits_{t}^{x_i} dx\int\limits_{y_{j-1}}^{y_j} k_3(x,y)dy =$$

$$= \frac{1}{h}\int\limits_{x_{i-1}}^{x_i} dt\int\limits_{t}^{x_i} dx\int\limits_{y_{j-1}}^{y_j} k_3(x,y)dy,$$

$$_3c_{n,j}^{i,j} = \frac{1}{h^2}\int\limits_{y_{j-1}}^{y_j}\int\limits_{x_{n-1}}^{x_n} m_3(x_{i-1},s,0)dtds = \frac{1}{h^2}\int\limits_{y_{j-1}}^{y_j} ds\int\limits_{x_{n-1}}^{x_n} dt\int\limits_{x_{i-1}}^{x_i} dx\int\limits_{s}^{y_j} k_3(x,y)dy =$$

$$= \frac{1}{h}\int\limits_{y_{j-1}}^{y_j} ds\int\limits_{x_{i-1}}^{x_i} dx\int\limits_{s}^{y_j} k_3(x,y)dy,$$

$$_3d_{i,j}^{i,j} = \frac{1}{h^2}\int\limits_{x_{i-1}}^{x_i}\int\limits_{y_{j-1}}^{y_j} m_3(t,s,0)dsdt = \frac{1}{h^2}\int\limits_{x_{i-1}}^{x_i} dt\int\limits_{y_{j-1}}^{y_j} ds\int\limits_{t}^{x_i} dx\int\limits_{s}^{y_j} k_3(x,y)dy.$$

THEOREM 3.23.– *Let the conditions of Theorem 3.22 be fulfilled and let the coefficients $k_1(x,y)$, $k_2(x,y)$, $k_3(x,y)$ satisfy the inequality*

$$q \equiv \frac{\|k_1\|_{C(\bar{\Omega})}}{\Gamma(2-\alpha)} T^{1-\alpha}\left(T+\frac{h}{2}\right) + \frac{\|k_2\|_{C(\bar{\Omega})}}{\Gamma(2-\alpha)} T^{1-\alpha}\left(T+\frac{h}{2}\right) + \|k_3\|_{C(\bar{\Omega})}\left(T+\frac{h}{2}\right)^2 < 1.$$

Then, the accuracy of the mesh scheme [3.122] is characterized by the estimate

$$\|z_{\bar{x}\bar{y}}\|_{T,T} \leq \frac{M}{1-q} h^{1-\alpha} \left|\frac{\partial^2 u}{\partial x \partial y}\right|_{C^{0,1-\alpha}(\bar{\Omega})} \qquad [3.123]$$

with

$$M = \frac{\|k_1\|_{C(\bar{\Omega})}}{\Gamma(2-\alpha)} T^{1-\alpha}(2T+h) + \frac{\|k_2\|_{C(\bar{\Omega})}}{\Gamma(2-\alpha)} T^{1-\alpha}(2T+h) + \frac{1}{2}\|k_3\|_{C(\bar{\Omega})}(2T+h)^2.$$

PROOF.– For the accuracy $z_{i,j} = v_{i,j} - u_{i,j}$, we have the following mesh scheme:

$$Lz_{i,j} \equiv z_{\bar{x}\bar{y},i,j} +$$

$$+ \sum_{r=1}^{3}\left\{\sum_{l=1}^{j-1}\sum_{n=1}^{i-1} {}_r a_{n,l}^{i,j} z_{\bar{x}\bar{y},n,l} + \sum_{l=1}^{j-1} {}_r b_{i,l}^{i,j} z_{\bar{x}\bar{y},i,l} + \sum_{n=1}^{i-1} {}_r c_{n,j}^{i,j} z_{\bar{x}\bar{y},n,j} + {}_r d_{i,j}^{i,j} z_{\bar{x}\bar{y},i,j}\right\} =$$

$$= \psi_{i,j}, \quad i,j = 1,2,\ldots,N,$$

where $\psi_{i,j} = \sum_{r=1}^{3} \eta_r = \sum_{r=1}^{3}\sum_{k=1}^{4} \eta_{r,k}$ is the approximation error.

This gives the inequality

$$|z_{\bar{x}\bar{y},i,j}| \leq \qquad [3.124]$$

$$\leq \sum_{r=1}^{3}\left\{\sum_{l=1}^{j-1}\sum_{n=1}^{i-1}|{}_r a_{n,l}^{i,j}| + \sum_{l=1}^{j-1}|{}_r b_{i,l}^{i,j}| + \sum_{n=1}^{i-1}|{}_r c_{n,j}^{i,j}| + |{}_r d_{i,j}^{i,j}|\right\}\|z_{\bar{x}\bar{y}}\|_{x_i,y_j} + |\psi_{i,j}|,$$

$i,j = 1,2,\ldots,N.$

Next, we will estimate each sum in the curly brackets:

$$\sum_{l=1}^{j-1}\sum_{n=1}^{i-1}\left|{}_1 a_{n,l}^{i,j}\right| = \sum_{l=1}^{j-1}\sum_{n=1}^{i-1}\left|\frac{1}{h^2}\int_{y_{l-1}}^{y_l}\int_{x_{n-1}}^{x_n}\int_{y_{j-1}}^{y_j}\int_{x_{i-1}}^{x_i}\frac{k_1(x,y)}{\Gamma(1-\alpha)}\frac{dxdy}{(x-t)^\alpha}dtds\right| \le$$

$$\le \frac{\|k_1\|_{C(\bar\Omega)} hy_{j-1}}{\Gamma(1-\alpha)h^2}\int_0^{x_{i-1}}\int_{x_{i-1}}^{x_i}\frac{dx}{(x-t)^\alpha}dt =$$

$$= \frac{\|k_1\|_{C(\bar\Omega)} hy_{j-1}}{\Gamma(1-\alpha)h^2}\int_0^{x_{i-1}}\frac{(x_i-t)^{1-\alpha}-(x_{i-1}-t)^{1-\alpha}}{1-\alpha}dt =$$

$$= \frac{\|k_1\|_{C(\bar\Omega)} hy_{j-1}}{\Gamma(2-\alpha)h^2}\frac{x_i^{2-\alpha}-h^{2-\alpha}-x_{i-1}^{2-\alpha}}{2-\alpha},$$

$$\left|{}_1 a_{n,l}^{i,j}\right| = \left|\frac{1}{h^2}\int_{y_{l-1}}^{y_l}\int_{x_{n-1}}^{x_n}\int_{y_{j-1}}^{y_j}\int_{x_{i-1}}^{x_i}\frac{k_1(x,y)}{\Gamma(1-\alpha)}\frac{dxdy}{(x-t)^\alpha}dtds\right| \le$$

$$\le \frac{\|k_1\|_{C(\bar\Omega)} h^2}{\Gamma(1-\alpha)h^2}\int_{x_{n-1}}^{x_n}\int_{x_{i-1}}^{x_i}\frac{dx}{(x-t)^\alpha}dt = \frac{\|k_1\|_{C(\bar\Omega)} h^2}{\Gamma(1-\alpha)h^2}\int_{x_{n-1}}^{x_n}\frac{(x_i-t)^{1-\alpha}-(x_{i-1}-t)^{1-\alpha}}{1-\alpha}dt \le$$

$$\le \frac{\|k_1\|_{C(\bar\Omega)} h^2}{\Gamma(2-\alpha)h^2} hh^{1-\alpha} = \frac{\|k_1\|_{C(\bar\Omega)} h^{2-\alpha}}{\Gamma(2-\alpha)},$$

$$\sum_{l=1}^{j-1}\left|{}_1 b_{i,l}^{i,j}\right| = \sum_{l=1}^{j-1}\left|\frac{1}{h^2}\int_{y_{l-1}}^{y_l}\int_{x_{i-1}}^{x_i}\int_{y_{j-1}}^{y_j}\int_t^{x_i}\frac{k_1(x,y)}{\Gamma(1-\alpha)}\frac{dxdy}{(x-t)^\alpha}dtds\right| \le$$

$$\le \frac{\|k_1\|_{C(\bar\Omega)} hy_{j-1}}{\Gamma(1-\alpha)h^2}\int_{x_{i-1}}^{x_i}\int_t^{x_i}\frac{dx}{(x-t)^\alpha}dt = \frac{\|k_1\|_{C(\bar\Omega)} hy_{j-1}}{\Gamma(1-\alpha)h^2}\int_{x_{i-1}}^{x_i}\frac{(x_i-t)^{1-\alpha}}{1-\alpha}dt =$$

$$= \frac{\|k_1\|_{C(\bar\Omega)} hy_{j-1}}{\Gamma(2-\alpha)h^2}\frac{h^{2-\alpha}}{2-\alpha},$$

$$\left|{}_1 b_{i,l}^{i,j}\right| = \left|\frac{1}{h^2}\int_{y_{l-1}}^{y_l}\int_{x_{i-1}}^{x_i}\int_{y_{j-1}}^{y_j}\int_t^{x_i}\frac{k_1(x,y)}{\Gamma(1-\alpha)}\frac{dxdy}{(x-t)^\alpha}dtds\right| \le \frac{\|k_1\|_{C(\bar\Omega)} h^2}{\Gamma(1-\alpha)h^2}\int_{x_{i-1}}^{x_i}\int_t^{x_i}\frac{dx}{(x-t)^\alpha}dt =$$

$$= \frac{\|k_1\|_{C(\bar{\Omega})}}{\Gamma(1-\alpha)h^2} \frac{h^2}{\omega} \int_{x_{i-1}}^{x_i} \frac{(x_i-t)^{1-\alpha}}{1-\alpha} dt = \frac{\|k_1\|_{C(\bar{\Omega})}}{\Gamma(2-\alpha)h^2} \frac{h^2}{2-\alpha} \frac{h^{2-\alpha}}{2-\alpha} = \frac{\|k_1\|_{C(\bar{\Omega})}}{\Gamma(3-\alpha)} h^{2-\alpha},$$

$$\sum_{n=1}^{i-1} \left|{}_1c_{n,j}^{i,j}\right| = \sum_{n=1}^{i-1} \left| \frac{1}{h^2} \int_{y_{j-1}}^{y_j} ds \int_{x_{n-1}}^{x_n} dt \int_s^{y_j} dy \int_{x_{i-1}}^{x_i} \frac{k_1(x,y)}{\Gamma(1-\alpha)} \frac{dx}{(x-t)^\alpha} \right| \le$$

$$\le \frac{\|k_1\|_{C(\bar{\Omega})}}{\Gamma(1-\alpha)h^2} \int_{y_{j-1}}^{y_j} (y_j - s)ds \int_0^{x_{i-1}} \frac{(x_i-t)^{1-\alpha} - (x_{i-1}-t)^{1-\alpha}}{1-\alpha} dt =$$

$$= \frac{\|k_1\|_{C(\bar{\Omega})}}{\Gamma(2-\alpha)h^2} \frac{h^2}{2} \frac{x_i^{2-\alpha} - h^{2-\alpha} - x_{i-1}^{2-\alpha}}{2-\alpha},$$

$$\left|{}_1c_{n,j}^{i,j}\right| = \left| \frac{1}{h^2} \int_{y_{j-1}}^{y_j} ds \int_{x_{n-1}}^{x_n} dt \int_s^{y_j} dy \int_{x_{i-1}}^{x_i} \frac{k_1(x,y)}{\Gamma(1-\alpha)} \frac{dx}{(x-t)^\alpha} \right| \le$$

$$\le \frac{\|k_1\|_{C(\bar{\Omega})}}{\Gamma(1-\alpha)h^2} \int_{y_{j-1}}^{y_j} (y_j - s)ds \int_{x_{n-1}}^{x_n} \frac{(x_i-t)^{1-\alpha} - (x_{i-1}-t)^{1-\alpha}}{1-\alpha} dt \le$$

$$\le \frac{\|k_1\|_{C(\bar{\Omega})}}{\Gamma(2-\alpha)h^2} \frac{h^2}{2} hh^{1-\alpha} = \frac{\|k_1\|_{C(\bar{\Omega})}}{2\Gamma(2-\alpha)} h^{2-\alpha},$$

$$\left|{}_1d_{i,j}^{i,j}\right| = \left| \frac{1}{h^2} \int_{x_{i-1}}^{x_i} dt \int_{y_{j-1}}^{y_j} ds \int_s^{y_j} dy \int_t^{x_i} \frac{k_1(x,y)}{\Gamma(1-\alpha)} \frac{dx}{(x-t)^\alpha} \right| \le$$

$$\le \frac{\|k_1\|_{C(\bar{\Omega})}}{\Gamma(1-\alpha)} \frac{1}{h^2} \int_{y_{j-1}}^{y_j} (y_j - s)ds \int_{x_{i-1}}^{x_i} \frac{(x_i-t)^{1-\alpha}}{1-\alpha} dt =$$

$$= \frac{\|k_1\|_{C(\bar{\Omega})}}{\Gamma(2-\alpha)h^2} \frac{h^2}{2} \frac{h^{2-\alpha}}{2-\alpha} = \frac{\|k_1\|_{C(\bar{\Omega})}}{2\Gamma(3-\alpha)} h^{2-\alpha}.$$

Then, we have

$$\sum_{l=1}^{j-1}\sum_{n=1}^{i-1}\left|{}_1 a_{n,l}^{i,j}\right| + \sum_{l=1}^{j-1}\left|{}_1 b_{i,l}^{i,j}\right| + \sum_{n=1}^{i-1}\left|{}_1 c_{n,j}^{i,j}\right| + \left|{}_1 d_{i,j}^{i,j}\right| \leq \qquad [3.125]$$

$$\leq \frac{\|k_1\|_{C(\bar{\Omega})}\, hy_{j-1}}{\Gamma(2-\alpha)h^2} \frac{x_i^{2-\alpha} - h^{2-\alpha} - x_{i-1}^{2-\alpha}}{2-\alpha} + \frac{\|k_1\|_{C(\bar{\Omega})}\, hy_{j-1}}{\Gamma(2-\alpha)h^2} \frac{h^{2-\alpha}}{2-\alpha} +$$

$$+ \frac{\|k_1\|_{C(\bar{\Omega})}}{\Gamma(2-\alpha)h^2} \frac{h^2}{2} \frac{x_i^{2-\alpha} - h^{2-\alpha} - x_{i-1}^{2-\alpha}}{2-\alpha} + \frac{\|k_1\|_{C(\bar{\Omega})}}{\Gamma(2-\alpha)h^2} \frac{h^2}{2} \frac{h^{2-\alpha}}{2-\alpha} =$$

$$= \frac{\|k_1\|_{C(\bar{\Omega})}\, hy_{j-1}}{\Gamma(2-\alpha)h^2} \frac{x_i^{2-\alpha} - x_{i-1}^{2-\alpha}}{2-\alpha} + \frac{\|k_1\|_{C(\bar{\Omega})}}{\Gamma(2-\alpha)h^2} \frac{h^2}{2} \frac{x_i^{2-\alpha} - x_{i-1}^{2-\alpha}}{2-\alpha} \leq$$

$$\leq \frac{\|k_1\|_{C(\bar{\Omega})}\, hy_{j-1}}{\Gamma(2-\alpha)h^2} x_i^{1-\alpha} h + \frac{\|k_1\|_{C(\bar{\Omega})}}{\Gamma(2-\alpha)h^2} \frac{h^2}{2} x_i^{1-\alpha} h = \frac{\|k_1\|_{C(\bar{\Omega})}}{\Gamma(2-\alpha)} x_i^{1-\alpha} \left(y_{j-1} + \frac{h}{2}\right) \leq$$

$$\leq \frac{\|k_1\|_{C(\bar{\Omega})}}{\Gamma(2-\alpha)} T^{1-\alpha}\left(T + \frac{h}{2}\right).$$

Similarly, we obtain the estimates

$$\sum_{l=1}^{j-1}\sum_{n=1}^{i-1}\left|{}_2 a_{n,l}^{i,j}\right| = \sum_{l=1}^{j-1}\sum_{n=1}^{i-1}\left|\frac{1}{h^2}\int_{y_{l-1}}^{y_l}\int_{x_{n-1}}^{x_n}\int_{x_{i-1}}^{x_i}\int_{y_{j-1}}^{y_j} \frac{k_2(x,y)}{\Gamma(1-\alpha)} \frac{dydx}{(y-s)^\alpha} dtds\right| \leq$$

$$\leq \frac{\|k_2\|_{C(\bar{\Omega})}\, hx_{i-1}}{\Gamma(1-\alpha)h^2} \int_0^{y_{j-1}}\int_{y_{j-1}}^{y_j} \frac{dy}{(y-s)^\alpha} ds =$$

$$= \frac{\|k_2\|_{C(\bar{\Omega})}\, hy_{j-1}}{\Gamma(1-\alpha)h^2} \int_0^{y_{j-1}} \frac{(y_j-s)^{1-\alpha} - (y_{j-1}-s)^{1-\alpha}}{1-\alpha} ds =$$

$$= \frac{\|k_2\|_{C(\bar{\Omega})}\, hx_{i-1}}{\Gamma(2-\alpha)h^2} \frac{y_j^{2-\alpha} - h^{2-\alpha} - y_{j-1}^{2-\alpha}}{2-\alpha},$$

$$\left|{}_2 a_{n,l}^{i,j}\right| = \left|\frac{1}{h^2}\int_{y_{l-1}}^{y_l}\int_{x_{n-1}}^{x_n}\int_{x_{i-1}}^{x_i}\int_{y_{j-1}}^{y_j} \frac{k_2(x,y)}{\Gamma(1-\alpha)} \frac{dydx}{(y-s)^\alpha} dtds\right| \leq$$

$$\leq \frac{\|k_2\|_{C(\bar{\Omega})}}{\Gamma(1-\alpha)h^2} \frac{h^2}{} \int_{y_{l-1}}^{y_l} \int_{y_{j-1}}^{y_{j-1}} \frac{dy}{(y-s)^\alpha} ds = \frac{\|k_2\|_{C(\bar{\Omega})}}{\Gamma(1-\alpha)h^2} \frac{h^2}{} \int_{y_{l-1}}^{y_l} \frac{(x_i-t)^{1-\alpha} - (x_{i-1}-t)^{1-\alpha}}{1-\alpha} ds \leq$$

$$\leq \frac{\|k_2\|_{C(\bar{\Omega})}}{\Gamma(2-\alpha)h^2} h^2 \, hh^{1-\alpha} = \frac{\|k_2\|_{C(\bar{\Omega})}}{\Gamma(2-\alpha)} h^{2-\alpha},$$

$$\sum_{l=1}^{j-1} \left|2 b_{i,l}^{i,j}\right| = \sum_{l=1}^{j-1} \left| \frac{1}{h^2} \int_{y_{l-1}}^{y_l} \int_{x_{i-1}}^{x_i} \int_{t}^{x_i} \int_{y_{j-1}}^{y_j} \frac{k_2(x,y)}{\Gamma(1-\alpha)} \frac{dy dx}{(y-s)^\alpha} dt ds \right| \leq$$

$$\leq \frac{\|k_2\|_{C(\bar{\Omega})}}{\Gamma(1-\alpha)h^2} \int_{y_{l-1}}^{y_l} \frac{(y_j-s)^{1-\alpha} - (y_{j-1}-s)^{1-\alpha}}{1-\alpha} ds \int_{x_{i-1}}^{x_i} (x_i-t) dt =$$

$$\leq \frac{\|k_2\|_{C(\bar{\Omega})}}{\Gamma(2-\alpha)h^2} h^{1-\alpha} \frac{h^2}{2} h = \frac{\|k_2\|_{C(\bar{\Omega})}}{2\Gamma(2-\alpha)} h^{2-\alpha},$$

$$\left|2 b_{i,l}^{i,j}\right| = \left| \frac{1}{h^2} \int_{y_{l-1}}^{y_l} \int_{x_{i-1}}^{x_i} \int_{t}^{x_i} \int_{y_{j-1}}^{y_j} \frac{k_2(x,y)}{\Gamma(1-\alpha)} \frac{dy dx}{(y-s)^\alpha} dt ds \right| \leq$$

$$\leq \frac{\|k_2\|_{C(\bar{\Omega})}}{\Gamma(1-\alpha)h^2} \int_{x_{i-1}}^{x_i} (x_i-t) dt \int_{y_{l-1}}^{y_l} \frac{(y_j-s)^{1-\alpha} - (y_{j-1}-s)^{1-\alpha}}{1-\alpha} ds \leq$$

$$\leq \frac{\|k_2\|_{C(\bar{\Omega})}}{\Gamma(2-\alpha)h^2} \frac{h^2}{2} hh^{1-\alpha} = \frac{\|k_2\|_{C(\bar{\Omega})}}{2\Gamma(2-\alpha)} h^{2-\alpha},$$

$$\sum_{n=1}^{i-1} \left|2 c_{n,j}^{i,j}\right| = \sum_{n=1}^{i-1} \left| \frac{1}{h^2} \int_{y_{j-1}}^{y_j} ds \int_{x_{n-1}}^{x_n} dt \int_{x_{i-1}}^{x_i} dx \int_{s}^{y_j} \frac{k_2(x,y)}{\Gamma(1-\alpha)} \frac{dy}{(y-s)^\alpha} \right| \leq$$

$$\leq \frac{\|k_2\|_{C(\bar{\Omega})} h x_{i-1}}{\Gamma(1-\alpha)h^2} \int_{y_{j-1}}^{y_j} \int_{s}^{y_j} \frac{dy}{(y-s)^\alpha} ds = \frac{\|k_2\|_{C(\bar{\Omega})} h x_{i-1}}{\Gamma(1-\alpha)h^2} \int_{y_{j-1}}^{y_j} \frac{(y_j-s)^{1-\alpha}}{1-\alpha} ds =$$

$$= \frac{\|k_2\|_{C(\bar{\Omega})}}{\Gamma(2-\alpha)h^2} \frac{hx_{i-1}}{2-\alpha} \frac{h^{2-\alpha}}{2-\alpha},$$

$$\left|{}_2 c_{n,j}^{i,j}\right| = \left| \frac{1}{h^2} \int\limits_{y_{j-1}}^{y_j} ds \int\limits_{x_{n-1}}^{x_n} dt \int\limits_{x_{i-1}}^{x_i} dx \int\limits_{s}^{y_j} \frac{k_2(x,y)}{\Gamma(1-\alpha)} \frac{dy}{(y-s)^\alpha} \right| \le$$

$$\le \frac{\|k_2\|_{C(\bar{\Omega})}}{\Gamma(1-\alpha)h^2} h^2 \int\limits_{y_{j-1}}^{y_j} \frac{(y_j-s)^{1-\alpha}}{1-\alpha} ds = \frac{\|k_2\|_{C(\bar{\Omega})}}{\Gamma(2-\alpha)h^2} \frac{h^2}{2-\alpha} \frac{h^{2-\alpha}}{2-\alpha} = \frac{\|k_2\|_{C(\bar{\Omega})}}{\Gamma(3-\alpha)} h^{2-\alpha},$$

$$\left|{}_2 d_{i,j}^{i,j}\right| = \left| \frac{1}{h^2} \int\limits_{y_{j-1}}^{y_j} ds \int\limits_{x_{i-1}}^{x_i} dt \int\limits_{t}^{x_i} dx \int\limits_{s}^{y_j} \frac{k_2(x,y)}{\Gamma(1-\alpha)} \frac{dy}{(y-s)^\alpha} \right| \le$$

$$\le \frac{\|k_2\|_{C(\bar{\Omega})}}{\Gamma(1-\alpha)} \frac{1}{h^2} \int\limits_{y_{j-1}}^{y_j} \frac{(y_j-s)^{1-\alpha}}{1-\alpha} ds \int\limits_{x_{i-1}}^{x_i} (x_i-t)dt =$$

$$= \frac{\|k_2\|_{C(\bar{\Omega})}}{\Gamma(2-\alpha)h^2} \frac{h^{2-\alpha}}{2-\alpha} \frac{h^2}{2} = \frac{\|k_2\|_{C(\bar{\Omega})}}{2\Gamma(3-\alpha)} h^{2-\alpha},$$

then

$$\sum_{l=1}^{j-1}\sum_{n=1}^{i-1} \left|{}_2 a_{n,l}^{i,j}\right| + \sum_{l=1}^{j-1}\left|{}_2 b_{i,l}^{i,j}\right| + \sum_{n=1}^{i-1}\left|{}_2 c_{n,j}^{i,j}\right| + \left|{}_2 d_{i,j}^{i,j}\right| \le \qquad [3.126]$$

$$\le \frac{\|k_2\|_{C(\bar{\Omega})}}{\Gamma(2-\alpha)} y_j^{1-\alpha}\left(x_{i-1}+\frac{h}{2}\right) \le \frac{\|k_2\|_{C(\bar{\Omega})}}{\Gamma(2-\alpha)} T^{1-\alpha}\left(T+\frac{h}{2}\right).$$

Now we find

$$\sum_{l=1}^{j-1}\sum_{n=1}^{i-1}\left|{}_3 a_{n,l}^{i,j}\right| = \sum_{l=1}^{j-1}\sum_{n=1}^{i-1}\left|\frac{1}{h^2}\int\limits_{y_{l-1}}^{y_l} ds \int\limits_{x_{n-1}}^{x_n} dt \int\limits_{x_{i-1}}^{x_i} dx \int\limits_{y_{j-1}}^{y_j} k_3(x,y)dy\right| \le$$

$$\le \frac{\|k_3\|_{C(\bar{\Omega})}}{h^2} x_{i-1} y_{j-1} h^2 = \|k_3\|_{C(\bar{\Omega})} x_{i-1} y_{j-1},$$

$$\left|{}_3a_{n,l}^{i,j}\right| = \left|\frac{1}{h^2}\int\limits_{y_{l-1}}^{y_l} ds \int\limits_{x_{n-1}}^{x_n} dt \int\limits_{x_{i-1}}^{x_i} dx \int\limits_{y_{j-1}}^{y_j} k_3(x,y)dy\right| \leq \frac{\|k_3\|_{C(\bar{\Omega})}}{h^2}h^4 = \|k_3\|_{C(\bar{\Omega})}h^2,$$

$$\sum_{l=1}^{j-1}\left|{}_3b_{i,l}^{i,j}\right| = \sum_{l=1}^{j-1}\left|\frac{1}{h^2}\int\limits_{y_{l-1}}^{y_l} ds \int\limits_{x_{i-1}}^{x_i} dt \int\limits_{t}^{x_i} dx \int\limits_{y_{j-1}}^{y_j} k_3(x,y)dy\right| \leq$$

$$\leq \frac{\|k_3\|_{C(\bar{\Omega})}}{h^2}y_{j-1}h\int\limits_{x_{i-1}}^{x_i}(x_i-t)dt = \frac{\|k_3\|_{C(\bar{\Omega})}}{h^2}y_{j-1}h\frac{h^2}{2} = \frac{1}{2}\|k_3\|_{C(\bar{\Omega})}y_{j-1}h,$$

$$\left|{}_3b_{i,l}^{i,j}\right| = \left|\frac{1}{h^2}\int\limits_{y_{l-1}}^{y_l} ds \int\limits_{x_{i-1}}^{x_i} dt \int\limits_{t}^{x_i} dx \int\limits_{y_{j-1}}^{y_j} k_3(x,y)dy\right| \leq$$

$$\leq \frac{\|k_3\|_{C(\bar{\Omega})}}{h^2}h^2\int\limits_{x_{i-1}}^{x_i}(x_i-t)dt = \frac{\|k_3\|_{C(\bar{\Omega})}}{h^2}h^2\frac{h^2}{2} = \frac{1}{2}\|k_3\|_{C(\bar{\Omega})}h^2,$$

$$\sum_{n=1}^{i-1}\left|{}_3c_{i,l}^{i,j}\right| = \sum_{n=1}^{i-1}\left|\frac{1}{h^2}\int\limits_{y_{j-1}}^{y_j} ds \int\limits_{x_{n-1}}^{x_n} dt \int\limits_{x_{i-1}}^{x_i} dx \int\limits_{s}^{y_j} k_3(x,y)dy\right| \leq$$

$$\leq \frac{\|k_3\|_{C(\bar{\Omega})}}{h^2}x_{i-1}h\int\limits_{y_{j-1}}^{y_j}(y_j-s)ds = \frac{\|k_3\|_{C(\bar{\Omega})}}{h^2}x_{i-1}h\frac{h^2}{2} = \frac{1}{2}\|k_3\|_{C(\bar{\Omega})}x_{i-1}h,$$

$$\left|{}_3c_{i,l}^{i,j}\right| = \left|\frac{1}{h^2}\int\limits_{y_{j-1}}^{y_j} ds \int\limits_{x_{n-1}}^{x_n} dt \int\limits_{x_{i-1}}^{x_i} dx \int\limits_{s}^{y_j} k_3(x,y)dy\right| \leq$$

$$\leq \frac{\|k_3\|_{C(\bar{\Omega})}}{h^2}h^2\int\limits_{y_{j-1}}^{y_j}(y_j-s)ds = \frac{\|k_3\|_{C(\bar{\Omega})}}{h^2}h^2\frac{h^2}{2} = \frac{1}{2}\|k_3\|_{C(\bar{\Omega})}h^2,$$

$$\left|{}_3d_{i,j}^{i,j}\right| = \left|\frac{1}{h^2}\int\limits_{x_{i-1}}^{x_i} dt \int\limits_{y_{j-1}}^{y_j} ds \int\limits_{t}^{x_i} dx \int\limits_{s}^{y_j} k_3(x,y)dy\right| \leq$$

$$\leq \frac{\|k_3\|_{C(\bar{\Omega})}}{h^2} \int_{x_{i-1}}^{x_i}(x_i-t)dt \int_{y_{j-1}}^{y_j}(y_j-s)ds = \frac{\|k_3\|_{C(\bar{\Omega})}}{h^2}\frac{h^2}{2}\frac{h^2}{2} = \frac{h^2}{4}\|k_3\|_{C(\bar{\Omega})},$$

which gives

$$\sum_{l=1}^{j-1}\sum_{n=1}^{i-1}\left|{}_3 a_{n,l}^{i,j}\right| + \sum_{l=1}^{j-1}\left|{}_3 b_{i,l}^{i,j}\right| + \sum_{n=1}^{i-1}\left|{}_3 c_{n,j}^{i,j}\right| + \left|{}_3 d_{i,j}^{i,j}\right| \leq \qquad [3.127]$$

$$\leq \|k_3\|_{C(\bar{\Omega})} x_{i-1}y_{j-1} + \frac{1}{2}\|k_3\|_{C(\bar{\Omega})} y_{j-1}h + \frac{1}{2}\|k_3\|_{C(\bar{\Omega})} x_{i-1}h + \frac{h^2}{4}\|k_3\|_{C(\bar{\Omega})} \leq$$

$$\leq \|k_3\|_{C(\bar{\Omega})} \left(T + \frac{h}{2}\right)^2.$$

Inequalities [3.125]–[3.127] yield the estimate

$$\sum_{r=1}^{3}\left\{\sum_{l=1}^{j-1}\sum_{n=1}^{i-1}\left|{}_r a_{n,l}^{i,j}\right| + \sum_{l=1}^{j-1}\left|{}_r b_{i,l}^{i,j}\right| + \sum_{n=1}^{i-1}\left|{}_r c_{n,j}^{i,j}\right| + \left|{}_r d_{i,j}^{i,j}\right|\right\} \leq \qquad [3.128]$$

$$\leq \frac{\|k_1\|_{C(\bar{\Omega})}}{\Gamma(2-\alpha)}T^{1-\alpha}\left(T+\frac{h}{2}\right) + \frac{\|k_2\|_{C(\bar{\Omega})}}{\Gamma(2-\alpha)}T^{1-\alpha}\left(T+\frac{h}{2}\right) + \|k_3\|_{C(\bar{\Omega})}\left(T+\frac{h}{2}\right)^2.$$

Now we consider the approximation error. We have

$$\eta_{1,1} = \sum_{l=1}^{j-1}\sum_{n=1}^{i-1}\frac{1}{h^2}\int_{x_{n-1}}^{x_n}\int_{y_{l-1}}^{y_l}\int_{y_{j-1}}^{y_j}\int_{x_{i-1}}^{x_i}\frac{k_1(x,y)}{\Gamma(1-\alpha)}\frac{dxdy}{(x-t)^\alpha}\left[\frac{\partial^2 u(t,s)}{\partial t\partial s} - u_{\bar{x}\bar{y},n,l}\right]dsdt =$$

$$= \sum_{l=1}^{j-1}\sum_{n=1}^{i-1}\frac{1}{h^2}\int_{x_{n-1}}^{x_n}\int_{y_{l-1}}^{y_l}\int_{y_{j-1}}^{y_j}\int_{x_{i-1}}^{x_i}\frac{k_1(x,y)}{\Gamma(1-\alpha)}\frac{dxdy}{(x-t)^\alpha}\times$$

$$\times\frac{1}{h^2}\int_{x_{n-1}}^{x_n}\int_{y_{l-1}}^{y_l}\left[\frac{\partial^2 u(t,s)}{\partial t\partial s} - \frac{\partial^2 u(\xi,s)}{\partial \xi\partial s} + \frac{\partial^2 u(\xi,s)}{\partial \xi\partial s} - \frac{\partial^2 u(\xi,\eta)}{\partial \xi\partial \eta}\right]d\eta d\xi dsdt,$$

$$\eta_{1,2} = \sum_{l=1}^{j-1} \frac{1}{h^2} \int_{y_{l-1}}^{y_l} \int_{x_{i-1}}^{x_i} \int_{y_{j-1}}^{y_j} \int_{t}^{x_i} \frac{k_1(x,y)}{\Gamma(1-\alpha)} \frac{dxdy}{(x-t)^\alpha} \left[\frac{\partial^2 u(t,s)}{\partial t \partial s} - u_{\overline{x}\,\overline{y},i,l} \right] dtds =$$

$$= \sum_{l=1}^{j-1} \frac{1}{h^2} \int_{y_{l-1}}^{y_l} \int_{x_{i-1}}^{x_i} \int_{y_{j-1}}^{y_j} \int_{t}^{x_i} \frac{k_1(x,y)}{\Gamma(1-\alpha)} \frac{dxdy}{(x-t)^\alpha} \times$$

$$\times \frac{1}{h^2} \int_{x_{i-1}}^{x_i} \int_{y_{l-1}}^{y_l} \left[\frac{\partial^2 u(t,s)}{\partial t \partial s} - \frac{\partial^2 u(\xi,s)}{\partial \xi \partial s} + \frac{\partial^2 u(\xi,s)}{\partial \xi \partial s} - \frac{\partial^2 u(\xi,\eta)}{\partial \xi \partial \eta} \right] d\eta d\xi dtds,$$

$$\eta_{1,3} = \sum_{n=1}^{i-1} \frac{1}{h^2} \int_{y_{j-1}}^{y_j} \int_{x_{n-1}}^{x_n} \int_{s}^{y_j} \int_{x_{i-1}}^{x_i} \frac{k_1(x,y)}{\Gamma(1-\alpha)} \frac{dxdy}{(x-t)^\alpha} \left[\frac{\partial^2 u(t,s)}{\partial t \partial s} - u_{\overline{x}\,\overline{y},n,j} \right] dtds =$$

$$= \sum_{n=1}^{i-1} \frac{1}{h^2} \int_{y_{j-1}}^{y_j} \int_{x_{n-1}}^{x_n} \int_{s}^{y_j} \int_{x_{i-1}}^{x_i} \frac{k_1(x,y)}{\Gamma(1-\alpha)} \frac{dxdy}{(x-t)^\alpha} \times$$

$$\times \frac{1}{h^2} \int_{x_{n-1}}^{x_n} \int_{y_{j-1}}^{y_j} \left[\frac{\partial^2 u(t,s)}{\partial t \partial s} - \frac{\partial^2 u(\xi,s)}{\partial \xi \partial s} + \frac{\partial^2 u(\xi,s)}{\partial \xi \partial s} - \frac{\partial^2 u(\xi,\eta)}{\partial \xi \partial \eta} \right] d\eta d\xi dtds,$$

$$\eta_{1,4} = \frac{1}{h^2} \int_{x_{i-1}}^{x_i} \int_{y_{j-1}}^{y_j} \int_{s}^{y_j} \int_{t}^{x_i} \frac{k_1(x,y)}{\Gamma(1-\alpha)} \frac{dxdy}{(x-t)^\alpha} \left[\frac{\partial^2 u(t,s)}{\partial t \partial s} - u_{\overline{x}\,\overline{y},i,j} \right] dsdt =$$

$$= \frac{1}{h^2} \int_{x_{i-1}}^{x_i} \int_{y_{j-1}}^{y_j} \int_{s}^{y_j} \int_{t}^{x_i} \frac{k_1(x,y)}{\Gamma(1-\alpha)} \frac{dxdy}{(x-t)^\alpha} \times$$

$$\times \frac{1}{h^2} \int_{x_{i-1}}^{x_i} \int_{y_{j-1}}^{y_j} \left[\frac{\partial^2 u(t,s)}{\partial t \partial s} - \frac{\partial^2 u(\xi,s)}{\partial \xi \partial s} + \frac{\partial^2 u(\xi,s)}{\partial \xi \partial s} - \frac{\partial^2 u(\xi,\eta)}{\partial \xi \partial \eta} \right] d\eta d\xi dsdt,$$

therefore

$$\sum_{k=1}^{4} |\eta_{1,k}| \leq \frac{\|k_1\|_{C(\bar{\Omega})}}{\Gamma(2-\alpha)} T^{1-\alpha} \left[T + \frac{h}{2} \right] \frac{1}{h^2} h^2 2h^{1-\alpha} \left| \frac{\partial^2 u(x,y)}{\partial x \partial y} \right|_{C^{0,1-\alpha}(\bar{\Omega})} = \qquad [3.129]$$

$$= h^{1-\alpha} \frac{\|k_1\|_{0,\infty}}{\Gamma(2-\alpha)} T^{1-\alpha} (2T+h) \left|\frac{\partial^2 u(x,y)}{\partial x \partial y}\right|_{C^{0,1-\alpha}(\bar{\Omega})}.$$

Similarly, we obtain the estimates

$$\sum_{k=1}^{4} |\eta_{2,k}| \leq h^{1-\alpha} \frac{\|k_2\|_{C(\bar{\Omega})}}{\Gamma(2-\alpha)} T^{1-\alpha} (2T+h) \left|\frac{\partial^2 u(x,y)}{\partial x \partial y}\right|_{C^{0,1-\alpha}(\bar{\Omega})}, \quad [3.130]$$

$$\sum_{k=1}^{4} |\eta_{3,k}| \leq h^{1-\alpha} \frac{1}{2} \|k_3\|_{C(\bar{\Omega})} (2T+h)^2 \left|\frac{\partial^2 u(x,y)}{\partial x \partial y}\right|_{C^{0,1-\alpha}(\bar{\Omega})}. \quad [3.131]$$

Inequalities [3.129]–[3.131] yield the estimate

$$\|\psi\|_{T,T} \leq h^{1-\alpha} \left[\frac{\|k_1\|_{C(\bar{\Omega})}}{\Gamma(2-\alpha)} T^{1-\alpha} (2T+h) + \frac{\|k_2\|_{C(\bar{\Omega})}}{\Gamma(2-\alpha)} T^{1-\alpha} (2T+h) + \right. \quad [3.132]$$

$$\left. + \frac{1}{2} \|k_3\|_{C(\bar{\Omega})} (2T+h)^2 \right] \left|\frac{\partial^2 u(x,y)}{\partial x \partial y}\right|_{C^{0,1-\alpha}(\bar{\Omega})}.$$

Taking into account estimates [3.128] and [3.132], we obtain estimate [3.123] from inequality [3.124] and therefore complete the proof of the theorem.

THEOREM 3.24.– *Let the conditions of Theorem 3.23 be fulfilled. Then, the accuracy of the mesh scheme [3.122] is characterized by the following weighted estimates:*

$$\max_{i,j=1,\ldots,N} \left|\frac{z(x_i,y_j)}{x_i y_j}\right| \leq \frac{M}{1-q} h^{1-\alpha} \left|\frac{\partial^2 u}{\partial x \partial y}\right|_{C^{0,1-\alpha}(\bar{\Omega})},$$

$$\max_{j=1,\ldots,N} \left|\frac{z_{\bar{x}}(x_i,y_j)}{y_j}\right| \leq \frac{M}{1-q} h^{1-\alpha} \left|\frac{\partial^2 u}{\partial x \partial y}\right|_{C^{0,1-\alpha}(\bar{\Omega})}, \quad i=1,\ldots,N,$$

$$\max_{i=1,\ldots,N} \left|\frac{z_{\bar{y}}(x_i,y_j)}{x_i}\right| \leq \frac{M}{1-q} h^{1-\alpha} \left|\frac{\partial^2 u}{\partial x \partial y}\right|_{C^{0,1-\alpha}(\bar{\Omega})}, \quad j=1,\ldots,N.$$

The proof follows from estimate [3.123] and the inequalities

$$|z(x_i,y_j)| = \left|\sum_{n=1}^{i}\sum_{l=1}^{j}h^2 z_{\bar{x}\bar{y}}(x_n,y_l)\right| \leq \|z_{\bar{x}\bar{y}}\|_{T,T} \sum_{n=1}^{i}h\sum_{l=1}^{j}h = x_i y_j \|z_{\bar{x}\bar{y}}\|_{T,T},$$

$$|z_{\bar{x}}(x_i,y_j)| = \left|\sum_{l=1}^{j}h z_{\bar{x}\bar{y}}(x_i,y_l)\right| \leq \|z_{\bar{x}\bar{y}}\|_{T,T} \sum_{l=1}^{j}h = y_j \|z_{\bar{x}\bar{y}}\|_{T,T},$$

$$|z_{\bar{y}}(x_i,y_j)| = \left|\sum_{n=1}^{i}h z_{\bar{x}\bar{y}}(x_n,y_j)\right| \leq \|z_{\bar{x}\bar{y}}\|_{T,T} \sum_{n=1}^{i}h = x_i \|z_{\bar{x}\bar{y}}\|_{T,T}.$$

The theorem is proved.

REMARK 3.9.– *The mesh scheme [3.122] is exact for problem [3.106] when the solution is* $u(x,y) = xy$.

REMARK 3.10.– *It seems possible to prove a stronger result than Theorem 3.22 if we use a discrete analogue of Wendroff's inequality (Qin 2016).*

3.5.3. Conclusion

In Theorems 3.19, 3.20 and 3.22, it is proved that for a weak solution $u(x,y)$ of the Goursat problem [3.106] we respectively have $u(x,y) \in C(\bar{\Omega})$, $\dfrac{\partial^2 u(x,y)}{\partial x \partial y} \in C(\bar{\Omega})$, $\dfrac{\partial^2 u(x,y)}{\partial x \partial y} \in C^{0,1-\alpha}(\bar{\Omega})$. Moreover, in Theorem 3.22 the solution is characterized in the norm $C(\bar{\Omega})$ by the weighted estimate with the weight function $\rho^{-1}(x,y)$, $\rho(x,y) = xy$.

The weighted estimates obtained in Theorem 3.24 indicate that in the norm $C(\omega)$ the accuracy of the mesh scheme [3.122] is $O(h^{2-\alpha})$ near the sides $x = 0$ and $y = 0$ of the square $\Omega = (0,T)^2$ and $O(h^{3-\alpha})$ near the square's vertices, whereas it is $O(h^{1-\alpha})$ far from the boundary of the square.

4

The Abstract Cauchy Problem

4.1. The approximation of the operator exponential function in a Hilbert space

Let H be a Hilbert space with the inner product (u,v) and the associated norm $\|u\| = \sqrt{(u,u)}$. In this section, we consider in H the Cauchy problem

$$\frac{dx(t)}{dt} + Ax(t) = 0, \quad t > 0,$$
$$x(0) = x_0,$$
[4.1]

where A is a self-adjoint positive definite operator with a dense domain $D(A) \subset H$.

We will recall now some important results obtained in Gavrilyuk and Makarov (1994) and Arov et al. (1995). It is proved that under the assumption $x_0 \in D(A^\sigma)$, $\sigma > 1$, which actually means finite smoothness of the initial vector x_0, the solution $x(t)$ of problem [4.1] can be represented as a series:

$$x(t) = e^{-tA} x_0 = e^{-\gamma t} \sum_{p=0}^{\infty} (-1)^p L_p^{(0)}(2\gamma t) T_\gamma^p (I + T_\gamma) x_0 ,$$
[4.2]

where $\gamma > 0$ is an arbitrary number, $L_p^{(0)}(t)$ is the Laguerre polynomial (Bateman and Erdélyi 1953b), I is an identity operator, and T_γ is the Cayley transformation of the operator A:

$$T_\gamma = (\gamma I + A)^{-1}(\gamma I - A). \qquad [4.3]$$

Note that series [4.2] converges uniformly in H for $t \geq 0$ with $x_0 \in D(A^\sigma)$, $\sigma > 0$. It implies that for $\sigma > 0$, the sum $x(t)$ of series [4.2] is a continuous function for all $t \geq 0$.

As the approximate solution of problem [4.1], we take the partial sum of series [4.2], i.e.

$$x_N(t) = e^{-\gamma t} \sum_{p=0}^{N} (-1)^p L_p^{(0)}(2\gamma t) T_\gamma^p (I + T_\gamma) x_0. \qquad [4.4]$$

For the accuracy of [4.4], the following estimate holds true:

$$\| x_N(t) - x(t) \| \leq \frac{C}{N^\sigma} \| A^\sigma x_0 \|, \quad t \geq 0, \qquad [4.5]$$

with a postitive constant C independent of N and x_0.

The study of the Cayley transform method [4.4] was further continued in Makarov et al. (2002). The estimate was obtained similar to [4.5] but of the integral type, namely:

$$z_N \equiv \left\{ \int_0^{+\infty} \| x(t) - x_N(t) \|^2 \, dt \right\}^{1/2} \leq \frac{C}{N^{\sigma+1/2}} \| A^\sigma x_0 \|, \qquad [4.6]$$

where $\sigma > 0$ and the constant $C > 0$ is independent of N and x_0.

In addition, it was shown that under the assumptions $\lambda_0 = \gamma \leq 1$, $\sigma > 0$, $\frac{2\gamma}{1+\sigma} \leq 1$, estimate [4.6] is almost (up to logarithm) unimprovable in the order N:

$$z_N^2 \geq \frac{4^{-11/2-\sigma}}{(N+\gamma/2)^{2\sigma+1} \ln^2(2N)}. \qquad [4.7]$$

We will now obtain estimate [4.6] in a different way and find the positive constant C in explicit form (Ryabichev and Mayko 2004).

THEOREM 4.1.– *Let the operator A and the initial vector x_0 in problem [4.1] satisfy the conditions*:

$$A = A^* \geq \lambda_0 I, \ \lambda_0 > 0; \ x_0 \in D(A^\sigma), \ \sigma > 0, \ 0 < \gamma \leq \lambda_0.$$

Then, the accuracy of the approximate solution [4.4] is characterized by the estimate

$$z_N \equiv \left\{ \int_0^{+\infty} \| x(t) - x_N(t) \|^2 \, dt \right\}^{1/2} \leq \frac{C}{N^{\sigma+1/2}} \| A^\sigma x_0 \| \qquad [4.8]$$

with the positive constant $C = \dfrac{(1+\sigma)^{2(1+\sigma)}}{(2\sigma+1)(2\gamma)^{2\sigma+1}}$ *independent of N and x_0 for all*

$$N + 1 \geq \frac{\lambda_0(1+\sigma)}{2\gamma}.$$

PROOF.– It is known from Gavrilyuk and Makarov (1994) and Arov et al. (1995) that the continuous problem [4.1] generates the discrete problem

$$y_{\gamma,n+1} = T_\gamma y_{\gamma,n}, \ n = 0, 1, \ldots, \ y_{\gamma,0} = x_0.$$

The solution of the latter can be written through the discrete semi-group T_γ^n as follows:

$$y_{\gamma,n} = T_\gamma^n y_{\gamma,0} \equiv T_\gamma^n x_0, \ n = 0, 1, \ldots. \qquad [4.9]$$

Next, for the error, we have the series

$$x(t) - x_N(t) = e^{-\gamma t} \sum_{p=N+1}^{\infty} (-1)^p L_p^{(0)}(2\gamma t) \left(y_{\gamma,p} + y_{\gamma,p+1} \right), \ t \geq 0,$$

which leads to the relation

$$\| x(t) - x_N(t) \|^2 = \left(x(t) - x_N(t), x(t) - x_N(t) \right) = \qquad [4.10]$$

$$= \left(e^{-\gamma t} \sum_{p=N+1}^{\infty} (-1)^p L_p^{(0)}(2\gamma t) \left(y_{\gamma,p} + y_{\gamma,p+1} \right), \right.$$

$$\left. e^{-\gamma t} \sum_{p=N+1}^{\infty} (-1)^p L_p^{(0)}(2\gamma t) \left(y_{\gamma,p} + y_{\gamma,p+1} \right) \right) =$$

$$= e^{-2\gamma t} \sum_{j,k=N+1}^{\infty} (-1)^{j+k} L_j^{(0)}(2\gamma t) L_k^{(0)}(2\gamma t) \left(y_{\gamma,j} + y_{\gamma,j+1}, y_{\gamma,k} + y_{\gamma,k+1} \right), \quad t \geq 0.$$

The generalized Laguerre polynomial $L_n^{(\alpha)}(t)$, $\alpha > -1$, satisfy the orthogonality condition with the weight function (Bateman and Erdélyi 1953b):

$$\int_0^{+\infty} e^{-t} t^{\alpha} L_j^{(\alpha)}(t) L_k^{(\alpha)}(t) dt = \frac{\Gamma(\alpha+k+1)}{k!} \delta_{jk}, \quad j,k = 0,1,\ldots, \alpha > -1, \quad [4.11]$$

where δ_{jk} is the Kronecker delta symbol. In particular, for $\alpha = 0$ we have

$$\int_0^{+\infty} e^{-2\gamma t} L_j^{(0)}(2\gamma t) L_k^{(0)}(2\gamma t) dt = \frac{1}{2\gamma} \frac{\Gamma(k+1)}{k!} \delta_{jk} = \frac{1}{2\gamma} \delta_{jk}, \quad [4.12]$$

$$j,k = 0,1,\ldots.$$

Taking this formula into account while integrating relation [4.10], we obtain

$$z_N^2 \equiv \int_0^{+\infty} \| x(t) - x_N(t) \|^2 \, dt =$$

$$= \sum_{j,k=N+1}^{\infty} (-1)^{j+k} \int_0^{+\infty} e^{-2\gamma t} L_j^{(0)}(2\gamma t) L_k^{(0)}(2\gamma t) dt \left(y_{\gamma,j} + y_{\gamma,j+1}, y_{\gamma,k} + y_{\gamma,k+1} \right) =$$

$$= \frac{1}{2\gamma} \sum_{k=N+1}^{\infty} \left(y_{\gamma,k} + y_{\gamma,k+1}, y_{\gamma,k} + y_{\gamma,k+1} \right) = \frac{1}{2\gamma} \sum_{k=N+1}^{\infty} \| y_{\gamma,k} + y_{\gamma,k+1} \|^2. \quad [4.13]$$

Due to equations [4.3] and [4.9], we obtain

$$y_{\gamma,k} + y_{\gamma,k+1} = T_\gamma^k (I + T_\gamma) x_0 =$$

$$= (\gamma I + A)^{-k} (\gamma I - A)^k \left(I + (\gamma I + A)^{-1}(\gamma I - A) \right) x_0 = \quad [4.14]$$

$$= (\gamma I + A)^{-k} (\gamma I - A)^k 2\gamma (\gamma I + A)^{-1} A^{-\sigma} A^\sigma x_0 \quad (\sigma > 0).$$

Denoting by E_λ the corresponding resolution of the identity for the operator A, we have the representation

$$y_{\gamma,k} + y_{\gamma,k+1} = \int_{\lambda_0}^{+\infty} \left(\frac{\gamma - \lambda}{\gamma + \lambda} \right)^k \frac{2\gamma}{\gamma + \lambda} \frac{1}{\lambda^\sigma} dE_\lambda A^\sigma x_0 ,$$

therefore,

$$\| y_{\gamma,k} + y_{\gamma,k+1} \|^2 = \int_{\lambda_0}^{+\infty} \left(\frac{\gamma - \lambda}{\gamma + \lambda} \right)^{2k} \frac{(2\gamma)^2}{(\gamma + \lambda)^2} \frac{1}{\lambda^{2\sigma}} d\| E_\lambda A^\sigma x_0 \|^2 . \quad [4.15]$$

Due to [4.15], formula [4.13] takes the form

$$z_N^2 = 2\gamma \sum_{k=N+1}^{\infty} \int_{\lambda_0}^{+\infty} \left(\frac{\gamma - \lambda}{\gamma + \lambda} \right)^{2k} \frac{1}{(\gamma + \lambda)^2} \frac{1}{\lambda^{2\sigma}} d\| E_\lambda A^\sigma x_0 \|^2 . \quad [4.16]$$

Next, we consider the function

$$\varphi(\lambda) = \frac{1}{\lambda^{2\rho}} \left(\frac{\gamma - \lambda}{\gamma + \lambda} \right)^{2k} \quad (\rho = \sigma + 1).$$

The derivative

$$\varphi'(\lambda) = \frac{-2\rho}{\lambda^{2\rho+1}} \frac{(\lambda - \gamma)^{2k-1}}{(\lambda + \gamma)^{2k+1}} \left(\lambda^2 - \frac{2k\gamma}{\rho} \lambda - \gamma^2 \right)$$

has two positive roots $\lambda_1 = \gamma$ and $\lambda_2 = \frac{\gamma}{\rho}\left(k + \sqrt{k^2 + \rho^2} \right)$. It is obvious that $\lambda_1 < \lambda_2$ for all $k \in \mathbb{N}$. We chose the constant $\gamma > 0$ such that $\gamma \le \lambda_0$. Then, for all $k \in \mathbb{N}$ such that:

$$\lambda_0 < \lambda_2 = \frac{\gamma}{\rho}\left(k + \sqrt{k^2 + \rho^2}\right), \qquad [4.17]$$

we have $\lambda_1 = \gamma \leq \lambda_0 < \lambda_2$. This yields the estimate

$$\max_{\lambda \geq \lambda_0} \varphi(\lambda) = \varphi(\lambda_2) =$$

$$= \frac{\rho^{2\rho}}{\gamma^{2\rho}\left(k + \sqrt{k^2 + \rho^2}\right)^{2\rho}} \left(1 - \frac{2\rho}{k + \sqrt{k^2 + \rho^2} + \rho}\right)^{2k} < \frac{\rho^{2\rho}}{\gamma^{2\rho}(2k)^{2\rho}}.$$

Applying this inequality to [4.16] for all $k \geq \dfrac{\lambda_0 \rho}{2\gamma} = \dfrac{\lambda_0(1+\sigma)}{2\gamma}$, we have

$$z_N^2 \leq 2\gamma \frac{(1+\sigma)^{2(1+\sigma)}}{(2\gamma)^{2(1+\sigma)}} \| A^\sigma x_0 \|^2 \sum_{k=N+1}^{\infty} \frac{1}{k^{2(1+\sigma)}} \leq$$

$$\leq \frac{(1+\sigma)^{2(1+\sigma)}}{(2\gamma)^{2\sigma+1}} \| A^\sigma x_0 \|^2 \int_N^{+\infty} \frac{ds}{s^{2(1+\sigma)}} = \frac{(1+\sigma)^{2(1+\sigma)}}{(2\sigma+1)(2\gamma)^{2\sigma+1}} \frac{1}{N^{2\sigma+1}} \| A^\sigma x_0 \|^2.$$

The theorem is proved.

We will now study whether estimate [4.8] is unimprovable in order N. That is, we will obtain an estimate similar to [4.7] but without the logarithm.

THEOREM 4.2.– *Estimate [4.8] is unimprovable in order N.*

PROOF.– To prove the proposition, it is sufficient to find an operator A and an initial vector x_0 such that the assumptions of Theorem 4.1 are fulfilled and the accuracy of approximation [4.4] is estimated from below in the same way as in [4.8].

Let the operator A have a countable set of eigenvalues

$$0 < \lambda_0 \leq \lambda_1 \leq \ldots \leq \lambda_k \leq \ldots$$

and the corresponding eigenvectors e_k, $k = 0,1,2,\ldots$, forming an orthonormal basis in H: $(e_j, e_k) = \delta_{jk}$, $j,k = 0,1,2,\ldots$, where δ_{jk} is the Kronecker delta symbol.

For any $x_0 \in D(A^\sigma)$, $\sigma > 0$, we have the representation $A^\sigma x_0 = \sum_{i=0}^{\infty} (A^\sigma x_0, e_i) e_i$.

Taking into account [4.3] and [4.14], we obtain the relation

$$y_{\gamma,k} + y_{\gamma,k+1} = \sum_{i=0}^{\infty} \left(\frac{\gamma - \lambda_i}{\gamma + \lambda_i}\right)^k \frac{2\gamma}{\gamma + \lambda_i} \frac{1}{\lambda_i^\sigma} (A^\sigma x_0, e_i) e_i,$$

and therefore,

$$\| y_{\gamma,k} + y_{\gamma,k+1} \|^2 = \sum_{i=0}^{\infty} \left(\frac{\gamma - \lambda_i}{\gamma + \lambda_i}\right)^{2k} \left(\frac{2\gamma}{\gamma + \lambda_i}\right)^2 \frac{1}{\lambda_i^{2\sigma}} (A^\sigma x_0, e_i)^2.$$

Then, formula [4.12] can be written as follows:

$$z_N^2 = 2\gamma \sum_{k=N+1}^{\infty} \sum_{i=0}^{\infty} \left(\frac{\gamma - \lambda_i}{\gamma + \lambda_i}\right)^{2k} \frac{1}{(\gamma + \lambda_i)^2} \frac{1}{\lambda_i^{2\sigma}} (A^\sigma x_0, e_i)^2. \quad [4.18]$$

Denoting $x_0^\sigma = A^\sigma x_0$ and putting $\lambda_i = i$ for all $i \in \mathbb{N}$, we have

$$z_N^2 \geq 2\gamma \sum_{i=1}^{2N+2} \sum_{k=N+1}^{\infty} \left(\frac{i - \gamma}{i + \gamma}\right)^{2k} \frac{1}{(i+\gamma)^2 i^{2\sigma}} (x_0^\sigma, e_i)^2 =$$

$$= \frac{1}{2} \sum_{i=1}^{2N+2} \frac{1}{i^{2\sigma+1}} \left(\frac{i-\gamma}{i+\gamma}\right)^{2N+2} (x_0^\sigma, e_i)^2 \geq$$

$$\geq \frac{1}{2^{2\sigma+2}(N+1)^{2\sigma+1}} \sum_{i=1}^{2N+2} \left(\frac{i-\gamma}{i+\gamma}\right)^{2N+2} (x_0^\sigma, e_i)^2. \quad [4.19]$$

We will now establish that for $\gamma = \dfrac{1}{2N+2}$ and $(x_0^\sigma, e_i)^2 = \dfrac{1}{i^2}$, $i \in \mathbb{N}$, the limit

$$\lim_{N \to \infty} \sum_{i=1}^{2N+2} \frac{1}{i^2} \left(\frac{i - \frac{1}{2N+2}}{i + \frac{1}{2N+2}}\right)^{2N+2} = 0.50...$$

exists and is a positive number. For convenience, we consider the sequence:

$$S_N = \sum_{i=1}^{N} \frac{1}{i^2} \left(\frac{i - \frac{1}{N}}{i + \frac{1}{N}} \right)^N, \quad N \in \mathbb{N},$$

and prove its boundedness and monotonicity. Indeed, since

$$S_N = \sum_{i=1}^{N} \frac{1}{i^2} \left(\frac{i - \frac{1}{N}}{i + \frac{1}{N}} \right)^N < \sum_{i=1}^{\infty} \frac{1}{i^2} = \frac{\pi^2}{6} = 1.644934068\ldots$$

or even more precisely

$$S_N = \sum_{i=1}^{N} \frac{1}{i^2} \left(\frac{i - \frac{1}{N}}{i + \frac{1}{N}} \right)^N < \sum_{i=1}^{\infty} \frac{e^{-2/i}}{i^2} = 0.5012176777\ldots,$$

the sequence (S_N) of positive numbers is bounded from above.

Next, we consider the difference:

$$S_{N+1} - S_N =$$

$$= \sum_{i=1}^{N} \frac{1}{(i+1)^2} \left\{ \left(\frac{i - \frac{1}{N+1}}{i + \frac{1}{N+1}} \right)^{N+1} - \left(\frac{i - \frac{1}{N}}{i + \frac{1}{N}} \right)^N \right\} + \frac{1}{(N+1)^2} \left(\frac{N+1 - \frac{1}{N+1}}{N+1 + \frac{1}{N+1}} \right)^{N+1}$$

and show that it is positive for all $N \in \mathbb{N}$. To this end, we introduce the function

$$g(N) = \left(\frac{i - \frac{1}{N}}{i + \frac{1}{N}} \right)^N, \quad N > 0, \text{ and find the derivative}$$

$$g'(N) = \left(\frac{iN-1}{iN+1}\right)^N \left[\ln\frac{iN-1}{iN+1} + \frac{2iN}{(iN)^2 - 1}\right].$$

To study the sign of the expression in the square brackets, we denote $(iN)^{-1} = x$ and consider the auxiliary function

$$\varphi(x) = \ln\frac{1-x}{1+x} + \frac{2x}{1-x^2}, \quad x \in [0,1).$$

Since $\varphi'(x) = \dfrac{4x^2}{(1-x^2)^2} > 0$ for all $x \in (0,1)$, i.e. $\varphi(x)$ is strictly increasing on the interval $[0,1)$, and $\varphi(0) = 0$, then we have $\varphi(x) > 0$ for all $x \in (0,1)$.

It means that $g'(N) > 0$ for all $N > 0$, i.e. $g(N)$ strictly increases, and therefore $S_{N+1} - S_N > 0$ for all $N \in \mathbb{N}$. This yields that (S_N) is a strictly increasing number sequence.

Combining monotonicity and boundedness of (S_N), we prove its convergence. The limit $\lim\limits_{N \to \infty} S_N$ is obviously positive. Then, there exist $N^* \in \mathbb{N}$ and $\tilde{C} > 0$ such that $S_N \geq \tilde{C}$ for all $N \geq N^*$. Consequently, inequality [4.19] leads to the estimate

$$z_N^2 \geq \frac{C}{(N+1)^{2\sigma+1}} \quad \text{for all } N \geq N^*$$

with the constant $C > 0$ independent of N. The theorem is proved.

Let E be a Banach space. We denote by $\mathfrak{R}(A,\mu,L)$ the set of elements $x \in E$ such that

$$\|x\|_{\mathfrak{R}(A,\mu,L)} = \sup_{0 \leq n < \infty} \frac{\|A^n x\|}{L^n M_n} < \infty,$$

where $0 < L < +\infty$, $\mu = (M_n)_{n=0}^{\infty}$ is a non-decreasing sequence of positive numbers.

It is known from Gorbachuk and Gorbachuk (1984) that $\mathfrak{R}(A,\mu,L)$ is continuously embedded in E.

In Makarov et al. (2002), it is proved that in the case of the discrete spectrum of the operator A and infinite smoothness of the initial vector x_0, i.e. $x_0 \in D(A^n)$ for all $n \in \mathbb{N}$, the Cayley transform method [4.4] is exponentially convergent and its accuracy is characterized by the estimate

$$z_N \leq \frac{\exp\left(-(N+1)\ln\left(1+\sqrt{2}\right)\right)}{\left(\gamma\sqrt{2}(2+\sqrt{2})\right)^{1/2}} \| x_0 \|_{\Re(A,(1),\lambda_0)}, \qquad [4.20]$$

which is unimprovable in order N.

As stated in Makarov et al. (2002), an estimate similar to [4.20] can be obtained without assuming the discreteness of the spectrum of the operator A. We will now prove it in the following proposition.

THEOREM 4.3.– *Let the operator A and the initial vector x_0 in problem [4.1] satisfy the conditions*

$$A = A^* \geq \lambda_0 I, \ \lambda_0 > 0; \ x_0 \in \Re(A,(1),\lambda_0), \ \gamma = \frac{\lambda_0}{1+\sqrt{2}}.$$

Then, the accuracy of the approximate solution [4.4] is characterized by the estimate

$$z_N \leq \left(\frac{1+\sqrt{2}}{2\gamma}\right)^{1/2} \exp\left(-(N+2)\ln(1+\sqrt{2})\right) \| x_0 \|_{\Re(A,(1),\lambda_0)}, \ N = 0,1,\ldots, \quad [4.21]$$

which is unimprovable in order N.

PROOF.– For all $n \in \mathbb{N}$, formula [4.14] takes the form

$$y_{\gamma,k} + y_{\gamma,k+1} = (\gamma I + A)^{-k}(\gamma I - A)^k 2\gamma(\gamma I + A)^{-1} A^{-n} A^n x_0 =$$

$$= \int_{\lambda_0}^{+\infty} \left(\frac{\gamma-\lambda}{\gamma+\lambda}\right)^k \frac{2\gamma}{\gamma+\lambda} \frac{\lambda_0^n}{\lambda^n} dE_\lambda \frac{A^n x_0}{\lambda_0^n},$$

therefore,

$$\|y_{\gamma,k}+y_{\gamma,k+1}\|^2 = \int_{\lambda_0}^{+\infty}\left(\frac{\gamma-\lambda}{\gamma+\lambda}\right)^{2k}\frac{(2\gamma)^2}{(\gamma+\lambda)^2}\frac{\lambda_0^{2n}}{\lambda^{2n}}d\left\|E_\lambda\frac{A^n x_0}{\lambda_0^n}\right\|^2.$$

In particular, for $n = k$, we have the relation

$$\|y_{\gamma,k}+y_{\gamma,k+1}\|^2 = \int_{\lambda_0}^{+\infty}\left(\frac{\gamma-\lambda}{\gamma+\lambda}\right)^{2k}\frac{(2\gamma)^2}{(\gamma+\lambda)^2}\left(\frac{\lambda_0}{\lambda}\right)^{2k}d\left\|E_\lambda\frac{A^k x_0}{\lambda_0^k}\right\|^2,$$

and formula [4.13] turns into the following:

$$z_N^2 = 2\gamma\sum_{k=N+1}^{\infty}\int_{\lambda_0}^{+\infty}\left(\frac{\gamma-\lambda}{\gamma+\lambda}\right)^{2k}\frac{1}{(\gamma+\lambda)^2}\left(\frac{\lambda_0}{\lambda}\right)^{2k}d\left\|E_\lambda\frac{A^k x_0}{\lambda_0^k}\right\|^2. \qquad [4.22]$$

Next, we consider the function $\varphi(\lambda) = \dfrac{1}{(\lambda+\gamma)^{k+1}}\left(\dfrac{\lambda-\gamma}{\lambda}\right)^k$ and find the derivative

$$\varphi'(\lambda) = \frac{-(\lambda-\gamma)^{k-1}}{(\lambda+\gamma)^{k+2}\lambda^{k+1}}\left((k+1)\lambda^2 - \gamma(2k+1)\lambda - k\gamma^2\right).$$

For all $\gamma > 0$ and $k \in \mathbb{N}$, the roots of the latter satisfy the inequality

$$\lambda_1 = \gamma < \lambda_2 = \frac{\gamma(2k+1)+\gamma 2\sqrt{2}\sqrt{(k+1/2)^2-1/8}}{2(k+1)} < \frac{\gamma(2k+1)(1+\sqrt{2})}{2k+2} < \gamma(1+\sqrt{2}).$$

If $\gamma = \dfrac{\lambda_0}{1+\sqrt{2}}$, then the function $\varphi(\lambda)$ strictly decreases on the ray $[\lambda_0,+\infty)$, and relation [4.22] finally gives the estimate

$$z_N^2 \leq 2\gamma\sum_{k=N+1}^{\infty}\int_{\lambda_0}^{+\infty}\left(\frac{\gamma-\lambda_0}{\gamma+\lambda_0}\right)^{2k}\frac{1}{(\gamma+\lambda_0)^2}\left(\frac{\lambda_0}{\lambda_0}\right)^{2k}d\left\|E_\lambda\frac{A^k x_0}{\lambda_0^k}\right\|^2 \leq$$

$$\leq \frac{1}{\gamma}\|x_0\|^2_{\Re(A,(1),\lambda_0)}\sum_{k=N+1}^{\infty}\left(\frac{1}{1+\sqrt{2}}\right)^{2(k+1)} =$$

$$= \frac{1+\sqrt{2}}{2\gamma}(1+\sqrt{2})^{-2(N+2)} \|x_0\|^2_{\Re(A,(1),\lambda_0)}.$$

As shown in Makarov et al. (2002), for the operator A with the discrete spectrum $\lambda_0 = \gamma(1+\sqrt{2}) \leq \lambda_1 \leq \ldots$ and for the initial vector $x_0 \in \Re(A,(1),\lambda_0)$, the following estimate holds true:

$$z_N \geq \frac{|(x_0, e_0)|}{\gamma^{1/2}} \exp(-(N+2)\ln(1+\sqrt{2})), \quad N = 0, 1, \ldots, \qquad [4.23]$$

where e_0 is a normalized eigenvector corresponding to the eigenvalue λ_0.

Inequality [4.23] indicates that estimate [4.21] is unimprovable in order N. This completes the proof of the theorem.

Now we will obtain another integral estimate (similar to [4.8] but with a weight function) and then study its unimprovability in order. First, we prove the following proposition.

THEOREM 4.4.– *Let the operator A and the initial vector x_0 in problem [4.1] satisfy the conditions*

$$A = A^* \geq \lambda_0 I, \ \lambda_0 > 0; \ x_0 \in D(A^\sigma), \ \sigma > 0, \ 0 < \gamma \leq \lambda_0.$$

Then, the accuracy of the approximate solution [4.4] is characterized by the estimate

$$z_N \equiv \left\{ \int_0^{+\infty} t^{-1} \|x(t) - x_N(t)\|^2 \, dt \right\}^{1/2} \leq \frac{C}{N^\sigma} \|A^\sigma x_0\| \qquad [4.24]$$

for all $N \geq \max\{\lambda_0\sigma/(2\gamma), \sigma\} - 1$ with the constant $C = \dfrac{\sigma^{2\sigma-1}}{e^\sigma 2^{2\sigma+1} \gamma^{2\sigma}}$ independent of N and x_0.

PROOF.– Making use of the well-known properties of the Laguerre polynomials, we transform the solution $x(t)$ as follows:

$$x(t) = e^{-\gamma t}\sum_{n=0}^{\infty}(-1)^n L_n^{(0)}(2\gamma t)\left(y_{\gamma,n}+y_{\gamma,n+1}\right)=$$

$$= e^{-\gamma t}y_{\gamma,0}+e^{-\gamma t}\sum_{n=1}^{\infty}(-1)^{n+1}\frac{2\gamma t}{n}L_{n-1}^{(1)}(2\gamma t)y_{\gamma,n}$$

with T_γ and $y_{\gamma,n}$ defined in [4.3] and [4.9] respectively. Then, we have

$$\|x(t)-x_N(t)\|^2 =$$

$$= e^{-2\gamma t}(2\gamma t)^2 \sum_{i,j=N+1}^{\infty}\frac{(-1)^{i+j}}{ij}L_{i-1}^{(1)}(2\gamma t)L_{j-1}^{(1)}(2\gamma t)\left(y_{\gamma,i},y_{\gamma,j}\right).$$

Taking into account the orthogonality condition [4.11], we obtain

$$z_N^2 = \int_0^\infty \frac{1}{t}\|x(t)-x_N(t)\|^2\,dt = \sum_{n=N+1}^{\infty}\frac{1}{n}\|y_{\gamma,n}\|^2 =$$

$$= \sum_{n=N+1}^{\infty}\frac{1}{n}\int_{\lambda_0}^{\infty}\left(\frac{\gamma-\lambda}{\gamma+\lambda}\right)^{2n}\frac{1}{\lambda^{2\sigma}}d\|E_\lambda A^\sigma x_0\|^2.$$

[4.25]

Now, we consider the function $f(\lambda)=\left(\dfrac{\lambda-\gamma}{\lambda+\gamma}\right)^{2n}\dfrac{1}{\lambda^{2\sigma}}$. Choosing the parameter γ such that $\gamma \le \lambda_0$ and making some calculations, we come to the inequality:

$$\max_{\lambda\ge\lambda_0} f(\lambda) = f\left(\frac{\gamma(n+\sqrt{n^2+\sigma^2})}{\sigma}\right) \le$$

$$\le \left(\frac{n+\sqrt{n^2+\sigma^2}-\sigma}{n+\sqrt{n^2+\sigma^2}+\sigma}\right)^{2n}\frac{\sigma^{2\sigma}}{\gamma^{2\sigma}\left(n+\sqrt{n^2+\sigma^2}\right)^{2\sigma}} \le \frac{\sigma^{2\sigma}}{e^\sigma \gamma^{2\sigma}(2n)^{2\sigma}}$$

for all $n \ge \max\{\lambda_0\sigma/(2\gamma),\sigma\}$.

Then, representation [4.25] implies the estimate

$$z_N^2 \le \frac{\sigma^{2\sigma}}{e^\sigma \gamma^{2\sigma} 2^{2\sigma}} \| A^\sigma x_0 \|^2 \sum_{n=N+1}^{\infty} \frac{1}{n^{2\sigma+1}} \le$$

$$\le \frac{\sigma^{2\sigma}}{e^\sigma \gamma^{2\sigma} 2^{2\sigma}} \| A^\sigma x_0 \|^2 \int_N^{+\infty} \frac{ds}{s^{2\sigma+1}} = \frac{\sigma^{2\sigma-1}}{e^\sigma \gamma^{2\sigma} 2^{2\sigma+1} N^{2\sigma}} \| A^\sigma x_0 \|^2$$

for all $N+1 \ge \max\{\lambda_0 \sigma/(2\gamma), \sigma\}$. This completes the proof of the theorem.

Next, we address the question of the unimprovability of estimate [4.24].

THEOREM 4.5.– *Estimate [4.24] is unimprovable in order N.*

PROOF.– To prove the proposition, we will show that there is an operator A and an initial vector x_0 such that the accuracy of the Cayley transform method [4.4] is estimated from below in order N in the same way as in [4.24].

Let the operator A have a countable set of eigenvalues

$$0 < \lambda_0 \le \lambda_1 \le \ldots \le \lambda_k \le \ldots$$

with the corresponding eigenvectors e_k, $k = 0,1,2,\ldots$, forming an orthonormal basis in H: $(e_j, e_k) = \delta_{jk}$, $j,k = 0,1,2,\ldots$, where δ_{jk} is the Kronecker delta symbol. Then

$$y_{\gamma,n} = T_\gamma^n x_0 = \sum_{i=0}^{\infty} \left(\frac{\gamma - \lambda_i}{\gamma + \lambda_i}\right)^n \frac{1}{\lambda_i^\sigma} (A^\sigma x_0, e_i) e_i,$$

and the error squared takes the form

$$z_N^2 = \sum_{n=N+1}^{\infty} \frac{1}{n} \| y_{\gamma,n} \|^2 = \sum_{n=N+1}^{\infty} \frac{1}{n} \sum_{i=0}^{\infty} \left(\frac{\gamma - \lambda_i}{\gamma + \lambda_i}\right)^{2n} \frac{1}{\lambda_i^{2\sigma}} (A^\sigma x_0, e_i)^2, \qquad [4.26]$$

$$N = 0,1,\ldots \quad (\gamma > 0).$$

Let $\lambda_i = i$, $i = 1,2,\ldots$, $0 < \lambda_0 \le 1$, and let $x_0 \in D(A^\sigma)$ satisfy the condition $(A^\sigma x_0, e_i)^2 = \frac{1}{i^2}$, $i = 2,3,\ldots$. Then, for $N > \gamma$, we have

$$z_N^2 \geq \sum_{n=N+1}^{2N} \frac{1}{n} \sum_{i=N}^{\infty} \left(\frac{i-\gamma}{i+\gamma}\right)^{2n} \frac{1}{i^{2+2\sigma}} \geq$$

$$\geq \sum_{n=N+1}^{2N} \frac{1}{n} \left(\frac{N-\gamma}{N+\gamma}\right)^{2n} \sum_{i=N}^{\infty} \frac{1}{i^{2+2\sigma}} \geq \left(\frac{N-\gamma}{N+\gamma}\right)^{4N} \frac{N}{2N} \int_{N}^{+\infty} \frac{dx}{x^{2+2\sigma}} =$$

$$= \left(\frac{N-\gamma}{N+\gamma}\right)^{4N} \frac{1}{2(1+2\sigma)} \frac{1}{N^{1+2\sigma}}.$$

Making use of the inequality $\ln(1+r) \geq \frac{r}{r+1}$ $(r > -1)$ with $r = -\frac{2\gamma}{N+\gamma}$, where $N > \gamma$, we come to the inequality

$$\left(\frac{N-\gamma}{N+\gamma}\right)^{4N} = e^{4N \ln\left(1 - \frac{2\gamma}{N+\gamma}\right)} \geq e^{-\frac{8\gamma N}{N-\gamma}} > e^{-\frac{8\gamma}{1-\gamma}} \quad (N \geq 1).$$

This finally gives the estimate

$$z_N^2 \geq \frac{e^{-\frac{8\gamma}{1-\gamma}}}{2(1+2\sigma)} \frac{1}{N^{1+2\sigma}}.$$

The theorem is proved.

We continue to study the unimprovability of estimate [4.24] in the following statement.

THEOREM 4.6.– *Let $(c_N, N \in \mathbb{N})$ be an unbounded sequence of positive numbers. Then, for some self-adjoint positive definite operator A and some initial vector $x_0 \in D(A^\sigma)$, $\sigma > 0$, there is no constant $C > 0$ such that*

$$z_N^2 \leq \frac{C}{N^{2\sigma} c_N}.$$

PROOF.– Let the vectors $e_n, n \in \mathbb{N}$, form an orthogonal basis in H. We define the operator A acting in H as follows:

$$D(A) = \left\{ x = \sum_{n=1}^{\infty} x_n e_n \in H : \sum_{n=1}^{\infty} |x_n|^2 n^2 < \infty \right\},$$

$$Ax = \sum_{n=1}^{\infty} x_n n e_n \quad \text{for all } x \in D(A).$$

It is clear that $\overline{D(A)} = H$ and A is an unbounded linear operator on $D(A)$. In addition, A is a symmetric operator with the range $R(A) = H$. It implies that A is a self-adjoint operator.

Assume that the subsequence $(c_{n_k}, k \in \mathbb{N})$ of the sequence $(c_N, N \in \mathbb{N})$ satisfies the condition

$$c_{n_k} \geq k^3, \quad n_1 < n_2 < \cdots < n_k < \cdots. \qquad [4.27]$$

We choose:

$$x_0 = \sum_{k=1}^{\infty} \frac{1}{k n_k^{\sigma}} e_{n_k} \in D(A^{\sigma}) = \left\{ x = \sum_{n=1}^{\infty} x_n e_n \in H : \sum_{n=1}^{\infty} |x_n|^2 n^{2\sigma} < \infty \right\},$$

which gives $A^{\sigma} x_0 = \sum_{k=1}^{\infty} \frac{1}{k} e_{n_k}$. Then, we have

$$z_{n_k}^2 = \int_0^{+\infty} t^{-1} \| x(t) - x_{n_k}(t) \|^2 \, dt = \qquad [4.28]$$

$$= \sum_{j=n_k+1}^{\infty} \sum_{p=1}^{\infty} \frac{1}{j} \left(\frac{n_p - \gamma}{n_p + \gamma} \right)^{2j} \frac{1}{n_p^{2\sigma}} |(A^{\sigma} x_0, e_{n_p})|^2 \geq \sum_{j=n_k+1}^{\infty} \frac{1}{j} \left(\frac{n_k - \gamma}{n_k + \gamma} \right)^{2j} \frac{1}{n_k^{2\sigma}} \frac{1}{k^2} \geq$$

$$\geq \sum_{j=n_k+1}^{2n_k} \frac{1}{j} \left(\frac{n_k - \gamma}{n_k + \gamma} \right)^{2j} \frac{1}{n_k^{2\sigma}} \frac{1}{k^2} \geq n_k \frac{1}{2n_k} \left(\frac{n_k - \gamma}{n_k + \gamma} \right)^{4n_k} \frac{1}{n_k^{2\sigma}} \frac{1}{k^2} =$$

$$= \frac{1}{2} \left(\frac{2\gamma - \gamma}{2\gamma + \gamma} \right)^{8\gamma} \frac{1}{n_k^{2\sigma}} \frac{1}{k^2} \geq \frac{1}{2} 3^{-8\gamma} \frac{1}{n_k^{2\sigma} k^2} \quad \text{for all } n_k \geq 2\gamma.$$

We will prove the theorem by contradiction. Suppose that, on the contrary, there is a positive constant C and a number $N' \in \mathbb{N}$ such that for $n_k \geq N'$

$$z_{n_k}^2 \leq \frac{C}{n_k^{2\sigma} c_{n_k}}. \qquad [4.29]$$

Estimates [4.28] and [4.29] yield the inequality $\dfrac{c_{n_k}}{k^2} \leq 3^{8\gamma} 2c$ for all $n_k \geq N'$, which contradicts condition [4.27] and therefore completes the proof of the theorem.

We will now address the case of the vector $x_0 \in \Re(A,(1),\lambda_0)$.

THEOREM 4.7.– *Let the operator A in problem [4.1] have a discrete spectrum*

$$0 < \gamma(1+\sqrt{2}) = \lambda_0 \leq \lambda_1 \leq \lambda_2 \leq \ldots$$

with the corresponding eigenvectors e_i, $i = 0,1,\ldots$, forming the orthonormal basis in H, and let $x_0 \in \Re(A,(1),\lambda_0)$. Then, the accuracy of the approximate solution [4.4] is characterized by the estimate

$$z_N^2 \equiv \int_0^{+\infty} t^{-1} \|x(t) - x_N(t)\|^2 \, dt \leq \qquad [4.30]$$

$$\leq \frac{1+\sqrt{2}}{2} e^{-(2N+2)\ln(1+\sqrt{2}) - \ln(N+1)} \| x \|^2_{\Re(A,(1),\lambda_0)} \quad \textit{for all } N = 0,1,\ldots,$$

which is unimprovable in order N.

PROOF.– Due to relation [4.26], we have

$$z_N^2 = \sum_{n=N+1}^{\infty} \frac{1}{n} \sum_{i=0}^{\infty} \left(\frac{\gamma - \lambda_i}{\gamma + \lambda_i} \right)^{2n} \left(\frac{\lambda_0}{\lambda_i} \right)^{2n} \left[\frac{(A^n x_0, e_i)}{\lambda_0^n} \right]^2. \qquad [4.31]$$

Since the function $\varphi(\lambda) = \dfrac{\lambda - \gamma}{\lambda(\lambda + \gamma)}$ is strictly decreasing for $\lambda \in [\gamma(1+\sqrt{2}), +\infty)$, then [4.31] gives us the upper inequality

$$z_N^2 \leq \sum_{n=N+1}^{\infty} \frac{1}{n}\left(\frac{1}{1+\sqrt{2}}\right)^{2n} \|x\|_{\Re(A,(1),\lambda_0)}^2 \leq$$
$$\leq \frac{(1+\sqrt{2})^{-2N-2}}{N+1} \frac{1+\sqrt{2}}{2} \|x\|_{\Re(A,(1),\lambda_0)}^2 .$$
[4.32]

On the contrary, formula [4.31] leads to the lower estimate

$$z_{t,N}^2 \geq \frac{1}{N+1}\left(\frac{\gamma-\lambda_0}{\gamma+\lambda_0}\right)^{2(N+1)}\left[\frac{(A^{N+1}x_0,e_0)}{\lambda_0^{N+1}}\right]^2 =$$
$$= \frac{(1+\sqrt{2})^{-2N-2}}{N+1}(x_0,e_0)^2 .$$
[4.33]

Combining [4.32] and [4.33], we come to estimate [4.30] and establish its unimprovability in order N. The theorem is proved.

4.2. Inverse theorems for the operator sine and cosine functions

This section is dedicated to some complementary results. In particular, we will now prove the inverse approximation theorem for problem [4.1].

THEOREM 4.8.– *Let for some operator $A = A^* \geq \lambda_0 I$, $\lambda_0 > 0$, some constant $\sigma > 0$ and some non-decreasing sequence of positive numbers $(c_N, N \in \mathbb{N})$ with the convergent series $\sum_{k=0}^{\infty} \frac{1}{c_{2^k}}$, the following inequality holds true:*

$$z_N^2 \equiv \int_0^{+\infty} t^{-1} \|x(t) - x_N(t)\|^2 dt \leq \frac{1}{N^{2\sigma} c_N} .$$
[4.34]

Then, $x_0 \in D(A^\sigma)$.

PROOF.– Making use of representation [4.25] and estimate [4.34], we have

$$z_N^2 \cdot N^{2\sigma} = \int_{\lambda_0}^{+\infty} N^{2\sigma} \sum_{j=N+1}^{\infty} \frac{1}{j}\left(\frac{\lambda-\gamma}{\lambda+\gamma}\right)^{2j} d(E_\lambda x_0, x_0) \leq \frac{1}{c_N} .$$
[4.35]

Next, we introduce the function

$$\psi_N(\lambda) = N^{2\sigma} \sum_{j=N+1}^{\infty} \frac{1}{j}\left(\frac{\lambda-\gamma}{\lambda+\gamma}\right)^{2j}$$

and consider a non-decreasing sequence $(\varphi_n, n \in \mathbb{N})$ of the functions

$$\varphi_n(\lambda) = \max\{\psi_1(\lambda), \psi_2(\lambda), \ldots, \psi_n(\lambda)\}, \quad \lambda \in [\lambda_0, +\infty).$$

For each $k \in \{1, 2, \ldots, n\}$, we can find $r \in \mathbb{N}_0$ such that $2^r \le k < 2^{r+1}$. Then

$$\frac{\psi_k(\lambda)}{\psi_{2^r}(\lambda)} = \frac{k^{2\sigma} \sum_{j=k+1}^{\infty} \frac{1}{j}\left(\frac{\lambda-\gamma}{\lambda+\gamma}\right)^{2j}}{(2^r)^{2\sigma} \sum_{j=2^r+1}^{\infty} \frac{1}{j}\left(\frac{\lambda-\gamma}{\lambda+\gamma}\right)^{2j}} < \left(\frac{k}{2^r}\right)^{2\sigma} < 2^{2\sigma},$$

and therefore, $\varphi_n(\lambda) < 2^{2\sigma} \sum_{r=0}^{\lfloor \log_2 n \rfloor} \psi_{2^r}(\lambda)$, where $\lfloor x \rfloor$ is the floor function. Taking into account [4.35], we then have for all $n \in \mathbb{N}$

$$\int_{\lambda_0}^{+\infty} \varphi_n(\lambda) d(E_\lambda x_0, x_0) < 2^{2\sigma} \sum_{r=0}^{\lfloor \log_2 n \rfloor} \int_{\lambda_0}^{+\infty} \psi_{2^r}(\lambda) d(E_\lambda x_0, x_0) < 2^{2\sigma} \sum_{r=0}^{\lfloor \log_2 n \rfloor} \frac{1}{c_{2^r}} < M,$$

where M is a positive constant.

Applying here Beppo Levi's lemma, we obtain the same inequality for the function $\varphi(\lambda) = \lim_{n\to\infty} \varphi_n(\lambda)$, namely:

$$\int_{\lambda_0}^{+\infty} \varphi(\lambda) d(E_\lambda x_0, x_0) \le M.$$

We will now show that $\varphi(\lambda) \ge C\lambda^{2\sigma}$ for $\lambda \ge \lambda' = \max\{\lambda_0, 2, \gamma\}$ with some positive constant C. This will prove the convergence of the integral

$$\int_{\lambda_0}^{+\infty} \lambda^{2\sigma} d(E_\lambda x_0, x_0) < \infty,$$

and therefore provide $x_0 \in D(A^\sigma)$.

To this end, for each $\lambda \geq \lambda' = \max\{\lambda_0, 2, \gamma\}$, we find

$$\psi_{\lfloor \lambda \rfloor}(\lambda) = \lfloor \lambda \rfloor^{2\sigma} \sum_{j=\lfloor \lambda \rfloor+1}^{\infty} \frac{1}{j}\left(\frac{\lambda-\gamma}{\lambda+\gamma}\right)^{2j} > \lfloor \lambda \rfloor^{2\sigma} \sum_{j=\lfloor \lambda \rfloor+1}^{2\lfloor \lambda \rfloor} \frac{1}{j}\left(\frac{\lambda-\gamma}{\lambda+\gamma}\right)^{2j} >$$

$$> \lfloor \lambda \rfloor^{2\sigma} \lfloor \lambda \rfloor \frac{1}{2\lfloor \lambda \rfloor}\left(\frac{\lambda-\gamma}{\lambda+\gamma}\right)^{4\lfloor \lambda \rfloor} \geq (\lambda-1)^{2\sigma} \frac{1}{2}\left(\frac{\lambda-\gamma}{\lambda+\gamma}\right)^{4\lambda} \geq$$

$$\geq (\lambda-1)^{2\sigma} \frac{1}{2}\left(\frac{2\gamma-\gamma}{2\gamma+\gamma}\right)^{8\lambda} \geq \frac{\lambda^{2\sigma}}{2^{2\sigma+1}} 3^{-8\gamma},$$

which finally gives

$$\varphi(\lambda) \geq \varphi_{\lfloor \lambda \rfloor}(\lambda) \geq \psi_{\lfloor \lambda \rfloor}(\lambda) \geq \frac{1}{2^{2\sigma+1}3^{8\gamma}} \lambda^{2\sigma}$$

and completes the proof of the theorem.

We now address the initial value problem for the following second-order differential equation in a Hilbert space:

$$\frac{d^2 x(t)}{dt^2} + Ax(t) = 0, \quad t \in (0, T],$$
$$x(0) = x_0, \quad x'(0) = 0,$$
[4.36]

where $A = A^* \geq \lambda_0 I$, $\lambda_0 > 0$, is a self-adjoint positive definite operator with the dense domain $D(A) \subset H$.

It is constructed and proved in Gavrilyuk and Makarov (1999) that for $x_0 \in D(A^\sigma)$, $\sigma > 5/4$, the solution $x(t)$ of problem [4.36] can be represented by the series

$$x(t) = e^{-\delta t} \sum_{n=0}^{\infty} \left(L_n^{(0)}(t) - L_{n-1}^{(0)}(t) \right) u_n, \quad L_{-1}^{(0)}(t) \equiv 0, \qquad [4.37]$$

where $\delta < 1/2$ is an arbitrary number, $L_n^{(0)}(t)$ are the Laguerre polynomials, and the sequence $(u_n, n = 0, 1, \ldots)$ satisfies the recurrence relation

$$\left(A + (\delta-1)^2 I\right) u_{n+1} = 2\left(A + \delta(\delta-1)I\right) u_n - \left(A + \delta^2 I\right) u_{n-1}, \quad n = 1, 2, \ldots,$$

$$u_0 = x_0, \quad u_1 = \left(A + \delta(\delta-1)I\right)\left(A + (\delta-1)^2 I\right)^{-1} x_0.$$

The partial sum of series [4.37] is taken as the approximation of the exact solution $x(t)$, namely:

$$x_N(t) = e^{-\delta t} \sum_{n=0}^{N} \left(L_n^{(0)}(t) - L_{n-1}^{(0)}(t) \right) u_n. \qquad [4.38]$$

The accuracy of that approximation is characterized in Makarov and Ryabichev (2002) by the estimate of the integral type

$$z_N^2 \equiv \int_0^{+\infty} t^{-1} e^{-(1-2\delta)t} \| x(t) - x_N(t) \|^2 dt \leq \frac{2^{4\sigma-1} \sigma^{2\sigma-1}}{(1-2\delta)^{2\sigma}} \| A^\sigma x_0 \|^2 N^{-2\sigma}$$

with $x_0 \in D(A^\sigma)$, $\sigma > 0$, $0 \leq \delta < 1/2$ and for all integer numbers $N \geq N'(\lambda_0, \delta, \sigma)$. Furthermore, it is shown that this estimate in unimprovable (up to logarithm) in order N.

We will now prove the inverse proposition.

THEOREM 4.9.– *Let for some operator $A = A^* \geq \lambda_0 I$, $\lambda_0 > 0$, some constant $\sigma > 0$ and some non-decreasing sequence of positive numbers $(c_N, N \in \mathbb{N})$ with the convergent series $\sum_{k=0}^{\infty} \frac{1}{c_{2^k}}$, the following inequality holds true:*

$$z_N^2 \equiv \int_0^{+\infty} t^{-1} e^{-(1-2\delta)t} \| x(t) - x_N(t) \|^2 dt \leq \frac{1}{N^{2\sigma} c_N}. \qquad [4.39]$$

Then, $x_0 \in D(A^{\sigma/2})$.

PROOF.– Using assumption [4.39], we obtain the estimate

$$z_N^2 \cdot N^{2\sigma} = \sum_{j=N+1}^{+\infty} \frac{1}{j} \int_{\lambda_0}^{+\infty} N^{2\sigma} \chi_1^j(\lambda) T_j^2(\chi_2(\lambda)) d(E_\lambda x_0, x_0) \le \frac{1}{c_N}, \qquad [4.40]$$

where E_λ is the corresponding resolution of the identity for the operator A, $T_n(x) = \cos(n \arccos x)$, $n = 0, 1, \ldots$, are the Chebyshev polynomials of the first kind

$$\chi_1(\lambda) = \frac{\lambda + \delta^2}{\lambda + (\delta-1)^2}, \quad \chi_2(\lambda) = \frac{\lambda + \delta(\delta-1)}{\sqrt{(\lambda+\delta^2)(\lambda+(\delta-1)^2)}}.$$

Next, we introduce the notation:

$$\psi_N(\lambda) = N^{2\sigma} \sum_{j=N+1}^{\infty} \frac{1}{j} \chi_1^j(\lambda) T_j^2(\chi_2(\lambda))$$

and consider the non-decreasing sequence $(\varphi_n, n \in \mathbb{N})$ of functions

$$\varphi_n(\lambda) = \max\{\psi_1(\lambda), \psi_2(\lambda), \ldots, \psi_n(\lambda)\}, \quad \lambda \in [\lambda_0, +\infty).$$

For each $k \in \{1, 2, \ldots, n\}$, we find $r \in \mathbb{N}_0$ such that $2^r \le k < 2^{r+1}$. Then

$$\frac{\psi_k(\lambda)}{\psi_{2^r}(\lambda)} = \frac{k^{2\sigma} \sum_{j=k+1}^{\infty} \frac{1}{j} \chi_1^j(\lambda) T_j^2(\chi_2(\lambda))}{(2^r)^{2\sigma} \sum_{j=2^r+1}^{\infty} \frac{1}{j} \chi_1^j(\lambda) T_j^2(\chi_2(\lambda))} < \left(\frac{k}{2^r}\right)^{2\sigma} < 2^{2\sigma},$$

and therefore $\varphi_n(\lambda) < 2^{2\sigma} \sum_{r=0}^{\lfloor \log_2 n \rfloor} \psi_{2^r}(\lambda)$.

Taking into account inequality [4.40], we have for all $n \in \mathbb{N}$

$$\int_{\lambda_0}^{+\infty} \varphi_n(\lambda) d(E_\lambda x_0, x_0) < 2^{2\sigma} \sum_{r=0}^{\lfloor \log_2 n \rfloor} \int_{\lambda_0}^{+\infty} \psi_{2^r}(\lambda) d(E_\lambda x_0, x_0) < 2^{2\sigma} \sum_{r=0}^{\lfloor \log_2 n \rfloor} \frac{1}{c_{2^r}} < M,$$

where M is a positive constant.

By Beppo Levi's lemma, we then obtain for the function $\varphi(\lambda) = \lim\limits_{n \to \infty} \varphi_n(\lambda)$, the estimate

$$\int_{\lambda_0}^{+\infty} \varphi(\lambda) d(E_\lambda x_0, x_0) \leq M.$$

We will show now that $\varphi(\lambda) \geq C\lambda^\sigma$ for $\lambda \geq \lambda'$, $\lambda' \geq \lambda_0$, with some number λ' and some positive constant C. Since this inequality implies the convergence of the integral

$$\int_{\lambda_0}^{+\infty} \lambda^\sigma d(E_\lambda x_0, x_0) < \infty,$$

then it means $x_0 \in D(A^{\sigma/2})$.

First, for each fixed $\lambda \in [\lambda_0, +\infty)$, we find N such that

$$\cos\frac{\pi}{6N} \leq \frac{\lambda + \delta(\delta-1)}{\sqrt{(\lambda + \delta(\delta-1))^2 + \lambda}}.$$

Using the inequalities

$$\cos\frac{\pi}{6N} = 1 - 2\sin^2\frac{\pi}{12N} < 1 - 2\left(\frac{2}{\pi}\frac{\pi}{12N}\right)^2,$$

$$\frac{\lambda + \delta(\delta-1)}{\sqrt{(\lambda + \delta(\delta-1))^2 + \lambda}} = \left(1 + \frac{\lambda}{(\lambda + \delta(\delta-1))^2}\right)^{-1/2} > 1 - \frac{\lambda}{2(\lambda + \delta(\delta-1))^2},$$

we obtain $N^2 < \dfrac{(\lambda+\delta(\delta-1))^2}{9\lambda}$. For all such N, we have

$$T_j\left(\frac{\lambda+\delta(\delta-1)}{\sqrt{(\lambda+\delta(\delta-1))^2+\lambda}}\right) \geq T_j\left(\cos\frac{\pi}{6N}\right) \geq T_{2N}\left(\cos\frac{\pi}{6N}\right) = \frac{1}{2}, \quad j=0,1,\ldots,2N.$$

Next, we take $n_0 = \left[\dfrac{\lambda+\delta(\delta-1)}{3\sqrt{\lambda}}\right]$, then

$$\psi_{n_0}(\lambda) \geq n_0^{2\sigma} \sum_{j=n_0+1}^{2n_0} \frac{1}{j}\left(\frac{\lambda+\delta^2}{\lambda+(\delta-1)^2}\right)^j T_j^2\left(\frac{\lambda+\delta(\delta-1)}{\sqrt{(\lambda+\delta^2)(\lambda+(\delta-1)^2)}}\right) \geq$$

$$\geq n_0^{2\sigma} n_0 \frac{1}{2n_0}\left(\frac{\lambda+\delta^2}{\lambda+(\delta-1)^2}\right)^{2n_0} \frac{1}{4} \geq$$

$$\geq \left(\frac{\lambda+\delta(\delta-1)}{3\sqrt{\lambda}}-1\right)^{2\sigma} \frac{1}{8}\left(\frac{\lambda+\delta^2}{\lambda+(\delta-1)^2}\right)^{2\frac{\lambda+\delta(\delta-1)}{3\sqrt{\lambda}}} \geq C\lambda^{\sigma},$$

and therefore $\varphi(\lambda) \geq \varphi_{n_0}(\lambda) \geq \psi_{n_0}(\lambda) \geq C\lambda^{\sigma}$. This proves the theorem.

Note (Torba 2007) that the assumptions of the theorem are fulfilled if we take the sequence $c_N = \ln^{1+\varepsilon} N$, $\varepsilon > 0$.

4.3. The approximation of the operator exponential function in a Banach space

We consider now the initial value problem for the homogeneous first-order differential equation in a Banach space E:

$$\frac{dx(t)}{dt} + Ax(t) = 0, \quad t > 0,$$

$$x(0) = x_0,$$

[4.41]

where A is a closed (generally speaking, unbounded) linear operator with the dense domain $D(A) \subset E$ and the resolvent set $\rho(A)$.

A solution of problem [4.41], by definition, is a continuous on $[0,+\infty)$ and continuously differentiable on $(0,+\infty)$ E-valued function that satisfies the equation and the initial condition in [4.41].

We recall (Gavrilyuk and Makarov 1996) that a closed linear operator A is called *strongly positive* if there exist $\varphi \in (0, \pi/2)$, $\gamma > 0$ and $L > 0$ such that

$$\Sigma \equiv \{z \in \mathbb{C} : \varphi \le \arg z \le 2\pi - \varphi\} \cup \{z \in \mathbb{C} : |z| \le \gamma\} \subset \rho(A),$$

$$\|(zI - A)^{-1}\| \le \frac{L}{1+|z|} \quad \forall z \in \Sigma.$$

The constructive representation for the solution $x(t)$ of problem [4.41] is proposed and studied in Gavrilyuk and Makarov (1996, 2004). That is, it is proved that for a strongly positive operator A and the initial vector $x_0 \in D(A^\sigma)$, $\sigma > 1$, the solution of problem [4.41] exists, is unique and can be represented by the series

$$x(t) = e^{-\gamma t} \sum_{k=0}^{\infty} (-1)^k L_k^{(0)}(2\gamma t)(y_{\gamma,k} + y_{\gamma,k+1}) =$$

$$= e^{-\gamma t} \sum_{k=0}^{\infty} (-1)^k L_k^{(0)}(2\gamma t) T_\gamma^k (I + T_\gamma) x_0,$$

[4.42]

where $\gamma > 0$ is an arbitrary number, $L_k^{(0)}(t)$ are the Laguerre polynomials, I is the identity operator, and T_γ is the Cayley transformation of the operator A:

$$T_\gamma = (\gamma I - A)(\gamma I + A)^{-1}.$$

[4.43]

In the case $0 < \sigma \le 1$, the sum $x(t)$ of series [4.42] can be considered a generalized solution of problem [4.41].

As the approximation of the exact solution $x(t)$, it is proposed to take the partial sum of series [4.42]:

$$x_N(t) = e^{-\gamma t} \sum_{k=0}^{N} (-1)^k L_k^{(0)}(2\gamma t) T_\gamma^k (I + T_\gamma) x_0.$$

[4.44]

The accuracy of the approximate solution [4.44] is characterized by the estimate

$$\sup_{t\in[0,+\infty)} \| x(t) - x_N(t) \| \leq \frac{C}{N^{\sigma-\delta}} \| A^\sigma x_0 \|, \qquad [4.45]$$

where C is a positive constant independent of N and x_0, $\delta \in (0,\sigma)$ is an arbitrary number, and $x_0 \in D(A^\sigma)$, $\sigma > 0$.

Below we will obtain the representation and the estimate similar to formulas [4.42] and [4.45] respectively in the case of a logarithmically sectorial operator A and give an explicit form of the constant C in the estimate of type [4.45].

Recall that a closed linear operator A with a dense domain $D(A)$ in a Banach space E is called a *logarithmically sectorial operator* if there exists $\varphi \in (0, \pi/2)$, $\delta > 0$ and $L > 0$ such that

$$\Sigma \equiv \{z \in \mathbb{C}: \varphi \leq \arg z \leq 2\pi - \varphi\} \cup \{z \in \mathbb{C}: |z| \leq \delta\} \subset \rho(A),$$

$$\| (zI - A)^{-1} \| \leq L \frac{\ln(1+|z|)}{|z|} \quad \forall z \in \Sigma \setminus \{0\}. \qquad [4.46]$$

The class of such operators is discussed in Gorodnii et al. (2004) where, in addition, an example of a logarithmically sectorial but not a sectorial operator is given.

Formally differentiating series [4.42] and taking into account the relation (Bateman and Erdélyi 1953b)

$$\frac{d}{dt}\left(L_k^{(0)}(t) - L_{k-1}^{(0)}(t)\right) = -L_{k-1}^{(0)}(t),$$

we obtain the series

$$x^{(1)}(t) = \gamma e^{-\gamma t} \sum_{k=0}^{\infty} (-1)^k L_k^{(0)}(2\gamma t)\left(y_{\gamma,k+1} - y_{\gamma,k}\right), \quad L_{-1}^{(0)} \equiv 0, \qquad [4.47]$$

which will be needed further together with series [4.42].

We start with the following auxiliary proposition.

LEMMA 4.1.– Let A in [4.41] be a sectorial operator. Then:

1) if $\sigma > 0$, $0 < \gamma \leq \delta$, then series [4.42] converges uniformly in $t \in [0, +\infty)$, and therefore represents a continuous for $t \in [0, +\infty)$ function $x(t)$;

2) if $\sigma > 1$, $0 < \gamma \leq \delta$, then series [4.47] converges uniformly in $t \in [0, +\infty)$, and therefore represents a continuous for $t \in [0, +\infty)$ function $x^{(1)}(t)$, which, in addition, satisfies the relation $x^{(1)}(t) = x'(t)$ for all $t \in [0, +\infty)$;

3) if $\sigma > 1$, $0 < \gamma \leq \delta$, then $x(t) \in D(A)$ for all $t \in [0, +\infty)$.

PROOF.– We use the notation $x_0^\sigma = A^\sigma x_0$ and the integration curve

$$\Gamma = \Gamma_1 \cup \Gamma_2 \cup \Gamma_3,$$

$$\Gamma_1 = \{z \in \mathbb{C}: z = -\rho e^{i\varphi}, -\infty < \rho \leq -\gamma\}, \quad \Gamma_2 = \{z \in \mathbb{C}: z = \rho e^{-i\varphi}, \gamma \leq \rho < +\infty\},$$

$$\Gamma_3 = \{z \in \mathbb{C}: z = \gamma e^{-i\theta}, -\varphi \leq \theta \leq \varphi\}.$$

Now, we prove Proposition 1). Based on formula [4.43], we have

$$y_{\gamma,k} + y_{\gamma,k+1} = T_\gamma^k (I + T_\gamma) A^{-\sigma} x_0^\sigma =$$

$$= \frac{1}{2\pi i} \int_\Gamma \left(\frac{\gamma - z}{\gamma + z} \right)^k \left(1 + \frac{\gamma - z}{\gamma + z} \right) z^{-\sigma} (zI - A)^{-1} x_0^\sigma dz =$$

$$= \frac{\gamma}{\pi i} \int_\Gamma \left(\frac{\gamma - z}{\gamma + z} \right)^k \frac{1}{(\gamma + z) z^\sigma} (zI - A)^{-1} x_0^\sigma dz.$$

Then, taking [4.46] into account, we obtain the inequality

$$\| y_{\gamma,k} + y_{\gamma,k+1} \| \leq \frac{M\gamma}{\pi} \int_\Gamma \left| \frac{\gamma - z}{\gamma + z} \right|^k \frac{1}{|\gamma + z|} \frac{\ln(1 + |z|)}{|z|^{\sigma+1}} |dz| \, \| x_0^\sigma \|,$$

which, given the parametrization for $\Gamma_i, i = 1, 2, 3$, can be written as follows:

$$\| y_{\gamma,k} + y_{\gamma,k+1} \| \leq \frac{L\gamma}{\pi} \int_\Gamma \left| \frac{\gamma - z}{\gamma + z} \right|^k \frac{1}{|\gamma + z|} \frac{\ln(1 + |z|)}{|z|^{\sigma+1}} |dz| \, \| x_0^\sigma \| =$$

$$= \frac{L\gamma}{\pi} \left\{ \int_{-\infty}^{-\gamma} \left| \frac{\gamma + \rho e^{i\varphi}}{\gamma - \rho e^{i\varphi}} \right|^k \frac{1}{|\gamma - \rho e^{i\varphi}|} \frac{\ln(1-\rho)}{(-\rho)^{\sigma+1}} d\rho + \int_{\gamma}^{+\infty} \left| \frac{\gamma - \rho e^{i\varphi}}{\gamma + \rho e^{i\varphi}} \right|^k \frac{1}{|\gamma + \rho e^{i\varphi}|} \frac{\ln(1+\rho)}{\rho^{\sigma+1}} d\rho + \right.$$

$$\left. + \int_{-\varphi}^{\varphi} \left| \frac{\gamma - \gamma e^{-i\theta}}{\gamma + \gamma e^{-i\theta}} \right|^k \frac{1}{|\gamma + \gamma e^{-i\theta}|} \frac{\ln(1+\gamma)}{\gamma^{\sigma+1}} \gamma d\theta \right\} \| x_0^{\sigma} \| = \qquad [4.48]$$

$$= \frac{2L\gamma}{\pi} \left\{ \int_{\gamma}^{+\infty} \left| \frac{\gamma - \rho e^{i\varphi}}{\gamma + \rho e^{i\varphi}} \right|^k \frac{1}{|\gamma + \rho e^{i\varphi}|} \frac{\ln(1+\rho)}{\rho^{\sigma+1}} d\rho + \right.$$

$$\left. + \frac{\ln(1+\gamma)}{2\gamma^{\sigma+1}} \int_{0}^{\varphi} \operatorname{tg}^k \frac{\theta}{2} \cos^{-1} \frac{\theta}{2} d\theta \right\} \| x_0^{\sigma} \| .$$

Now, we simplify the first integral using the relations

$$\left| \frac{\gamma - \rho e^{i\varphi}}{\gamma + \rho e^{i\varphi}} \right|^2 = \frac{\gamma^2 + \rho^2 - 2\gamma\rho\cos\varphi}{\gamma^2 + \rho^2 + 2\gamma\rho\cos\varphi} =$$

$$= \frac{\rho - \gamma\cos\varphi}{\rho + \gamma\cos\varphi} \cdot \frac{(\rho(\gamma^2 + \rho^2) - 2\gamma^2\rho\cos\varphi) - \gamma\cos\varphi(\rho^2 - \gamma^2)}{(\rho(\gamma^2 + \rho^2) - 2\gamma^2\rho\cos\varphi) + \gamma\cos\varphi(\rho^2 - \gamma^2)} \leq \frac{\rho - \gamma\cos\varphi}{\rho + \gamma\cos\varphi},$$

$$\frac{1}{|\gamma + \rho e^{i\varphi}|} = \frac{1}{\sqrt{\gamma^2 + \rho^2 + 2\gamma\rho\cos\varphi}} \leq \frac{1}{\sqrt{\gamma^2 + \rho^2}} \leq \frac{1}{\rho}$$

$$(\rho \geq \gamma, \ 0 < \varphi < \pi/2),$$

and we estimate the second integral using the formula

$$\int_{0}^{\varphi} \cos^{-1} \frac{\theta}{2} d\theta = 2 \ln \left| \operatorname{tg}\left(\frac{\varphi}{4} + \frac{\pi}{4} \right) \right|.$$

Then, inequality [4.48] takes the form

$$\| y_{\gamma,k} + y_{\gamma,k+1} \| \leq$$

$$\leq \frac{2L\gamma}{\pi}\left\{\int_{\gamma}^{+\infty}\left|\frac{\rho-\gamma\cos\varphi}{\rho+\gamma\cos\varphi}\right|^{k/2}\frac{\ln(1+\rho)}{\rho^{\sigma+2}}d\rho+\frac{\ln(1+\gamma)}{\gamma^{\sigma+1}}\ln\left|\mathrm{tg}\left(\frac{\varphi}{4}+\frac{\pi}{4}\right)\right|\mathrm{tg}^{k}\frac{\varphi}{2}\right\}\|x_{0}^{\sigma}\|\leq$$

$$\leq \frac{2L\gamma}{\pi}\left\{\max_{\rho\in[\gamma,+\infty)}\left[\left(\frac{\rho-\gamma\cos\varphi}{\rho+\gamma\cos\varphi}\right)^{k/2}\rho^{-(\sigma+1-\varepsilon)}\right]\int_{\gamma}^{+\infty}\frac{\ln(1+\rho)}{\rho^{1+\varepsilon}}d\rho+\right.\quad[4.49]$$

$$\left.+\frac{1}{k^{1+\sigma-\varepsilon}}\frac{\ln(1+\gamma)}{\gamma^{\sigma+1}}\ln\left|\mathrm{tg}\left(\frac{\varphi}{4}+\frac{\pi}{4}\right)\right|\cdot\max_{k\geq 0}\left[k^{1+\sigma-\varepsilon}\mathrm{ctg}^{-k}\frac{\varphi}{2}\right]\right\}\|x_{0}^{\sigma}\|$$

for an arbitrary number $\varepsilon \in (0, \sigma)$.

Next, we consider separately each of the two terms in the curly brackets.

For the function

$$f(\rho) = \left(\frac{\rho-\gamma\cos\varphi}{\rho+\gamma\cos\varphi}\right)^{s}\rho^{-\alpha},\ \rho\geq\gamma,\ \alpha>0,\ s>0,$$

we find the derivative

$$f'(\rho) = -\frac{(\rho-\gamma\cos\varphi)^{s-1}}{(\rho+\gamma\cos\varphi)^{s+1}}(\alpha\rho^{2}-2\gamma s\cos\varphi\rho-\alpha\gamma^{2}\cos^{2}\varphi).$$

For the positive root of the derivative, we have

$$\rho^{*} = \gamma\cos\varphi\frac{s+\sqrt{s^{2}+\alpha^{2}}}{\alpha} > \gamma\cos\varphi.$$

We choose:

$$s \geq \frac{\alpha}{2\cos\varphi},\quad\quad\quad [4.50]$$

so that the inequality $\rho^{*} > \gamma$ holds true. Then, we have

$$\max_{\rho\in[\gamma,+\infty)} f(\rho) = f(\rho^*) = \left(\frac{s+\sqrt{s^2+\alpha^2}-\alpha}{s+\sqrt{s^2+\alpha^2}+\alpha}\right)^s \left(\frac{\alpha}{\gamma\cos\varphi}\right)^\alpha \left(s+\sqrt{s^2+\alpha^2}\right)^{-\alpha} =$$

$$= \exp\left\{s\ln\left(1-\frac{2\alpha}{s+\sqrt{s^2+\alpha^2}+\alpha}\right)\right\} \left(\frac{\alpha}{\gamma\cos\varphi}\right)^\alpha \left(s+\sqrt{s^2+\alpha^2}\right)^{-\alpha}.$$

Since $\ln(1-x) < -x$ for $0 \neq x < 1$, then due to assumption [4.50], we have

$$s\ln\left(1-\frac{2\alpha}{s+\sqrt{s^2+\alpha^2}+\alpha}\right) < \frac{-2\alpha s}{s+\sqrt{s^2+\alpha^2}+\alpha} =$$

$$= \frac{-2\alpha}{1+\sqrt{1+\left(\frac{\alpha}{s}\right)^2}+\frac{\alpha}{s}} \leq \frac{-2\alpha}{1+\sqrt{1+4\cos^2\varphi}+2\cos\varphi}.$$

Thus, for each natural $s \geq \dfrac{\alpha}{2\cos\varphi}$, we obtain the inequality

$$\max_{\rho\in[\gamma,+\infty)} f(\rho) < \exp\left\{\frac{-2\alpha}{1+\sqrt{1+4\cos^2\varphi}+2\cos\varphi}\right\} \left(\frac{\alpha}{\gamma\cos\varphi}\right)^\alpha \frac{1}{(2s)^\alpha},$$

which for $s = k/2$, $\alpha = 1+\sigma-\varepsilon$ and all integer $k \geq \dfrac{1+\sigma-\varepsilon}{\cos\varphi}$ leads to the estimate

$$\max_{\rho\in[\gamma,+\infty)} \left(\frac{\rho-\gamma\cos\varphi}{\rho+\gamma\cos\varphi}\right)^{k/2} \rho^{-(\sigma+1-\varepsilon)} <$$

$$< \frac{1}{k^{1+\sigma-\varepsilon}} \left(\frac{1+\sigma-\varepsilon}{\gamma\cos\varphi}\right)^{1+\sigma-\varepsilon} \exp\left\{\frac{-2(1+\sigma-\varepsilon)}{1+\sqrt{1+4\cos^2\varphi}+2\cos\varphi}\right\}. \qquad [4.51]$$

The function $\varphi(x) = x^\beta a^{-x}$, $x \geq 0$ $(a > 1, \beta > 0)$, takes its maximum at the point $x = \beta/\ln a$, therefore

$$\max_{x\geq 0} \varphi(x) = \varphi(\beta/\ln a) = (\beta/\ln a)^\beta a^{-\beta/\ln a}.$$

For $\beta = 1+\sigma-\varepsilon$, $a = \operatorname{ctg}\frac{\varphi}{2}$, we obtain from here

$$\max_{k \geq 0}\left(k^{1+\sigma-\varepsilon}\operatorname{ctg}^{-k}\frac{\varphi}{2}\right) = \left(\frac{1+\sigma-\varepsilon}{\ln\operatorname{ctg}\frac{\varphi}{2}}\right)^{1+\sigma-\varepsilon}\left(\operatorname{tg}\frac{\varphi}{2}\right)^{(1+\sigma-\varepsilon)\ln^{-1}\operatorname{ctg}\frac{\varphi}{2}}. \qquad [4.52]$$

Due to [4.51] and [4.52], estimate [4.49] can be written in the form

$$\| y_{\gamma,k} + y_{\gamma,k+1} \| \leq \frac{c_1}{k^{1+\sigma-\varepsilon}} \| x_0^\sigma \| \qquad [4.53]$$

with the constant $c_1 = c_1(\varphi,\gamma,L,\sigma,\varepsilon) = \frac{2L\gamma}{\pi}\max\{c_2,c_3\}$, where

$$c_2 = \exp\left\{\frac{-2(1+\sigma-\varepsilon)}{1+\sqrt{1+4\cos^2\varphi}+2\cos\varphi}\right\}\left(\frac{1+\sigma-\varepsilon}{\gamma\cos\varphi}\right)^{1+\sigma-\varepsilon}\int_\gamma^{+\infty}\frac{\ln(1+\rho)}{\rho^{1+\varepsilon}}d\rho,$$

$$c_3 = \frac{\ln(1+\gamma)}{\gamma^{\sigma+1}}\ln\left|\operatorname{tg}\left(\frac{\varphi}{4}+\frac{\pi}{4}\right)\right|\left(\frac{1+\sigma-\varepsilon}{\ln\operatorname{ctg}\frac{\varphi}{2}}\right)^{1+\sigma-\varepsilon}\left(\operatorname{tg}\frac{\varphi}{2}\right)^{(1+\sigma-\varepsilon)\ln^{-1}\operatorname{ctg}\frac{\varphi}{2}}$$

for all $k \geq \frac{1+\sigma-\varepsilon}{\cos\varphi}$ and an arbitrary number $\varepsilon \in (0,\sigma)$, $\sigma > 0$.

Estimate [4.53] together with the inequality from Bateman and Erdélyi (1953b):

$$e^{-t/2}|L_n^{(0)}(t)| \leq 1 \quad (t \geq 0) \qquad [4.54]$$

indicates that series [4.42] is majorized by the convergent number series $\sum_{k=k_1}^{\infty}\frac{c_1\|x_0^\sigma\|}{k^{1+\sigma-\delta}}$ and therefore converges in E uniformly in $t \in [0,+\infty)$. Therefore, the sum $x(t)$ is continuous for all $t \in [0,+\infty)$.

Now we move on to Proposition 2). By reasoning, as just before, we obtain the representation

$$y_{\gamma,k+1} - y_{\gamma,k} = T_\gamma^k (I - T_\gamma) A^{-\sigma} x_0^\sigma =$$

$$= \frac{1}{2\pi i} \int_\Gamma \left(\frac{\gamma - z}{\gamma + z}\right)^k \left(\frac{\gamma - z}{\gamma + z} - 1\right) z^{-\sigma} (zI - A)^{-1} x_0^\sigma \, dz =$$

$$= -\frac{1}{\pi i} \int_\Gamma \left(\frac{\gamma - z}{\gamma + z}\right)^k \frac{1}{(\gamma + z) z^{\sigma - 1}} (zI - A)^{-1} x_0^\sigma \, dz.$$

Then

$$\| y_{\gamma,k+1} - y_{\gamma,k} \| \leq \frac{L}{\pi} \int_{\Gamma_1 \cup \Gamma_2 \cup \Gamma_3} \left|\frac{\gamma - z}{\gamma + z}\right|^k \frac{1}{|\gamma + z|} \frac{\ln(1 + |z|)}{|z|^\sigma} |dz| \, \| x_0^\sigma \| \leq$$

$$\leq \frac{2L}{\pi} \left\{ \max_{\rho \in [\gamma, +\infty)} \left[\left(\frac{\rho - \gamma \cos \varphi}{\rho + \gamma \cos \varphi}\right)^{k/2} \rho^{-(\sigma - \delta)} \right] \int_\gamma^{+\infty} \frac{\ln(1 + \rho)}{\rho^{1+\delta}} d\rho + \right. \quad [4.55]$$

$$\left. + \frac{1}{k^{\sigma - \varepsilon}} \frac{\ln(1 + \gamma)}{\gamma^\sigma} \ln\left|\text{tg}\left(\frac{\varphi}{4} + \frac{\pi}{4}\right)\right| \max_{k \geq 0} \left[k^{\sigma - \varepsilon} \text{ctg}^{-k} \frac{\varphi}{2} \right] \right\} \| x_0^\sigma \| \leq \frac{c_4}{k^{\sigma - \varepsilon}} \| x_0^\sigma \|$$

with the constant $c_4 = c_4(\varphi, \gamma, L, \sigma, \varepsilon) = \dfrac{2L}{\pi} \max\{c_5, c_6\}$, where

$$c_5 = \exp\left\{\frac{-2(\sigma - \varepsilon)}{1 + \sqrt{1 + 4\cos^2 \varphi} + 2\cos \varphi}\right\} \left(\frac{\sigma - \varepsilon}{\gamma \cos \varphi}\right)^{\sigma - \varepsilon} \int_\gamma^{+\infty} \frac{\ln(1 + \rho)}{\rho^{1 + \varepsilon}} d\rho,$$

$$c_6 = \frac{\ln(1 + \gamma)}{\gamma^\sigma} \ln\left|\text{tg}\left(\frac{\varphi}{4} + \frac{\pi}{4}\right)\right| \left(\frac{\sigma - \varepsilon}{\ln \text{ctg} \dfrac{\varphi}{2}}\right)^{\sigma - \varepsilon} \left(\text{tg} \frac{\varphi}{2}\right)^{(\sigma - \varepsilon) \ln^{-1} \text{ctg} \frac{\varphi}{2}}$$

for all $k \geq \dfrac{\sigma - \varepsilon}{\cos \varphi}$ and an arbitrary number $\varepsilon \in (0, \sigma)$, $\sigma > 1$.

As a consequence of [4.55] and [4.54], series [4.47] converges in E uniformly in $t \in [0,+\infty)$ and therefore its sum $x^{(1)}(t)$ is continuous for all $t \in [0,+\infty)$ and satisfies the relation $x^{(1)}(t) = x'(t)$.

Finally, we prove Proposition 3). Given the relation

$$AT_\gamma^k (I+T_\gamma) x_0 = \frac{\gamma}{\pi i} \int_\Gamma \left(\frac{\gamma-z}{\gamma+z}\right)^k \frac{1}{(\gamma+z)z^{\sigma-1}} (zI-A)^{-1} x_0^\sigma dz,$$

we have for $k \geq \dfrac{\sigma - \varepsilon}{\cos \varphi}$ and an arbitrary number $\varepsilon \in (0, \sigma)$, $\sigma > 1$, that

$$\| AT_\gamma^k (I+T_\gamma) x_0 \| \leq$$

$$\leq \frac{L\gamma}{\pi} \int_{\Gamma_1 \cup \Gamma_2 \cup \Gamma_3} \left|\frac{\gamma-z}{\gamma+z}\right|^k \frac{1}{|\gamma+z|} \frac{\ln(1+|z|)}{|z|^\sigma} |dz| \, \| x_0^\sigma \| \leq \frac{c_6}{k^{\sigma-\delta}} \| x_0^\sigma \|, \quad [4.56]$$

where $c_7 = c_7(\varphi, \gamma, L, \sigma, \varepsilon) = \gamma c_4(\varphi, \gamma, L, \sigma, \varepsilon)$.

Estimates [4.56] and [4.54] imply a uniform in $t \in [0,+\infty)$ convergence in E of the series

$$e^{-\gamma t} \sum_{k=0}^{\infty} (-1)^k L_k^{(0)}(2\gamma t) AT_\gamma^k (I+T_\gamma) x_0 =$$

$$= \lim_{N \to \infty} A \left(e^{-\gamma t} \sum_{k=0}^{N} (-1)^k L_k^{(0)}(2\gamma t) T_\gamma^k (I+T_\gamma) x_0 \right).$$
[4.57]

Since the operator A is closed, formula [4.57] indicates that $x(t) \in D(A)$ for all $t \geq 0$. The lemma is proved.

Lemma 4.1 easily leads to the following result.

THEOREM 4.10.– *Let the operator A in problem [4.41] be logarithmically sectorial and let $x_0 \in D(A^\sigma)$, $\sigma > 1$. Then, problem [4.41] has a unique solution, and for this solution, representation [4.42] holds true.*

PROOF.– In view of Lemma 4.1, it remains only to verify that the series [4.42] is a solution of problem [4.41]. Indeed, due to [4.47] and [4.57], we have for each $t > 0$

$$\frac{dx(t)}{dt} + Ax(t) =$$

$$= \gamma e^{-\gamma t} \sum_{k=0}^{\infty} (-1)^k L_k^{(0)}(2\gamma t) T_\gamma^k \left(T_\gamma - I\right) x_0 + e^{-\gamma t} \sum_{k=0}^{\infty} (-1)^k L_k^{(0)}(2\gamma t) A T_\gamma^k \left(I + T_\gamma\right) x_0 =$$

$$= e^{-\gamma t} \sum_{k=0}^{\infty} (-1)^k L_k^{(0)}(2\gamma t) \times$$

$$\times \frac{1}{2\pi i} \int_\Gamma \left(\frac{\gamma-z}{\gamma+z}\right)^k \left(\gamma \frac{\gamma-z}{\gamma+z} - \gamma + z + z \frac{\gamma-z}{\gamma+z}\right) \frac{1}{z^\sigma} (zI - A)^{-1} x_0^\sigma dz = 0,$$

and therefore $x(t)$ satisfies the equation in [4.41].

The initial condition in [4.41] is also fulfilled, namely:

$$\lim_{t \to +0} x(t) = x(0) = \sum_{k=0}^{\infty} (-1)^k (y_{\gamma,k} + y_{\gamma,k+1}) =$$

$$= \lim_{N \to +\infty} \sum_{k=0}^{N} (-1)^k (y_{\gamma,k} + y_{\gamma,k+1}) = \lim_{N \to +\infty} \left(x_0 + (-1)^N y_{\gamma,N+1}\right) = x_0,$$

due to the estimate and the limit

$$\| y_{\gamma,k} \| = \left\| \frac{1}{2\pi i} \int_\Gamma \left(\frac{\gamma-z}{\gamma+z}\right)^k \frac{1}{z^\sigma} (zI - A)^{-1} x_0^\sigma dz \right\| \le$$

$$\le \frac{L}{2\pi} \int_\Gamma \left|\frac{\gamma-z}{\gamma+z}\right|^k \frac{\ln(1+|z|)}{|z|^{\sigma+1}} |dz| \, \| x_0^\sigma \| \le \frac{c_8}{k^{\sigma-\delta}} \| x_0^\sigma \| \to 0 \text{ as } k \to +\infty,$$

where $c_8 = c_8(\varphi, \gamma, L, \sigma, \varepsilon)$. The theorem is proved.

Finally, we will obtain an estimate similar to [4.45].

THEOREM 4.11.– *Let the operator A in problem [4.41] be logarithmically sectorial, and let* $x_0 \in D(A^\sigma)$, $\sigma > 1$. *Then, the accuracy of the approximate solution [4.44] is characterized by the estimate*

$$\sup_{t \in [0,+\infty)} \| x(t) - x_N(t) \| \leq \frac{c}{N^{\sigma-\varepsilon}} \| A^\sigma x_0 \|, \qquad [4.58]$$

with the positive constant $c = \dfrac{c_1(\varphi, \gamma, M, \sigma, \varepsilon)}{\sigma - \varepsilon}$ *defined in [4.53]*, $N + 1 \geq \dfrac{1 + \sigma - \varepsilon}{\cos \varphi}$, $\varepsilon \in (0, \sigma)$.

PROOF.– Based on [4.42], [4.44], [4.53] and [4.54], we have for all $t \in [0, +\infty)$

$$\| x(t) - x_N(t) \| = \left\| e^{-\gamma t} \sum_{k=N+1}^{\infty} (-1)^k L_k^{(0)}(2\gamma t) T_\gamma^k (I + T_\gamma) x_0 \right\| \leq$$

$$\leq c_1(\varphi, \gamma, L, \sigma, \varepsilon) \| x_0^\sigma \| \sum_{k=N+1}^{\infty} \frac{1}{k^{1+\sigma-\varepsilon}} \leq c_1(\varphi, \gamma, L, \sigma, \varepsilon) \| x_0^\sigma \| \int_N^{+\infty} \frac{ds}{s^{1+\sigma-\varepsilon}} =$$

$$= \frac{c_1(\varphi, \gamma, M, \sigma, \varepsilon)}{\sigma - \varepsilon} N^{-(\sigma-\delta)} \| x_0^\sigma \|,$$

which proves the theorem.

4.4. Conclusion

We will now summarize the results obtained above.

Estimate [4.8] obtained in Theorem 4.1 and estimate [4.24] obtained in Theorem 4.4 indicate that the rate of convergence of the Cayley transform method [4.4] for the approximate solution of the abstract Cauchy problem in a Hilbert space automatically depends on the smoothness of the initial vector x_0 (that is on the powers of the parameter σ), and therefore this is a method without saturation of accuracy in the sense of Babenko (2002). Unimprovability (in order N) of estimate [4.8] is proved in Theorem 4.2 and of estimate [4.24] in Theorems 4.5 and 4.6.

In the case of infinite smoothness of the initial vector x_0, estimate [4.21] obtained in Theorem 4.3 and estimate [4.30] obtained in Theorem 4.7 show the exponential rate of convergence of the Cayley transformation method [4.4] and are unimproved in order N.

In Theorems 4.8 and 4.9, we studied the smoothness of the initial vector x_0 in terms of the accuracy order of the Cayley transform method for the operator exponential and operator cosine functions respectively.

In Theorem 4.10, we proposed and justified the constructive representation (in the form of a series) of the solution of the Cauchy problem for a homogeneous equation in a Banach space with a logarithmically sectorial operator. In Theorem 4.11, we obtained the uniform estimate [4.58] which demonstrates that the accuracy order of the approximate solution [4.44] is automatically dependent on the parameter σ (i.e. on smoothness of the initial vector x_0).

5

The Cayley Transform Method for Abstract Differential Equations

5.1. Exact and approximate solutions of the BVP in a Hilbert space

Let H be a Hilbert space with the inner product (u,v) and the associated norm $\|u\| = \sqrt{(u,u)}$. In this section, we consider in H the Dirichlet boundary value problem for a second-order inhomogeneous differential equation with an unbounded operator coefficient, namely:

$$\frac{d^2 u(x)}{dx^2} - Au(x) = -f(x), \quad x \in (0,1), \qquad [5.1]$$
$$u(0) = 0, \quad u(1) = 0,$$

where $u : [0,1] \to H$ is an unknown solution, $f : (0,1) \to H$ is a given function, A is a self-adjoint positive definite operator with the dense domain $D(A) \subset H$ and the spectral set $\Sigma(A) \subset [\lambda_0, +\infty)$, $\lambda_0 > 0$.

To illustrate problem [5.1], we consider the boundary value problem for the two-dimensional Poisson equation

$$\frac{\partial^2 u}{\partial x^2} + \frac{\partial^2 u}{\partial y^2} = -f(x,y), \quad (x,y) \in \Omega = (0,1)^2, \qquad [5.2]$$
$$u(x,y) = 0, \quad (x,y) \in \partial\Omega,$$

with a given function $f(x, y)$. In a Hilbert space $H = L_2(0,1)$ with the inner product and associated norm

$$(v, w) = \int_0^1 v(x)w(x)\,dx, \quad \|v\| = \left(\int_0^1 v^2(x)dx\right)^{1/2}, \quad v, w \in L_2(0,1),$$

we define the operator

$$Au = -\frac{d^2u}{dy^2}, \quad D(A) = H^2(0,1) \cap \overset{\circ}{H}{}^1(0,1).$$

Then, problem [5.2] can be written in the abstract setting [5.1].

To solve problem [5.1] numerically, we represent the solution $u(x)$ as an infinite series in powers of the linear fractional transformation of the operator A with coefficients depending only on the variable x. To that end, below we use the operator Green's function $G(x, \xi; A)$, Meixner-type polynomials $v_k(x)$ and the Fourier series representation of the right-hand side $f(x)$. We start with some useful preliminaries.

5.1.1. *Auxiliary results*

Consider the following two BVPs for second-order ODEs:

$$u''(x) - au(x) = -f(x), \quad x \in (0,1),$$
$$u(0) = 0, \quad u(1) = 0, \qquad [5.3]$$

and

$$w''(x) - aw(x) = 0, \quad x \in (0,1),$$
$$w(0) = 0, \quad w(1) = w_1, \qquad [5.4]$$

with $a = \text{const} > 0$. The solution $u(x)$ of problem [5.3] can be presented in the form

$$u(x) = \int_0^1 G(x, \xi) f(\xi)\,d\xi, \quad x \in [0,1],$$

where

$$G(x,\xi) = (\sqrt{a}\,\text{sh}\sqrt{a})^{-1}\begin{cases}\text{sh}(\sqrt{a}x)\text{sh}(\sqrt{a}(1-\xi)), & x \le \xi, \\ \text{sh}(\sqrt{a}\xi)\text{sh}(\sqrt{a}(1-x)), & \xi \le x,\end{cases}$$

is the Green's function of the differential operator

$$Lu = -u''(x) + au(x) \quad (a > 0), \quad D(L) = H^2(0,1) \cap \overset{\circ}{H}{}^1(0,1).$$

To solve problem [5.4], we first investigate the BVP

$$v''(x) - \frac{z^2}{1-z^2}v(x) = 0, \quad x \in (0,1), \qquad [5.5]$$
$$v(0) = 0, \quad v(1) = v_1,$$

with a parameter $z \in (0,1)$. We try to find the solution $v(x)$ as a series expansion in even powers of z (e.g. Hartman (1964)):

$$v(x) = \sum_{k=0}^{\infty} v_k(x) z^{2k} v_1. \qquad [5.6]$$

Substituting this series in equation [5.5], we have

$$\sum_{k=0}^{\infty} v_k''(x) z^{2k} v_1 - \left(\sum_{k=0}^{\infty} z^{2k+2}\right)\left(\sum_{k=0}^{\infty} v_k(x) z^{2k} v_1\right) = 0,$$

$$\sum_{k=0}^{\infty} v_k''(x) z^{2k} - \sum_{k=1}^{\infty} z^{2k} \sum_{p=0}^{k-1} v_p(x) = 0,$$

which leads to the relations

$$v_0''(x) = 0, \quad v_k''(x) = \sum_{p=0}^{k-1} v_p(x), \quad k = 1, 2, \ldots. \qquad [5.7]$$

Now, given the boundary conditions in [5.5], we obtain

$$v(0) = \sum_{k=0}^{\infty} v_k(0) z^{2k} v_1 = 0, \quad v(1) = \sum_{k=0}^{\infty} v_k(1) z^{2k} v_1 = v_1,$$

and therefore

$$v_0(0) = 0, \quad v_0(1) = 1; \quad v_k(0) = 0, \quad v_k(1) = 0, \quad k = 1, 2, \dots. \qquad [5.8]$$

From formulas [5.7] and [5.8], we come to the recurrent sequence of problems

$$v_k''(x) = \sum_{p=0}^{k-1} v_p(x), \quad x \in (0,1), \quad v_k(0) = 0, \quad v_k(1) = 0, \quad k = 1, 2, \dots, \quad v_0(x) = x.$$

Here, we have the equations

$$\left(v_k(x) - v_{k-1}(x)\right)'' = v_{k-1}(x), \quad x \in (0,1), \quad v_k(0) = 0, \quad v_k(1) = 0, \quad k = 1, 2, \dots,$$
$$v_0(x) = x, \qquad [5.9]$$

which lead to the relations

$$v_k(x) = v_{k-1}(x) - \int_0^1 G_0(x,\xi) v_{k-1}(\xi) d\xi, \quad x \in [0,1], \quad k = 2, 3, \dots,$$
$$v_0(x) = x, \quad v_1(x) = -\frac{1}{3!} x(1-x^2), \qquad [5.10]$$

where $G_0(x,\xi) = \begin{cases} x(1-\xi), & x \leq \xi, \\ \xi(1-x), & \xi \leq x, \end{cases}$ is the Green's function of the differential operator

$$L_0 u = -u''(x), \quad D(L_0) = H^2(0,1) \cap \overset{\circ}{H}{}^1(0,1).$$

Using [5.10], we can easily find, for example

$$v_2(x) = \frac{1}{5!} x(1-x^2)\left(-x^2 - \frac{53}{3}\right), \quad v_3(x) = \frac{1}{7!} x(1-x^2)\left(-x^4 - 78x^2 - \frac{1963}{3}\right).$$

Note that polynomials $v_k(x)$ are closely related to the Meixner polynomials (Meixner 1934) and were recently studied in Makarov (2019). A short time later, we will need the representation of the functions $v_k(x)$ via the Fourier series.

We start with the following auxiliary proposition.

LEMMA 5.1.– *Let $m > 0$, $\alpha > 0$. Then, the following inequality holds true*:

$$\max_{t \geq 1} \left(1 - \frac{1}{t}\right)^m t^{-\alpha} \leq \left(\frac{\alpha}{e}\right)^{\alpha} m^{-\alpha}.$$

PROOF.– Consider the function

$$\varphi(t) = \left(1 - \frac{1}{t}\right)^m t^{-\alpha}, \quad t \geq 1,$$

and find the derivative $\varphi'(t) = (t-1)^{m-1} t^{-\alpha-m-1}(m + \alpha - \alpha t)$. Then, we have

$$\max_{t \geq 1} \varphi(t) = \varphi\left(\frac{m+\alpha}{\alpha}\right) = \left(1 - \frac{\alpha}{m+\alpha}\right)^m \left(\frac{m+\alpha}{\alpha}\right)^{-\alpha} =$$

$$= \left(1 - \frac{\alpha}{m+\alpha}\right)^m \left(1 + \frac{\alpha}{m}\right)^{-\alpha} \alpha^{\alpha} m^{-\alpha} = \left(\frac{m}{m+\alpha}\right)^{m+\alpha} \alpha^{\alpha} m^{-\alpha} \leq e^{-\alpha} \alpha^{\alpha} m^{-\alpha},$$

which proves the lemma.

We will now use Lemma 5.1 to establish some important inequalities for $v_k(x)$.

LEMMA 5.2.– *For the polynomials $v_k(x)$, the following estimates hold true*:

$$\left|\frac{v_k(x)}{\min(x, 1-x)}\right| \leq \frac{C_1}{k^{(1-\varepsilon_1)/2}}, \quad x \in [0,1], \quad k = 1, 2, \ldots, \qquad [5.11]$$

$$\left|\frac{v_k(x)}{\min(x, 1-x)}\right| \leq \frac{1}{3}, \quad x \in [0,1], \quad k = 1, 2, \ldots, \qquad [5.12]$$

where $C_1 = \left(\frac{1-\varepsilon_1}{e}\right)^{(1-\varepsilon_1)/2} \frac{2}{\pi^{1+\varepsilon_1}} \zeta(1+\varepsilon_1)$, $0 < \varepsilon_1 < 1$, *and $\zeta(\cdot)$ is the Riemann zeta function*.

PROOF.– First, we continue the polynomials $v_k(x)$ in an odd way onto the interval $[-1, 0]$ and then periodically along the whole real axis. Now we prove by induction that $v_k(x)$ can be presented as follows:

$$v_k(x) = \sum_{p=1}^{\infty} \sqrt{2} a_p^{(k)} \sin(p\pi x), \quad x \in [0,1], \quad k = 1, 2, \ldots, \quad [5.13]$$

with $a_p^{(k)} = \sqrt{2} \int_0^1 v_k(x) \sin(p\pi x) dx = \dfrac{\sqrt{2}(-1)^p}{(p\pi)^3} \left(1 - \dfrac{1}{(p\pi)^2}\right)^{k-1}$.

For $k = 1$, formula [5.13] holds true since

$$v_1(x) = \sum_{p=1}^{\infty} \sqrt{2} a_p^{(1)} \sin(p\pi x), \quad x \in [0,1],$$

$$a_p^{(1)} = \sqrt{2} \int_0^1 v_1(x) \sin(p\pi x) dx = -\dfrac{\sqrt{2}}{6} \int_0^1 x(1-x^2) \sin(p\pi x) dx = \dfrac{\sqrt{2}(-1)^p}{(p\pi)^3}.$$

Assuming that formula [5.13] is true for some $k \in \mathbb{N}$, we prove it for $k+1$. We have

$$v_{k+1}(x) = v_k(x) - \int_0^1 G_0(x, \xi) v_k(\xi) d\xi =$$

$$= \sum_{p=1}^{\infty} \sqrt{2} a_p^{(k)} \sin(p\pi x) - \sum_{p=1}^{\infty} \sqrt{2} a_p^{(k)} \int_0^1 G_0(x, \xi) \sin(p\pi\xi) d\xi =$$

$$= \sum_{p=1}^{\infty} \sqrt{2} a_p^{(k)} \sin(p\pi x) -$$

$$- \sum_{p=1}^{\infty} \sqrt{2} a_p^{(k)} \left[(1-x) \int_0^x \xi \sin(p\pi\xi) d\xi + x \int_x^1 (1-\xi) \sin(p\pi\xi) d\xi \right] =$$

$$= \sum_{p=1}^{\infty} \sqrt{2} a_p^{(k)} \sin(p\pi x) - \sum_{p=1}^{\infty} \sqrt{2} a_p^{(k)} \frac{1}{(p\pi)^2} \sin(p\pi x) =$$

$$= \sum_{p=1}^{\infty} \sqrt{2} a_p^{(k)} \left(1 - \frac{1}{(p\pi)^2}\right) \sin(p\pi x),$$

which leads to the relation

$$a_p^{(k+1)} = a_p^{(k)} \left(1 - \frac{1}{(p\pi)^2}\right) = \frac{\sqrt{2}(-1)^p}{(p\pi)^3} \left(1 - \frac{1}{(p\pi)^2}\right)^{k-1} \left(1 - \frac{1}{(p\pi)^2}\right) =$$

$$= \frac{\sqrt{2}(-1)^p}{(p\pi)^3} \left(1 - \frac{1}{(p\pi)^2}\right)^k$$

and therefore gives [5.13].

Now we are ready to prove estimate [5.11]. We have

$$\left| \frac{v_k(x)}{\min(x,1-x)} \right| = \left| \sum_{p=1}^{\infty} \sqrt{2} \frac{\sqrt{2}(-1)^p}{(p\pi)^3} \left(1 - \frac{1}{(p\pi)^2}\right)^{k-1} \frac{\sin(p\pi x)}{\min(x,1-x)} \right| \le$$

$$\le \sum_{p=1}^{\infty} \frac{2}{(p\pi)^3} \left(1 - \frac{1}{(p\pi)^2}\right)^{k-1} \frac{|\sin(p\pi x)|}{\min(x,1-x)} \le \sum_{p=1}^{\infty} \frac{2}{(p\pi)^2} \left(1 - \frac{1}{(p\pi)^2}\right)^{k-1} =$$

$$= \sum_{p=1}^{\infty} \frac{2}{(p\pi)^{1+\varepsilon_1}} \frac{1}{\left((p\pi)^2\right)^{(1-\varepsilon_1)/2}} \left(1 - \frac{1}{(p\pi)^2}\right)^{k-1}.$$

Applying here Lemma 5.1 with $m = k-1$, $\alpha = \dfrac{1-\varepsilon_1}{2}$, $k = 2,3,\ldots$, $0 < \varepsilon_1 < 1$, we obtain

$$\left| \frac{v_k(x)}{\min(x,1-x)} \right| \le \sup_{p \in \mathbb{N}} \left[\frac{1}{\left((p\pi)^2\right)^{(1-\varepsilon_1)/2}} \left(1 - \frac{1}{(p\pi)^2}\right)^{k-1} \right] \sum_{p=1}^{\infty} \frac{2}{(p\pi)^{1+\varepsilon_1}} \le$$

$$\leq \sup_{t \geq \pi^2} \left[\frac{1}{t^{(1-\varepsilon_1)/2}} \left(1-\frac{1}{t}\right)^{k-1} \right] \frac{2}{\pi^{1+\varepsilon_1}} \zeta(1+\varepsilon_1) \leq$$

$$\leq \sup_{t \geq 1} \left[\frac{1}{t^{(1-\varepsilon_1)/2}} \left(1-\frac{1}{t}\right)^{k-1} \right] \frac{2}{\pi^{1+\varepsilon_1}} \zeta(1+\varepsilon_1) \leq$$

$$\leq \left(\frac{1-\varepsilon_1}{2e}\right)^{(1-\varepsilon_1)/2} (k-1)^{-(1-\varepsilon_1)/2} \frac{2}{\pi^{1+\varepsilon_1}} \zeta(1+\varepsilon_1) =$$

$$= \left(\frac{1-\varepsilon_1}{2e}\right)^{(1-\varepsilon_1)/2} \left(1-\frac{1}{k}\right)^{-(1-\varepsilon_1)/2} \frac{2}{\pi^{1+\varepsilon_1}} \zeta(1+\varepsilon_1) k^{-(1-\varepsilon_1)/2} \leq$$

$$\leq \left(\frac{1-\varepsilon_1}{2e}\right)^{(1-\varepsilon_1)/2} \left(\frac{1}{2}\right)^{-(1-\varepsilon_1)/2} \frac{2}{\pi^{1+\varepsilon_1}} \zeta(1+\varepsilon_1) k^{-(1-\varepsilon_1)/2} =$$

$$= \left(\frac{1-\varepsilon_1}{e}\right)^{(1-\varepsilon_1)/2} \frac{2}{\pi^{1+\varepsilon_1}} \zeta(1+\varepsilon_1) k^{-(1-\varepsilon_1)/2}.$$

For $k=1$, inequality [5.11] takes the form

$$\left|\frac{v_1(x)}{\min(x,1-x)}\right| \leq C_1, \quad x \in [0,1],$$

and therefore is also correct since

$$\max_{0 \leq x \leq 1} \left|\frac{v_1(x)}{\min(x,1-x)}\right| = \max_{0 \leq x \leq 1} \left|\frac{-\frac{1}{3!}x(1-x^2)}{\min(x,1-x)}\right| = \frac{1}{3}$$

and

$$C_1 = \left(\frac{1-\varepsilon_1}{e}\right)^{(1-\varepsilon_1)/2} \frac{2}{\pi^{1+\varepsilon_1}} \zeta(1+\varepsilon_1) > \frac{1}{3} \text{ for } 0 < \varepsilon_1 < 1.$$

Thus, estimate [5.11] is true for all $k \in \mathbb{N}$.

We now move on to estimate [5.12] and find

$$\left|\frac{v_k(x)}{\min(x,1-x)}\right| = \left|\sum_{p=1}^{\infty} \sqrt{2}\frac{\sqrt{2}(-1)^p}{(p\pi)^3}\left(1-\frac{1}{(p\pi)^2}\right)^{k-1}\frac{\sin p\pi x}{\min(x,1-x)}\right| \le$$

$$\le \sum_{p=1}^{\infty}\frac{2}{(p\pi)^3}\left(1-\frac{1}{(p\pi)^2}\right)^{k-1}\frac{|\sin p\pi x|}{\min(x,1-x)} \le \sum_{p=1}^{\infty}\frac{2}{(p\pi)^2} = \frac{1}{3}, \quad k=1,2,\ldots,$$

which completes the proof of the lemma.

Comparing problems [5.4] and [5.5], and taking into account expansion [5.6], we can note that the solution $w(x)$ of problem [5.4] can be written as follows:

$$w(x) = \operatorname{sh}^{-1}(\sqrt{a})\operatorname{sh}(x\sqrt{a})w_1 = \sum_{k=0}^{\infty} v_k(x)y_k \qquad [5.14]$$

with $y_k = (1+a)^{-1}ay_{k-1} = \left[(1+a)^{-1}a\right]^k w_1$.

5.1.2. The exact solution of the BVP

Together with problem [5.1], we consider now the boundary value problem for the homogeneous equation:

$$\frac{d^2w(x)}{dx^2} - Aw(x) = 0, \quad x \in (0,1), \qquad [5.15]$$
$$w(0) = 0, \quad w(1) = w_1.$$

Similarly to [5.14], the solution $w(x)$ of problem [5.15] can be written in the form

$$w(x) = \operatorname{sh}^{-1}(\sqrt{A})\operatorname{sh}(x\sqrt{A})w_1 = \sum_{k=0}^{\infty} v_k(x)y_k, \qquad [5.16]$$

where $y_k = (I+A)^{-1}Ay_{k-1} = \left[(I+A)^{-1}A\right]^k w_1$, I is an identity operator, and the functions $v_k(x)$ are defined by the recurrent sequence of the integral equations [5.10].

We are now ready to discuss problem [5.1]. The function $f(x)$ on the right-hand side of the equation can be represented by the Fourier series:

$$f(x) = \sum_{k=1}^{\infty} \sqrt{2} \sin(2k\pi x) f_{s,k} + f_0 + \sum_{k=1}^{\infty} \sqrt{2} \cos(2k\pi x) f_{c,k} \qquad [5.17]$$

with

$$f_{s,k} = \int_0^1 f(x)\sqrt{2}\sin(2k\pi x)dx, \quad f_{c,k} = \int_0^1 f(x)\sqrt{2}\cos(2k\pi x)dx, \quad k=1,2,\ldots,$$

$$f_0 = \int_0^1 f(x)dx.$$

Next, making use of the operator Green's function

$$G(x,\xi;A) = \left(\sqrt{A}\,\mathrm{sh}\sqrt{A}\right)^{-1}\begin{cases} \mathrm{sh}(\sqrt{A}x)\mathrm{sh}(\sqrt{A}(1-\xi)), & x \leq \xi, \\ \mathrm{sh}(\sqrt{A}\xi)\mathrm{sh}(\sqrt{a}(1-x)), & \xi \leq x, \end{cases}$$

and performing some simple transformations, we come to the representation of the solution $u(x)$ as follows:

$$u(x) = \int_0^1 G(x,\xi;A)f(\xi)d\xi =$$

$$= \left(\sqrt{A}\,\mathrm{sh}\sqrt{A}\right)^{-1}\mathrm{sh}\left(\sqrt{A}(1-x)\right)\int_0^x \mathrm{sh}\left(\sqrt{A}\xi\right)f(\xi)d\xi +$$

$$+ \left(\sqrt{A}\,\mathrm{sh}\sqrt{A}\right)^{-1}\mathrm{sh}\left(\sqrt{A}x\right)\int_x^1 \mathrm{sh}\left(\sqrt{A}(1-\xi)\right)f(\xi)d\xi =$$

$$= \left(\sqrt{A}\,\mathrm{sh}\sqrt{A}\right)^{-1}\sum_{k=1}^{\infty}\sqrt{2}\left\{\mathrm{sh}\left(\sqrt{A}(1-x)\right)\int_0^x \mathrm{sh}\left(\sqrt{A}\xi\right)\sin(2k\pi\xi)d\xi +\right.$$

$$+\operatorname{sh}(\sqrt{A}x)\int_x^1 \operatorname{sh}(\sqrt{A}(1-\xi))\sin(2k\pi\xi)d\xi\bigg\}f_{s,k} +$$

$$+(\sqrt{A}\operatorname{sh}\sqrt{A})^{-1}\bigg\{\operatorname{sh}(\sqrt{A}(1-x))\int_0^x \operatorname{sh}(\sqrt{A}\xi)d\xi + \operatorname{sh}(\sqrt{A}x)\int_x^1 \operatorname{sh}(\sqrt{A}(1-\xi))d\xi\bigg\}f_0 +$$

$$+(\sqrt{A}\operatorname{sh}\sqrt{A})^{-1}\sum_{k=1}^{\infty}\sqrt{2}\bigg\{\operatorname{sh}(\sqrt{A}(1-x))\int_0^x \operatorname{sh}(\sqrt{A}\xi)\cos(2k\pi\xi)d\xi +$$

$$+\operatorname{sh}(\sqrt{A}x)\int_x^1 \operatorname{sh}(\sqrt{A}(1-\xi))\cos(2k\pi\xi)d\xi\bigg\}f_{c,k}.$$

This finally leads to the main representation, which we will use throughout the chapter:

$$u(x) = \sum_{k=1}^{\infty}\sqrt{2}\sin(2k\pi x)\big[(2k\pi)^2 I + A\big]^{-1}f_{s,k} +$$

$$+(A\operatorname{sh}\sqrt{A})^{-1}\big\{\operatorname{sh}\sqrt{A} - \operatorname{sh}(\sqrt{A}(1-x)) - \operatorname{sh}(\sqrt{A}x)\big\}f_0 + \qquad [5.18]$$

$$+\sum_{k=1}^{\infty}\sqrt{2}\big[(2k\pi)^2 I + A\big]^{-1}\operatorname{sh}^{-1}\sqrt{A}\times$$

$$\times\big\{\cos(2k\pi x)\operatorname{sh}\sqrt{A} - \operatorname{sh}(\sqrt{A}(1-x)) - \operatorname{sh}(\sqrt{A}x)\big\}f_{c,k}.$$

Next, we obtain some useful relations.

LEMMA 5.3.– *The following inequalities hold true:*

$$\frac{|\sin(2k\pi x)|}{\min(x,1-x)} \leq 2\pi k \quad \forall x \in (0,1), \quad k=1,2,\ldots; \qquad [5.19]$$

$$\frac{\left|\operatorname{sh}\sqrt{\lambda}-\operatorname{sh}\left(\sqrt{\lambda}(1-x)\right)-\operatorname{sh}\left(\sqrt{\lambda}x\right)\right|}{\min(x,1-x)\lambda\operatorname{sh}\sqrt{\lambda}} \le \frac{1}{2} \qquad [5.20]$$

$$\forall x \in (0,1) \quad \forall \lambda \in [\lambda_0, +\infty);$$

$$\frac{\left|\cos(2k\pi x)\operatorname{sh}\sqrt{\lambda}-\operatorname{sh}\left(\sqrt{\lambda}(1-x)\right)-\operatorname{sh}\left(\sqrt{\lambda}x\right)\right|}{\min(x,1-x)\left[(2k\pi)^2+\lambda\right]\operatorname{sh}\sqrt{\lambda}} \le \frac{C(\lambda_0)}{2\pi k} \qquad [5.21]$$

$$\forall x \in (0,1) \quad \forall \lambda \in [\lambda_0, +\infty), \quad k = 1, 2, \ldots,$$

with the positive constant $C(\lambda_0) = 1 + \dfrac{1}{2}\operatorname{cth}\sqrt{\lambda_0} + \dfrac{1}{4}\operatorname{sh}^{-1}\left(\dfrac{\sqrt{\lambda_0}}{2}\right)$.

PROOF.– To prove [5.19], we consider the function

$$h(x) = \frac{\sin(2k\pi x)}{\min(x,1-x)}, \quad x \in (0,1).$$

Making use of the relation $h(1-x) = -h(x)$ for all $x \in (0, 1/2]$, we obtain

$$\sup_{0<x<1}|h(x)| = \sup_{0<x\le 1/2}|h(x)| = \sup_{0<x\le 1/2}\frac{|\sin(2k\pi x)|}{x} \le 2\pi k, \quad k = 1, 2, \ldots,$$

which gives [5.19].

To prove [5.20], we consider the function

$$\varphi(x) = \frac{\operatorname{sh}\sqrt{\lambda}-\operatorname{sh}\left(\sqrt{\lambda}(1-x)\right)-\operatorname{sh}\left(\sqrt{\lambda}x\right)}{\min(x,1-x)\lambda\operatorname{sh}\sqrt{\lambda}} = \frac{2\operatorname{sh}\left(\dfrac{\sqrt{\lambda}(1-x)}{2}\right)\operatorname{sh}\left(\dfrac{\sqrt{\lambda}x}{2}\right)}{\min(x,1-x)\lambda\operatorname{ch}\left(\dfrac{\sqrt{\lambda}}{2}\right)}, \quad x \in (0,1).$$

Since $\varphi(x) > 0$ for all $x \in (0,1)$ and $\varphi(1-x) = \varphi(x)$ for all $x \in (0, 1/2]$, we have

$$\sup_{0<x<1}|\varphi(x)| = \sup_{0<x<1}\varphi(x) = \sup_{0<x\le 1/2}\varphi(x).$$

Due to the inequalities

$$\frac{\sqrt{\lambda}x}{2} < \operatorname{sh}\left(\frac{\sqrt{\lambda}x}{2}\right), \quad \operatorname{sh}\left(\frac{\sqrt{\lambda}(1-2x)}{2}\right) < \operatorname{sh}\left(\frac{\sqrt{\lambda}(1-x)}{2}\right), \quad x \in (0,1/2],$$

for the derivative $\varphi'(x)$, we obtain

$$\varphi'(x) = \frac{2}{\lambda \operatorname{ch}\left(\sqrt{\lambda}/2\right)} \times$$

$$\times \frac{\frac{\sqrt{\lambda}x}{2}\operatorname{sh}\left(\frac{\sqrt{\lambda}(1-2x)}{2}\right) - \operatorname{sh}\left(\frac{\sqrt{\lambda}x}{2}\right)\operatorname{sh}\left(\frac{\sqrt{\lambda}(1-x)}{2}\right)}{x^2} < 0, \quad x \in (0,1/2],$$

which means that $\varphi(x)$ is strictly decreasing on the interval $(0,1/2]$. Thus,

$$\sup_{0<x<1}|\varphi(x)| = \lim_{x\to+0}\varphi(x) = \lim_{x\to+0}\frac{2\operatorname{sh}\left(\frac{\sqrt{\lambda}(1-x)}{2}\right)\operatorname{sh}\left(\frac{\sqrt{\lambda}x}{2}\right)}{x\lambda\operatorname{ch}\left(\frac{\sqrt{\lambda}}{2}\right)} = \frac{\operatorname{th}\left(\frac{\sqrt{\lambda}}{2}\right)}{\sqrt{\lambda}} \le \frac{1}{2}.$$

To prove [5.21], we consider the function

$$g(x) = \frac{\cos(2k\pi x)\operatorname{sh}\sqrt{\lambda} - \operatorname{sh}\left(\sqrt{\lambda}(1-x)\right) - \operatorname{sh}(\sqrt{\lambda}\,x)}{\min(x,1-x)\left[(2k\pi)^2 + \lambda\right]\sinh\sqrt{\lambda}}, \quad x \in (0,1). \qquad [5.22]$$

Once again we have $g(1-x) = g(x)$ for all $x \in (0,1/2]$, and therefore

$$\sup_{0<x<1}|g(x)| = \sup_{0<x\le1/2}|g(x)|.$$

Noting that the numerator of [5.22] vanishes at $x = 0$, we apply the Lagrange mean theorem and perform some simple transformations:

$$|g(x)| = \frac{\left|\cos(2k\pi x)\operatorname{sh}\sqrt{\lambda} - \operatorname{sh}\left(\sqrt{\lambda}(1-x)\right) - \operatorname{sh}(\sqrt{\lambda}\,x)\right|}{x\left[(2k\pi)^2 + \lambda\right]\operatorname{sh}\sqrt{\lambda}} =$$

$$= \frac{\left|\left(\cos(2k\pi x)\operatorname{sh}\sqrt{\lambda} - \operatorname{sh}(\sqrt{\lambda}(1-x)) - \operatorname{sh}(\sqrt{\lambda}\,x)\right)'\Big|_{x=\xi\in(0,x)} \cdot x\right|}{x\left[(2k\pi)^2 + \lambda\right]\operatorname{sh}\sqrt{\lambda}} =$$

$$= \frac{\left|-2k\pi\sin(2k\pi\xi)\operatorname{sh}\sqrt{\lambda} + \sqrt{\lambda}\operatorname{ch}(\sqrt{\lambda}(1-\xi)) - \sqrt{\lambda}\operatorname{ch}(\sqrt{\lambda}\,\xi)\right|}{\left[(2k\pi)^2 + \lambda\right]\operatorname{sh}\sqrt{\lambda}} \le$$

$$\le \frac{2k\pi\operatorname{sh}\sqrt{\lambda} + \sqrt{\lambda}\operatorname{ch}\sqrt{\lambda} + \sqrt{\lambda}\operatorname{ch}\left(\dfrac{\sqrt{\lambda}}{2}\right)}{\left[(2k\pi)^2 + \lambda\right]\operatorname{sh}\sqrt{\lambda}} =$$

$$= \frac{2k\pi\operatorname{sh}\sqrt{\lambda}}{\left[(2k\pi)^2 + \lambda\right]\operatorname{sh}\sqrt{\lambda}} + \frac{\sqrt{\lambda}\operatorname{ch}\sqrt{\lambda} + \sqrt{\lambda}\operatorname{ch}\left(\dfrac{\sqrt{\lambda}}{2}\right)}{\left[(2k\pi)^2 + \lambda\right]\operatorname{sh}\sqrt{\lambda}} \le \frac{1}{2k\pi} + \frac{\operatorname{cth}\sqrt{\lambda} + \dfrac{1}{2}\operatorname{sh}^{-1}\left(\dfrac{\sqrt{\lambda}}{2}\right)}{4k\pi} \le$$

$$\le \frac{1}{2k\pi}\left(1 + \frac{\operatorname{cth}\sqrt{\lambda_0}}{2} + \frac{1}{4}\operatorname{sh}^{-1}\left(\frac{\sqrt{\lambda_0}}{2}\right)\right) = \frac{1}{2k\pi}\left(1 + \frac{\operatorname{ch}\left(\dfrac{3\sqrt{\lambda_0}}{4}\right)\operatorname{ch}\left(\dfrac{\sqrt{\lambda_0}}{4}\right)}{\operatorname{sh}\sqrt{\lambda_0}}\right),$$

$$x \in (0, 1/2],$$

which proves [5.21] and hence the whole lemma.

Now we are ready to obtain the weighted estimates for the solution $u(x)$. We begin with the case of finite (in a certain sense) smoothness of the function $f(x)$.

THEOREM 5.1.– Let the Fourier representation of the function $f(x)$ on the right-hand side of equation [5.17] satisfy the conditions

$$f_0 \in D(A^\sigma), \quad \sigma > 0,$$

$$\|f_s\|_\sigma \equiv \left(\sum_{k=1}^\infty k^{2\sigma}\|f_{s,k}\|^2\right)^{1/2} < \infty, \quad \|f_c\|_\sigma \equiv \left(\sum_{k=1}^\infty k^{2\sigma}\|f_{c,k}\|^2\right)^{1/2} < \infty.$$

Then, for the solution $u(x)$ of problem [5.1], the weighted estimate holds true:

$$\left\| \frac{u(x)}{\min(x,1-x)} \right\| \le C \left(\| f_s \|_\sigma + \| A^\sigma f_0 \| + \| f_c \|_\sigma \right),$$

where $C = C(\lambda_0, \sigma)$ is independent of $u(x)$:

$$C = \max \left\{ \frac{1}{2\lambda_0^\sigma}, \frac{\sqrt{2} C(\lambda_0)}{2\pi} \zeta^{1/2}(2+2\sigma) \right\},$$

$$C(\lambda_0) = 1 + \frac{1}{2} \operatorname{cth} \sqrt{\lambda_0} + \frac{1}{4} \operatorname{sh}^{-1}\left(\frac{\sqrt{\lambda_0}}{2} \right), \quad \zeta(2+2\sigma) = \sum_{k=1}^{\infty} \frac{1}{k^{2+2\sigma}}.$$

PROOF.– Let $E(\lambda)$ be the spectral family of the operator A. Using the spectral representation for an operator function of A and applying Lemma 5.3, we have

$$\left\| \frac{u(x)}{\min(x,1-x)} \right\| = \left\| \sum_{k=1}^{\infty} \frac{\sqrt{2} \sin(2k\pi x)}{\min(x,1-x)} \left[(2k\pi)^2 I + A \right]^{-1} f_{s,k} + \right.$$

$$+ \left(A \operatorname{sh} \sqrt{A} \right)^{-1} \frac{\operatorname{sh}\sqrt{A} - \operatorname{sh}\left(\sqrt{A}(1-x)\right) - \operatorname{sh}\left(\sqrt{A}x\right)}{\min(x,1-x)} f_0 +$$

$$+ \sum_{k=1}^{\infty} \sqrt{2} \left[(2k\pi)^2 I + A \right]^{-1} \operatorname{sh}^{-1} \sqrt{A} \times$$

$$\left. \times \frac{\cos(2k\pi x) \operatorname{sh}\sqrt{A} - \operatorname{sh}\left(\sqrt{A}(1-x)\right) - \operatorname{sh}\left(\sqrt{A}x\right)}{\min(x,1-x)} f_{c,k} \right\| \le$$

$$\le \sum_{k=1}^{\infty} \left\| \int_{\lambda_0}^{+\infty} \frac{\sqrt{2} \sin(2k\pi x)}{\min(x,1-x)\left[(2k\pi)^2 + \lambda\right]} dE(\lambda) f_{s,k} \right\| +$$

$$+ \left\| \int_{\lambda_0}^{+\infty} \frac{\operatorname{sh}\sqrt{\lambda} - \operatorname{sh}\left(\sqrt{\lambda}(1-x)\right) - \operatorname{sh}\left(\sqrt{\lambda}x\right)}{\min(x,1-x)\lambda^{\sigma+1} \operatorname{sh}\sqrt{\lambda}} dE(\lambda) A^\sigma f_0 \right\| +$$

$$+\sum_{k=1}^{\infty}\left\|\int_{\lambda_0}^{+\infty}\sqrt{2}\,\frac{\cos(2k\pi x)\operatorname{sh}\sqrt{\lambda}-\operatorname{sh}\left(\sqrt{\lambda}(1-x)\right)-\operatorname{sh}\left(\sqrt{\lambda}x\right)}{\min(x,1-x)\left[(2k\pi)^2+\lambda\right]\operatorname{sh}\sqrt{\lambda}}\,dE(\lambda)\,f_{c,k}\right\|\le$$

$$\le\sum_{k=1}^{\infty}\sup_{\substack{0<x<1\\ \lambda\ge\lambda_0}}\frac{\sqrt{2}\,|\sin(2k\pi x)|}{\min(x,1-x)\left[(2k\pi)^2+\lambda\right]}\|f_{s,k}\|+$$

$$+\sup_{\substack{0<x<1\\ \lambda\ge\lambda_0}}\frac{\left|\operatorname{sh}\sqrt{\lambda}-\operatorname{sh}\left(\sqrt{\lambda}(1-x)\right)-\operatorname{sh}\left(\sqrt{\lambda}x\right)\right|}{\min(x,1-x)\lambda^{\sigma+1}\operatorname{sh}\sqrt{\lambda}}\|A^{\sigma}f_0\|+$$

$$+\sum_{k=1}^{\infty}\sup_{\substack{0<x<1\\ \lambda\ge\lambda_0}}\frac{\sqrt{2}\left|\cos(2k\pi x)\operatorname{sh}\sqrt{\lambda}-\operatorname{sh}\left(\sqrt{\lambda}(1-x)\right)-\operatorname{sh}\left(\sqrt{\lambda}x\right)\right|}{\min(x,1-x)\left[(2k\pi)^2+\lambda\right]\operatorname{sh}\sqrt{\lambda}}\|f_{c,k}\|\le$$

$$\le\sum_{k=1}^{\infty}\frac{\sqrt{2}}{2k\pi}\|f_{s,k}\|+\frac{1}{2\lambda_0^{\sigma}}\|A^{\sigma}f_0\|+\sum_{k=1}^{\infty}\frac{\sqrt{2}C(\lambda_0)}{2\pi k}\|f_{c,k}\|=$$

$$=\frac{\sqrt{2}}{2\pi}\sum_{k=1}^{\infty}\frac{1}{k^{\sigma+1}}k^{\sigma}\|f_{s,k}\|+\frac{1}{2\lambda_0^{\sigma}}\|A^{\sigma}f_0\|+\frac{\sqrt{2}C(\lambda_0)}{2\pi}\sum_{k=1}^{\infty}\frac{1}{k^{\sigma+1}}k^{\sigma}\|f_{c,k}\|\le$$

$$\le\frac{\sqrt{2}}{2\pi}\left(\sum_{k=1}^{\infty}\frac{1}{k^{2+2\sigma}}\right)^{1/2}\left(\sum_{k=1}^{\infty}k^{2\sigma}\|f_{s,k}\|^2\right)^{1/2}+\frac{1}{2\lambda_0^{\sigma}}\|A^{\sigma}f_0\|+$$

$$+\frac{\sqrt{2}C(\lambda_0)}{2\pi}\left(\sum_{k=1}^{\infty}\frac{1}{k^{2+2\sigma}}\right)^{1/2}\left(\sum_{k=1}^{\infty}k^{2\sigma}\|f_{c,k}\|^2\right)^{1/2}=$$

$$=\frac{\sqrt{2}}{2\pi}\left(\sum_{k=1}^{\infty}\frac{1}{k^{2+2\sigma}}\right)^{1/2}\|f_s\|_{\sigma}+\frac{1}{2\lambda_0^{\sigma}}\|A^{\sigma}f_0\|+\frac{\sqrt{2}C(\lambda_0)}{2\pi}\left(\sum_{k=1}^{\infty}\frac{1}{k^{2+2\sigma}}\right)^{1/2}\|f_c\|_{\sigma}.$$

This proves the theorem.

In the next proposition, we obtain the weighted estimate for the solution $u(x)$ in the case of infinite (in a certain sense) smoothness of the function $f(x)$.

THEOREM 5.2.– *Let the Fourier representation of the function $f(x)$ on the right-hand side of equation [5.17] satisfy the conditions*

$$f_0 \in D(e^A), \quad \|f_s\|_\infty \equiv \left(\sum_{k=1}^\infty \|e^{\sqrt{k}} f_{s,k}\|^2\right)^{1/2} < \infty,$$

$$\|f_c\|_\infty \equiv \left(\sum_{k=1}^\infty \|e^{\sqrt{k}} f_{c,k}\|^2\right)^{1/2} < \infty.$$

Then for the solution $u(x)$ of problem [5.1], the weighted estimate holds true:

$$\left\|\frac{u(x)}{\min(x,1-x)}\right\| \leq C\left(\|f_s\|_\infty + \|e^A f_0\| + \|f_c\|_\infty\right), \quad x \in (0,1),$$

where $C = \max\left\{\dfrac{C(\lambda_0)}{2\sqrt{3e}}, \dfrac{1}{2e^{\lambda_0}}\right\}$ *and the constant $C(\lambda_0)$ is defined in [5.21].*

PROOF.– Let $E(\lambda)$ be the spectral family of the operator A. Using the spectral representation for an operator function of A and applying Lemma 5.3, we have

$$\left\|\frac{u(x)}{\min(x,1-x)}\right\| = \left\|\sum_{k=1}^\infty \frac{\sqrt{2}\sin(2k\pi x)}{\min(x,1-x)}\left[(2k\pi)^2 I + A\right]^{-1} f_{s,k} + \right.$$

$$+ \left(A\,\text{sh}\,\sqrt{A}\right)^{-1} \frac{\text{sh}\,\sqrt{A} - \text{sh}\left(\sqrt{A}(1-x)\right) - \text{sh}\left(\sqrt{A}x\right)}{\min(x,1-x)} f_0 +$$

$$+ \sum_{k=1}^\infty \sqrt{2}\left[(2k\pi)^2 I + A\right]^{-1} \text{sh}^{-1}\sqrt{A} \times$$

$$\left. \times \frac{\cos(2k\pi x)\,\text{sh}\,\sqrt{A} - \text{sh}\left(\sqrt{A}(1-x)\right) - \text{sh}\left(\sqrt{A}x\right)}{\min(x,1-x)} f_{c,k}\right\| \leq$$

$$\leq \sum_{k=1}^\infty \left\|\int_{\lambda_0}^{+\infty} \frac{\sqrt{2}\sin(2k\pi x)}{\min(x,1-x)\left[(2k\pi)^2 + \lambda\right]} dE(\lambda) f_{s,k}\right\| +$$

$$+\left\|\int_{\lambda_0}^{+\infty}\frac{\sh\sqrt{\lambda}-\sh(\sqrt{\lambda}(1-x))-\sh(\sqrt{\lambda}x)}{\min(x,1-x)\lambda^{\sigma+1}\sh\sqrt{\lambda}}dE(\lambda)A^\sigma f_0\right\|+$$

$$+\sum_{k=1}^{\infty}\left\|\int_{\lambda_0}^{+\infty}\sqrt{2}\frac{\cos(2k\pi x)\sh\sqrt{\lambda}-\sh(\sqrt{\lambda}(1-x))-\sh(\sqrt{\lambda}x)}{\min(x,1-x)\left[(2k\pi)^2+\lambda\right]\sh\sqrt{\lambda}}dE(\lambda)\,f_{c,k}\right\|\le$$

$$\le\sum_{k=1}^{\infty}\sup_{\substack{0<x<1\\ \lambda\ge\lambda_0}}\frac{\sqrt{2}\,|\sin(2k\pi x)|}{\min(x,1-x)\left[(2k\pi)^2+\lambda\right]}\,\|f_{s,k}\|+$$

$$+\sup_{\substack{0<x<1\\ \lambda\ge\lambda_0}}e^{-\lambda}\frac{\left|\sh\sqrt{\lambda}-\sh(\sqrt{\lambda}(1-x))-\sh(\sqrt{\lambda}x)\right|}{\min(x,1-x)\lambda\sh\sqrt{\lambda}}\,\|e^A f_0\|+$$

$$+\sum_{k=1}^{\infty}\sup_{\substack{0<x<1\\ \lambda\ge\lambda_0}}\frac{\sqrt{2}\left|\cos(2k\pi x)\sh\sqrt{\lambda}-\sh(\sqrt{\lambda}(1-x))-\sh(\sqrt{\lambda}x)\right|}{\min(x,1-x)\left[(2k\pi)^2+\lambda\right]\sh\sqrt{\lambda}}\,\|f_{c,k}\|\le$$

$$\le\sum_{k=1}^{\infty}\frac{\sqrt{2}}{2k\pi}\|f_{s,k}\|+\frac{1}{2e^{\lambda_0}}\|e^A f_0\|+\sum_{k=1}^{\infty}\frac{\sqrt{2}C(\lambda_0)}{2\pi k}\|f_{c,k}\|=$$

$$=\frac{\sqrt{2}}{2\pi}\sum_{k=1}^{\infty}\frac{e^{-\sqrt{k}}}{k}\|e^{\sqrt{k}}f_{s,k}\|+\frac{1}{2e^{\lambda_0}}\|e^A f_0\|+\frac{\sqrt{2}C(\lambda_0)}{2\pi}\sum_{k=1}^{\infty}\frac{e^{-\sqrt{k}}}{k}\|e^{\sqrt{k}}f_{c,k}\|\le$$

$$\le\frac{\sqrt{2}}{2\pi e}\left(\sum_{k=1}^{\infty}\frac{1}{k^2}\right)^{1/2}\left(\sum_{k=1}^{\infty}\|e^{\sqrt{k}}f_{s,k}\|^2\right)^{1/2}+\frac{1}{2e^{\lambda_0}}\|e^A f_0\|+$$

$$+\frac{\sqrt{2}C(\lambda_0)}{2\pi e}\left(\sum_{k=1}^{\infty}\frac{1}{k^2}\right)^{1/2}\left(\sum_{k=1}^{\infty}\|e^{\sqrt{k}}f_{c,k}\|^2\right)^{1/2}=$$

$$=\frac{1}{2\sqrt{3}e}\|f_s\|_\infty+\frac{1}{2e^{\lambda_0}}\|e^A f_0\|+\frac{C(\lambda_0)}{2\sqrt{3}e}\|f_c\|_\infty.$$

The theorem is proved.

5.1.3. *The approximate method without saturation of accuracy*

As an approximate solution of problem [5.1], we consider the partial sum of series [5.18]:

$$u_N(x) = \sum_{k=1}^{N} \sqrt{2} \sin(2k\pi x)\left[(2k\pi)^2 I + A\right]^{-1} f_{s,k} +$$ [5.23]

$$+ \left(A \operatorname{sh}\sqrt{A}\right)^{-1}\left\{\operatorname{sh}\sqrt{A} - \operatorname{sh}\left(\sqrt{A}(1-x)\right) - \operatorname{sh}\left(\sqrt{A}x\right)\right\} f_0 + \sum_{k=1}^{N} \sqrt{2}\left[(2k\pi)^2 I + A\right]^{-1} \times$$

$$\times \operatorname{sh}^{-1}\sqrt{A}\left\{\cos(2k\pi x)\operatorname{sh}\sqrt{A} - \operatorname{sh}\left(\sqrt{A}(1-x)\right) - \operatorname{sh}\left(\sqrt{A}x\right)\right\} f_{c,k}.$$

Next, we study the accuracy of this approximation.

THEOREM 5.3.– *Let the conditions of Theorem 5.1 be fulfilled. Then, the accuracy of the approximate solution [5.23] is characterized by the weighted estimate*

$$\left\|\frac{u(x) - u_N(x)}{\min(x, 1-x)}\right\| \leq \frac{C}{N^{\sigma+1/2}}\left(\|f_s\|_\sigma + \|f_c\|_\sigma + \|A^\sigma f_0\|\right),$$

where $C = C(\sigma, \lambda_0) = \dfrac{\sqrt{2}C(\lambda_0)}{2\pi(1+2\sigma)^{1/2}}$ *is independent of N, the constant* $C(\lambda_0)$ *is defined in [5.21].*

PROOF.– Making use of representations [5.18] and [5.23], analogously to the proof of Theorem 5.1, we obtain

$$\left\|\frac{u(x) - u_N(x)}{\min(x, 1-x)}\right\| \leq$$

$$\leq \frac{\sqrt{2}}{2\pi}\left(\sum_{k=N+1}^{\infty}\frac{1}{k^{2+2\sigma}}\right)^{1/2}\|f_s\|_\sigma + \frac{\sqrt{2}C(\lambda_0)}{2\pi}\left(\sum_{k=N+1}^{\infty}\frac{1}{k^{2+2\sigma}}\right)^{1/2}\|f_c\|_\sigma \leq$$

$$\leq \frac{\sqrt{2}}{2\pi}\left(\int_{N}^{+\infty}\frac{dx}{x^{2+2\sigma}}\right)^{1/2}\|f_s\|_\sigma + \frac{\sqrt{2}C(\lambda_0)}{2\pi}\left(\int_{N}^{+\infty}\frac{dx}{x^{2+2\sigma}}\right)^{1/2}\|f_c\|_\sigma =$$

$$= \frac{\sqrt{2}}{2\pi} \frac{1}{(1+2\sigma)^{1/2} N^{\sigma+1/2}} \| f_s \|_\sigma + \frac{\sqrt{2}C(\lambda_0)}{2\pi} \frac{1}{(1+2\sigma)^{1/2} N^{\sigma+1/2}} \| f_c \|_\sigma \le$$

$$\le \max \left\{ \frac{\sqrt{2}}{2\pi(1+2\sigma)^{1/2}} ; \frac{\sqrt{2}C(\lambda_0)}{2\pi(1+2\sigma)^{1/2}} \right\} \frac{1}{N^{\sigma+1/2}} (\| f_s \|_\sigma + \| f_c \|_\sigma).$$

The theorem is proved.

Now we consider another approximation of $u(x)$. Using representation [5.16], we can rewrite the partial sum [5.23] as follows:

$$u_N(x) = \sum_{k=1}^{N} \sqrt{2} \sin(2k\pi x) \left[(2k\pi)^2 I + A \right]^{-1} f_{s,k} +$$

$$+ A^{-1} \left\{ I - \sum_{j=0}^{\infty} \left[v_j(1-x) + v_j(x) \right] \left[(I+A)^{-1} A \right]^j \right\} f_0 + \sum_{k=1}^{N} \sqrt{2} \left[(2k\pi)^2 I + A \right]^{-1} \times$$

$$\times \left\{ \cos(2k\pi x) I - \sum_{j=0}^{\infty} \left[v_j(1-x) + v_j(x) \right] \left[(I+A)^{-1} A \right]^j \right\} f_{c,k}.$$

Then, we approximate the solution $u(x)$ of problem [5.1] by the partial sum

$$u_{N,M}(x) = \sum_{k=1}^{N} \sqrt{2} \sin(2k\pi x) \left[(2k\pi)^2 I + A \right]^{-1} f_{s,k} +$$

$$+ A^{-1} \left\{ I - \sum_{j=0}^{M} \left[v_j(1-x) + v_j(x) \right] \left[(I+A)^{-1} A \right]^j \right\} f_0 + \quad [5.24]$$

$$+ \sum_{k=1}^{N} \sqrt{2} \left[(2k\pi)^2 I + A \right]^{-1} \times$$

$$\times \left\{ (\cos(2k\pi x) - 1) I - \sum_{j=1}^{M} \left[v_j(1-x) + v_j(x) \right] \left[(I+A)^{-1} A \right]^j \right\} f_{c,k}.$$

To study the accuracy of $u_{N,M}(x)$, we need the following auxiliary proposition.

LEMMA 5.4.– *Under the assumptions $\alpha > 0$ and $\beta \geq \max\{2\alpha, (1+\lambda_0)\alpha\}$, the following inequality holds true:*

$$\max_{\lambda \geq \lambda_0} \lambda^{-\alpha} \left(\frac{\lambda}{\lambda+1}\right)^\beta \leq (2\alpha)^\alpha \beta^{-\alpha}. \qquad [5.25]$$

PROOF.– We consider the function

$$\varphi(\lambda) = \lambda^{-\alpha} \left(\frac{\lambda}{\lambda+1}\right)^\beta, \quad \lambda \geq \lambda_0,$$

and its derivative

$$\varphi'(\lambda) = \frac{\lambda^{-\alpha+\beta-1}}{(\lambda+1)^{\beta+1}} (\beta - \alpha - \alpha\lambda).$$

Due to the condition $\dfrac{\beta-\alpha}{\alpha} \geq \lambda_0$, we have

$$\max_{\lambda \geq \lambda_0} \varphi(\lambda) = \varphi\left(\frac{\beta-\alpha}{\alpha}\right) = \left(\frac{\beta}{\alpha}-1\right)^{-\alpha} \left(1-\frac{\alpha}{\beta}\right)^\beta.$$

Since $\beta \geq 2\alpha$, then $\dfrac{\beta}{\alpha} - 1 \geq \dfrac{\beta}{2\alpha}$, and therefore

$$\max_{\lambda \geq \lambda_0} \varphi(\lambda) \leq \left(\frac{\beta}{2\alpha}\right)^{-\alpha} = (2\alpha)^\alpha \beta^{-\alpha},$$

which proves the lemma.

Now we arrive at the proposition.

THEOREM 5.4.– *Let the following conditions be satisfied:*

$$M = N, \quad \sigma > 1, \quad M+1 \geq \max\{2\sigma, (1+\lambda_0)\sigma\},$$

$$f_0 \in D(A^{\sigma-1}), \quad \|f_s\|_{\sigma-3/2} \equiv \left(\sum_{k=1}^\infty k^{2\sigma-3} \|f_{s,k}\|^2\right)^{1/2} < \infty,$$

$$\|f_c\|_{\sigma-3/2} \equiv \left(\sum_{k=1}^{\infty} k^{2\sigma-3} \|f_{c,k}\|^2\right)^{1/2} < \infty,$$

$$f_{c,k} \in D(A^{\sigma-1/2}) \quad \forall k \in \mathbb{N} \quad \|f_c\|_{A^{\sigma-1/2}} \equiv \left(\sum_{k=1}^{\infty} \|A^{\sigma-1/2} f_{c,k}\|^2\right)^{1/2} < \infty.$$

Then, the accuracy of the approximate solution [5.24] is characterized by the weighted estimate

$$\left\|\frac{u(x)-u_{N,M}(x)}{\min(x,1-x)}\right\| \leq \frac{C}{N^{\sigma-1}}\left(\|f_c\|_{\sigma-3/2}+\|f_s\|_{\sigma-3/2}+\|f_c\|_{A^{\sigma-1/2}}+\|A^{\sigma-1}f_0\|\right),$$

$$x \in (0,1),$$

where the positive constant $C = C(\sigma)$ *is independent of* N *and* M.

PROOF.– We present the error of the approximate solution [5.24] as the sum of five terms:

$$u(x) - u_{N,M}(x) = D_1 + D_2 + D_3 + D_4 + D_5, \qquad [5.26]$$

where

$$D_1 = \sum_{k=N+1}^{\infty} \sqrt{2}\sin(2k\pi x)\left[(2k\pi)^2 I + A\right]^{-1} f_{s,k},$$

$$D_2 = -\sum_{j=M+1}^{\infty} \left[v_j(1-x)+v_j(x)\right] A^{-1}\left[(I+A)^{-1}A\right]^j f_0,$$

$$D_3 = \sum_{k=N+1}^{\infty} \sqrt{2}\left[(2k\pi)^2 I + A\right]^{-1}(\cos(2k\pi x)-1) f_{c,k},$$

$$D_4 = -\sum_{k=1}^{\infty} \sqrt{2}\left[(2k\pi)^2 I + A\right]^{-1}\sum_{j=M+1}^{\infty}\left[v_j(1-x)+v_j(x)\right]\left[(I+A)^{-1}A\right]^j f_{c,k},$$

$$D_5 = -\sum_{k=N+1}^{\infty} \sqrt{2}\left[(2k\pi)^2 I + A\right]^{-1} \sum_{j=1}^{M}\left[v_j(1-x) + v_j(x)\right]\left[(I+A)^{-1}A\right]^j f_{c,k}.$$

For the first summand, we have

$$\left\|\frac{D_1}{\min(x,1-x)}\right\| = \left\|\sum_{k=N+1}^{\infty} \sqrt{2}\,\frac{\sin(2k\pi x)}{\min(x,1-x)}\left[(2k\pi)^2 I + A\right]^{-1} f_{s,k}\right\| \le$$

$$\le \sum_{k=N+1}^{\infty} \frac{\sqrt{2}\,|\sin(2k\pi x)|}{\min(x,1-x)} \left\|\int_{\lambda_0}^{+\infty}\left[(2k\pi)^2 + \lambda\right]^{-1} dE(\lambda) f_{s,k}\right\| \le$$

$$\le \sum_{k=N+1}^{\infty} \frac{2k\pi\sqrt{2}}{(2k\pi)^2 + \lambda_0}\|f_{s,k}\| \le \sum_{k=N+1}^{\infty} \frac{\sqrt{2}}{2k\pi}\|f_{s,k}\| = \frac{\sqrt{2}}{2\pi}\sum_{k=N+1}^{\infty} \frac{1}{k^{\sigma-1/2}} k^{\sigma-3/2}\|f_{s,k}\| \le$$

$$\le \frac{\sqrt{2}}{2\pi}\left(\sum_{k=N+1}^{\infty} \frac{1}{k^{2\sigma-1}}\right)^{1/2} \left(\sum_{k=N+1}^{\infty} k^{2\sigma-3}\|f_{s,k}\|^2\right)^{1/2} =$$

$$= \frac{\sqrt{2}}{2\pi}\left(\sum_{k=N+1}^{\infty} \frac{1}{k^{2\sigma-1}}\right)^{1/2}\|f_s\|_{\sigma-3/2} \le \frac{\sqrt{2}}{2\pi}\left(\int_N^{+\infty} \frac{dx}{x^{2\sigma-1}}\right)^{1/2}\|f_s\|_{\sigma-3/2} =$$

$$= \frac{\|f_s\|_{\sigma-3/2}}{2\pi(\sigma-1)^{1/2} N^{\sigma-1}} \quad (\sigma > 1). \qquad [5.27]$$

Applying Lemmas 5.2.1 and 5.4, for the second summand, we obtain

$$\left\|\frac{D_2}{\min(x,1-x)}\right\| = \left\|\sum_{j=M+1}^{\infty} \frac{v_j(1-x) + v_j(x)}{\min(x,1-x)} A^{-1}\left[(I+A)^{-1}A\right]^j f_0\right\| \le$$

$$\le \sum_{j=M+1}^{\infty} \frac{|v_j(1-x)| + |v_j(x)|}{\min(x,1-x)} \left\|\int_{\lambda_0}^{+\infty} \lambda^{-\sigma}\left[(1+\lambda)^{-1}\lambda\right]^j dE(\lambda) A^{\sigma-1} f_0\right\| \le$$

$$\leq \sum_{j=M+1}^{\infty} \frac{2}{3} \frac{(2\sigma)^{\sigma}}{j^{\sigma}} \|A^{\sigma-1} f_0\| \leq \frac{2(2\sigma)^{\sigma}}{3} \int_{M}^{+\infty} \frac{dx}{x^{\sigma}} \|A^{\sigma-1} f_0\| \leq$$

$$\leq \frac{2(2\sigma)^{\sigma}}{3(\sigma-1)M^{\sigma-1}} \|A^{\sigma-1} f_0\| \qquad [5.28]$$

with $\sigma > 1$, $M+1 \geq \max\{2\sigma, (1+\lambda_0)\sigma\}$.

Next, we obtain

$$\left\|\frac{D_3}{\min(x,1-x)}\right\| = \left\|\sum_{k=N+1}^{\infty} \sqrt{2} \frac{\cos(2k\pi x)-1}{\min(x,1-x)} \left[(2k\pi)^2 I + A\right]^{-1} f_{c,k}\right\| \leq$$

$$\leq \sum_{k=N+1}^{\infty} \frac{2\sqrt{2}\sin^2(k\pi x)}{\min(x,1-x)} \left\|\int_{\lambda_0}^{+\infty} \left[(2k\pi)^2 + \lambda\right]^{-1} dE(\lambda) f_{c,k}\right\| \leq$$

$$\leq \sum_{k=N+1}^{\infty} \frac{2\sqrt{2}k\pi}{(2k\pi)^2 + \lambda_0} \|f_{c,k}\| \leq \sum_{k=N+1}^{\infty} \frac{\sqrt{2}}{2k\pi} \|f_{c,k}\| = \frac{\sqrt{2}}{2\pi} \sum_{k=N+1}^{\infty} \frac{1}{k^{\sigma-1/2}} k^{\sigma-3/2} \|f_{c,k}\| \leq$$

$$\leq \frac{\sqrt{2}}{2\pi} \left(\sum_{k=N+1}^{\infty} \frac{1}{k^{2\sigma-1}}\right)^{1/2} \left(\sum_{k=N+1}^{\infty} k^{2\sigma-3} \|f_{c,k}\|^2\right)^{1/2} \leq$$

$$\leq \frac{\sqrt{2}}{2\pi} \left(\sum_{k=N+1}^{\infty} \frac{1}{k^{2\sigma-1}}\right)^{1/2} \|f_c\|_{\sigma-3/2} \leq \frac{\sqrt{2}}{2\pi} \frac{1}{(2\sigma-2)^{1/2} N^{\sigma-1}} \|f_c\|_{\sigma-3/2} =$$

$$= \frac{\|f_c\|_{\sigma-3/2}}{2\pi(\sigma-1)^{1/2} N^{\sigma-1}} \quad (\sigma > 1). \qquad [5.29]$$

For the fourth summand, we have

$$\left\|\frac{D_4}{\min(x,1-x)}\right\| =$$

$$= \left\| -\sum_{k=1}^{\infty} \sqrt{2}\left[(2k\pi)^2 I + A\right]^{-1} \sum_{j=M+1}^{\infty} \frac{v_j(1-x)+v_j(x)}{\min(x,1-x)} \left[(I+A)^{-1}A\right]^j f_{c,k} \right\| \le$$

$$\le \sqrt{2} \sum_{k=1}^{\infty} \sum_{j=M+1}^{\infty} \frac{|v_j(1-x)|+|v_j(x)|}{\min(x,1-x)} \left\| \int_{\lambda_0}^{+\infty} \frac{\lambda^{-(\sigma-1/2)}\left[(I+\lambda)^{-1}\lambda\right]^j}{(2k\pi)^2+\lambda} dE(\lambda) A^{\sigma-1/2} f_{c,k} \right\| \le$$

$$\le \sqrt{2} \sum_{k=1}^{\infty} \sum_{j=M+1}^{\infty} \frac{2}{3} \left\| \int_{\lambda_0}^{+\infty} \frac{\lambda^{-(\sigma-1/2)}\left[(I+\lambda)^{-1}\lambda\right]^j}{4k\pi\sqrt{\lambda}} dE(\lambda) A^{\sigma-1/2} f_{c,k} \right\| \le$$

$$\le \sqrt{2} \sum_{k=1}^{\infty} \sum_{j=M+1}^{\infty} \frac{2}{3} \frac{(2\sigma)^{\sigma}}{4k\pi j^{\sigma}} \left\| A^{\sigma-1/2} f_{c,k} \right\| = \frac{\sqrt{2}(2\sigma)^{\sigma}}{6\pi} \sum_{k=1}^{\infty} \frac{\left\| A^{\sigma-1/2} f_{c,k} \right\|}{k} \sum_{j=M+1}^{\infty} \frac{1}{j^{\sigma}} \le$$

$$\le \frac{\sqrt{2}(2\sigma)^{\sigma}}{6\pi} \left(\sum_{k=1}^{\infty} \frac{1}{k^2} \right)^{1/2} \left(\sum_{k=1}^{\infty} \left\| A^{\sigma-1/2} f_{c,k} \right\|^2 \right)^{1/2} \int_{M}^{+\infty} \frac{dx}{x^{\sigma}} \le$$

$$\le \frac{(2\sigma)^{\sigma}}{6\sqrt{3}(\sigma-1)M^{\sigma-1}} \left\| f_c \right\|_{A^{\sigma-1/2}} \quad [5.30]$$

with $\sigma > 1$, $M+1 \ge \max\{2\sigma, (1+\lambda_0)\sigma\}$.

We can estimate the fifth summand as follows:

$$\left\| \frac{D_5}{\min(x,1-x)} \right\| =$$

$$= \left\| -\sum_{k=N+1}^{\infty} \sqrt{2}\left[(2k\pi)^2 I + A\right]^{-1} \sum_{j=1}^{M} \frac{v_j(1-x)+v_j(x)}{\min(x,1-x)} \left[(I+A)^{-1}A\right]^j f_{c,k} \right\| \le$$

$$\le \sqrt{2} \sum_{k=N+1}^{\infty} \sum_{j=1}^{M} \frac{|v_j(1-x)|+|v_j(x)|}{\min(x,1-x)} \left\| \int_{\lambda_0}^{+\infty} \frac{\left[(1+\lambda)^{-1}\lambda\right]^j}{(2k\pi)^2+\lambda} dE(\lambda) f_{c,k} \right\| \le$$

$$\leq \sqrt{2} \sum_{k=N+1}^{\infty} \sum_{j=1}^{M} \frac{2}{3} \frac{1}{(2k\pi)^2} \|f_{c,k}\| = \frac{2\sqrt{2}}{3} \frac{M}{4\pi^2} \sum_{k=N+1}^{\infty} \frac{\|f_{c,k}\|}{k^2} =$$

$$= \frac{2\sqrt{2}}{3} \frac{M}{4\pi^2} \sum_{k=N+1}^{\infty} \frac{1}{k^{\sigma+1/2}} k^{\sigma-3/2} \|f_{c,k}\| =$$

$$= \frac{2\sqrt{2}}{3} \frac{M}{4\pi^2} \left(\sum_{k=N+1}^{\infty} \frac{1}{k^{2\sigma+1}} \right)^{1/2} \left(\sum_{k=N+1}^{\infty} k^{2\sigma-3} \|f_{c,k}\|^2 \right)^{1/2} \leq$$

$$\leq \frac{2\sqrt{2}}{3} \frac{M}{4\pi^2} \left(\int_{N}^{+\infty} \frac{dx}{x^{2\sigma+1}} \right)^{1/2} \|f_c\|_{\sigma-3/2} = \frac{2\sqrt{2}}{3} \frac{M}{4\pi^2} \left(\frac{1}{2\sigma N^{2\sigma}} \right)^{1/2} \|f_c\|_{\sigma-3/2} \leq$$

$$\leq \frac{M}{6\pi^2 \sqrt{\sigma} N^{\sigma}} \|f_c\|_{\sigma-3/2} \leq \quad (\sigma > 0). \qquad [5.31]$$

Inequalities [5.27]–[5.31] yield the proof of the theorem.

5.1.4. The approximate method with a super-exponential rate of convergence

We will now study the accuracy of the approximate solution [5.23] under some other assumptions about the smoothness of the input data. We start with the following proposition.

THEOREM 5.5.– *Let the conditions of Theorem 5.2 be satisfied. Then, the accuracy of the approximate solution [5.23] is characterized by the weighted estimate*

$$\left\| \frac{u(x) - u_N(x)}{\min(x, 1-x)} \right\| \leq \frac{C(\lambda_0)}{\sqrt{2\pi}} \frac{e^{-\sqrt{N}}}{\sqrt{N}} (\|f_s\|_\infty + \|f_c\|_\infty), \quad x \in (0,1), \qquad [5.32]$$

with the constant $C(\lambda_0)$ *defined in [5.21].*

PROOF.– Analogously to the proof of Theorem 5.2, we have

$$\left\| \frac{u(x) - u_N(x)}{\min(x, 1-x)} \right\| \leq \sum_{k=N+1}^{\infty} \frac{\sqrt{2}}{2k\pi} \|f_{s,k}\| + \sum_{k=N+1}^{\infty} \frac{\sqrt{2} C(\lambda_0)}{2\pi k} \|f_{c,k}\| =$$

$$= \frac{\sqrt{2}}{2\pi} \sum_{k=N+1}^{\infty} \frac{e^{-\sqrt{k}}}{k} \left\| e^{\sqrt{k}} f_{s,k} \right\| + \frac{\sqrt{2} C(\lambda_0)}{2\pi} \sum_{k=N+1}^{\infty} \frac{e^{-\sqrt{k}}}{k} \left\| e^{\sqrt{k}} f_{c,k} \right\| \le$$

$$\le \frac{\sqrt{2}}{2\pi} e^{-\sqrt{N}} \left(\sum_{k=N+1}^{\infty} \frac{1}{k^2} \right)^{1/2} \left(\sum_{k=N+1}^{\infty} \left\| e^{\sqrt{k}} f_{s,k} \right\|^2 \right)^{1/2} +$$

$$+ \frac{\sqrt{2} C(\lambda_0)}{2\pi} e^{-\sqrt{N}} \left(\sum_{k=N+1}^{\infty} \frac{1}{k^2} \right)^{1/2} \left(\sum_{k=N+1}^{\infty} \left\| e^{\sqrt{k}} f_{c,k} \right\|^2 \right)^{1/2} \le$$

$$\le \frac{\sqrt{2}}{2\pi} e^{-\sqrt{N}} \left(\int_N^{+\infty} \frac{dx}{x^2} \right)^{1/2} \left(\sum_{k=1}^{\infty} \left\| e^{\sqrt{k}} f_{s,k} \right\|^2 \right)^{1/2} +$$

$$+ \frac{\sqrt{2} C(\lambda_0)}{2\pi} e^{-\sqrt{N}} \left(\int_N^{+\infty} \frac{dx}{x^2} \right)^{1/2} \left(\sum_{k=1}^{\infty} \left\| e^{\sqrt{k}} f_{c,k} \right\|^2 \right)^{1/2} =$$

$$= \frac{1}{\sqrt{2\pi}} \frac{e^{-\sqrt{N}}}{\sqrt{N}} \|f_s\|_\infty + \frac{C(\lambda_0)}{\sqrt{2\pi}} \frac{e^{-\sqrt{N}}}{\sqrt{N}} \|f_c\|_\infty,$$

which gives estimate [5.32]. The theorem is proved.

Next, we turn to the study of the approximate solution $u_{N,M}(x)$ defined by formula [5.24]. First, we prove the preliminary result.

LEMMA 5.5.– *The following two relations hold true:*

$$\max_{\lambda \ge \lambda_0} \lambda^{-1} e^{-\lambda} \left(\frac{\lambda}{1+\lambda} \right)^j = \frac{e^{1-\sqrt{j}}}{\sqrt{j}-1} \left(1 - \frac{1}{\sqrt{j}} \right)^j, \quad j \ge (\lambda_0+1)^2, \qquad [5.33]$$

$$\sup_{\lambda \ge \lambda_0} e^{-\lambda} \left(\frac{\lambda}{1+\lambda} \right)^j = e^{\frac{1-\sqrt{4j+1}}{2}} \left(1 - \frac{2}{1+\sqrt{4j+1}} \right)^j, \quad j \ge (\lambda_0+1)\lambda_0. \qquad [5.34]$$

PROOF.– To obtain [5.33], we find the derivative

$$\frac{d}{d\lambda}\left[\lambda^{-1}e^{-\lambda}\left(\frac{\lambda}{1+\lambda}\right)^j\right] = \frac{e^{-\lambda}\lambda^{j-2}}{(\lambda+1)^{j+1}}\left(j-(\lambda+1)^2\right).$$

Since $\sqrt{j}-1 \geq \lambda_0$, we have

$$\max_{\lambda \geq \lambda_0} \lambda^{-1}e^{-\lambda}\left(\frac{\lambda}{1+\lambda}\right)^j = \lambda^{-1}e^{-\lambda}\left(\frac{\lambda}{1+\lambda}\right)^j\bigg|_{\lambda=\sqrt{j}-1} = \frac{e^{1-\sqrt{j}}}{\sqrt{j}-1}\left(1-\frac{1}{\sqrt{j}}\right)^j.$$

To prove [5.34], we find the derivative

$$\frac{d}{d\lambda}\left[e^{-\lambda}\left(\frac{\lambda}{1+\lambda}\right)^j\right] = \frac{e^{-\lambda}\lambda^{j-1}}{(\lambda+1)^{j+1}}\left(-\lambda^2-\lambda+j\right).$$

Since $\sqrt{4j+1}-1 \geq 2\lambda_0$, we obtain

$$\sup_{\lambda \geq \lambda_0} e^{-\lambda}\left(\frac{\lambda}{1+\lambda}\right)^j = e^{-\lambda}\left(\frac{\lambda}{1+\lambda}\right)^j\bigg|_{\lambda=\frac{\sqrt{4j+1}-1}{2}} = e^{\frac{1-\sqrt{4j+1}}{2}}\left(1-\frac{2}{1+\sqrt{4j+1}}\right)^j.$$

The lemma is proved.

Now we come to the following assertion.

THEOREM 5.6.– *Let the following conditions be fulfilled:*

$$M = N, \quad M+1 \geq (\lambda_0+1)^2,$$

$$\|f_c\|_\infty \equiv \left(\sum_{k=1}^\infty \left\|e^{\sqrt{k}}f_{c,k}\right\|^2\right)^{1/2} < \infty, \quad \|f_s\|_\infty \equiv \left(\sum_{k=1}^\infty \left\|e^{\sqrt{k}}f_{s,k}\right\|^2\right)^{1/2} < \infty,$$

$$f_0 \in D(e^A), \quad f_{c,k} \in D(e^A) \ \forall k \in \mathbb{N}, \quad \|f_c\|_A \equiv \left(\sum_{k=1}^\infty \left\|e^A f_{c,k}\right\|^2\right)^{1/2} < \infty.$$

Then, the accuracy of the approximate solution [5.24] is characterized by the weighted estimate

$$\left\| \frac{u(x) - u_{N,M}(x)}{\min(x, 1-x)} \right\| \le C \frac{e^{-\sqrt{N}}}{\sqrt{N}} \left(\| f_s \|_\infty + \| e^A f_0 \| + \| f_c \|_\infty + \| f_c \|_A \right), \quad [5.35]$$

$$x \in (0,1),$$

where the constant C is independent of N:

$$C = \max\left\{ \frac{1}{\pi\sqrt{2}}, \frac{2e(1+\sqrt{2}) \cdot 0.62}{3}, \frac{3.94\sqrt{e}}{18\sqrt{5}}, \frac{1}{3\sqrt{6\pi^2}} \right\} = \frac{3.94\sqrt{e}}{18\sqrt{5}} = 2.712505313\ldots.$$

PROOF.– The error of the approximate solution [5.24] can be given as the sum of five terms:

$$u(x) - u_{N,M}(x) = D_1 + D_2 + D_3 + D_4 + D_5,$$

where

$$D_1 = \sum_{k=N+1}^{\infty} \sqrt{2} \sin(2k\pi x) \left[(2k\pi)^2 I + A \right]^{-1} f_{s,k},$$

$$D_2 = -\sum_{j=M+1}^{\infty} \left[v_j(1-x) + v_j(x) \right] A^{-1} \left[(I+A)^{-1} A \right]^j f_0,$$

$$D_3 = \sum_{k=N+1}^{\infty} \sqrt{2} \left[(2k\pi)^2 I + A \right]^{-1} (\cos(2k\pi x) - 1) f_{c,k},$$

$$D_4 = -\sum_{k=1}^{\infty} \sqrt{2} \left[(2k\pi)^2 I + A \right]^{-1} \sum_{j=M+1}^{\infty} \left[v_j(1-x) + v_j(x) \right] \left[(I+A)^{-1} A \right]^j f_{c,k},$$

$$D_5 = -\sum_{k=N+1}^{\infty} \sqrt{2} \left[(2k\pi)^2 I + A \right]^{-1} \sum_{j=1}^{M} \left[v_j(1-x) + v_j(x) \right] \left[(I+A)^{-1} A \right]^j f_{c,k}.$$

Next, we study each of these summands. For the first summand, we have

$$\left\|\frac{D_1}{\min(x,1-x)}\right\| = \left\|\sum_{k=N+1}^{\infty} \sqrt{2}\frac{\sin(2k\pi x)}{\min(x,1-x)}\left[(2k\pi)^2 I + A\right]^{-1} f_{s,k}\right\| \leq$$

$$\leq \sum_{k=N+1}^{\infty} \frac{\sqrt{2}|\sin(2k\pi x)|}{\min(x,1-x)} \left\|\int_{\lambda_0}^{+\infty} \left[(2k\pi)^2 + \lambda\right]^{-1} dE(\lambda) f_{s,k}\right\| \leq$$

$$\leq \sum_{k=N+1}^{\infty} \frac{2k\pi\sqrt{2}}{(2k\pi)^2 + \lambda_0} \|f_{s,k}\| \leq \sum_{k=N+1}^{\infty} \frac{\sqrt{2}}{2k\pi} e^{-\sqrt{k}} \|e^{\sqrt{k}} f_{s,k}\| \leq$$

$$\leq \frac{e^{-\sqrt{N}}}{\sqrt{2}\pi} \sum_{k=N+1}^{\infty} \frac{1}{k} \|e^{\sqrt{k}} f_{s,k}\| \leq \frac{e^{-\sqrt{N}}}{\sqrt{2}\pi} \left(\sum_{k=N+1}^{\infty} \frac{1}{k^2}\right)^{1/2} \left(\sum_{k=N+1}^{\infty} \|e^{\sqrt{k}} f_{s,k}\|^2\right)^{1/2} \leq$$

$$\leq \frac{e^{-\sqrt{N}}}{\sqrt{2}\pi} \left(\int_N^{+\infty} \frac{dx}{x^2}\right)^{1/2} \left(\sum_{k=1}^{\infty} \|e^{\sqrt{k}} f_{s,k}\|^2\right)^{1/2} = \frac{1}{\sqrt{2}\pi} \frac{e^{-\sqrt{N}}}{\sqrt{N}} \|f_s\|_\infty. \qquad [5.36]$$

To estimate the second summand, we apply Lemmas 5.2 and 5.5:

$$\left\|\frac{D_2}{\min(x,1-x)}\right\| = \left\|\sum_{j=M+1}^{\infty} \frac{v_j(1-x)+v_j(x)}{\min(x,1-x)} A^{-1}\left[(I+A)^{-1}A\right]^j f_0\right\| \leq$$

$$\leq \sum_{j=M+1}^{\infty} \frac{|v_j(1-x)|+|v_j(x)|}{\min(x,1-x)} \left\|\int_{\lambda_0}^{+\infty} \lambda^{-1} e^{-\lambda}\left[(1+\lambda)^{-1}\lambda\right]^j dE(\lambda) e^A f_0\right\| \leq$$

$$\leq \sum_{j=M+1}^{\infty} \frac{2}{3} \max_{\lambda \geq \lambda_0} \lambda^{-1} e^{-\lambda}\left[(1+\lambda)^{-1}\lambda\right]^j \|e^A f_0\| =$$

$$= \sum_{j=M+1}^{\infty} \frac{2}{3} \frac{e^{1-\sqrt{j}}}{\sqrt{j}-1}\left(1-\frac{1}{\sqrt{j}}\right)^j \|e^A f_0\| \leq \frac{2}{3} \frac{e^{1-\sqrt{M+1}}}{\sqrt{M+1}-1} \sum_{j=1}^{\infty}\left(1-\frac{1}{\sqrt{j}}\right)^j \|e^A f_0\| \leq$$

$$\leq \frac{2e(1+\sqrt{2}) \cdot 0.62}{3} \frac{e^{-\sqrt{M}}}{\sqrt{M}} \|e^A f_0\| \text{ for } M+1 \geq (\lambda_0+1)^2. \qquad [5.37]$$

Here, we used the inequality

$$\frac{1}{\sqrt{M+1}-1} = \frac{\sqrt{M+1}+1}{M} \leq \frac{\sqrt{2M}+\sqrt{M}}{M} = \frac{\sqrt{2}+1}{\sqrt{M}} \quad (M \geq 1)$$

and the series $\sum_{j=1}^{\infty}\left(1-\frac{1}{\sqrt{j}}\right)^j = 0.6159302120... < 0.62$, which is convergent due to the logarithmic test:

$$\frac{-\ln\left(1-\frac{1}{\sqrt{j}}\right)^j}{\ln j} = \frac{-j\ln\left(1-\frac{1}{\sqrt{j}}\right)}{\ln j} \geq l > 1 \quad \forall j \geq j_0.$$

For the third summand, we have

$$\left\|\frac{D_3}{\min(x,1-x)}\right\| = \left\|\sum_{k=N+1}^{\infty} \sqrt{2}\frac{\cos(2k\pi x)-1}{\min(x,1-x)}\left[(2k\pi)^2 I + A\right]^{-1} f_{c,k}\right\| \leq$$

$$\leq \sum_{k=N+1}^{\infty} \frac{2\sqrt{2}\sin^2(k\pi x)}{\min(x,1-x)} \left\|\int_{\lambda_0}^{+\infty}\left[(2k\pi)^2+\lambda\right]^{-1} dE(\lambda) f_{c,k}\right\| \leq$$

$$\leq \sum_{k=N+1}^{\infty} \frac{2\sqrt{2}k\pi}{(2k\pi)^2+\lambda_0} \|f_{c,k}\| \leq \sum_{k=N+1}^{\infty} \frac{\sqrt{2}}{2k\pi} e^{-\sqrt{k}} \|e^{\sqrt{k}} f_{c,k}\| \leq$$

$$\leq \frac{e^{-\sqrt{N}}}{\sqrt{2}\pi} \sum_{k=N+1}^{\infty} \frac{1}{k} \|e^{\sqrt{k}} f_{c,k}\| \leq \frac{e^{-\sqrt{N}}}{\sqrt{2}\pi}\left(\sum_{k=N+1}^{\infty} \frac{1}{k^2}\right)^{1/2}\left(\sum_{k=N+1}^{\infty} \|e^{\sqrt{k}} f_{c,k}\|^2\right)^{1/2} \leq$$

$$\leq \frac{e^{-\sqrt{N}}}{\sqrt{2}\pi}\left(\int_{N}^{+\infty}\frac{dx}{x^2}\right)^{1/2}\left(\sum_{k=1}^{\infty}\|e^{\sqrt{k}} f_{c,k}\|^2\right)^{1/2} = \frac{1}{\sqrt{2}\pi}\frac{e^{-\sqrt{N}}}{\sqrt{N}} \|f_c\|_{\infty}. \qquad [5.38]$$

The fourth summand can be estimated as follows:

$$\left\| \frac{D_4}{\min(x,1-x)} \right\| = \left\| -\sum_{k=1}^{\infty} \sqrt{2} \left[(2k\pi)^2 I + A \right]^{-1} \times \right.$$

$$\left. \times \sum_{j=M+1}^{\infty} \frac{v_j(1-x) + v_j(x)}{\min(x,1-x)} \left[(I+A)^{-1} A \right]^j f_{c,k} \right\| \le$$

$$\le \sqrt{2} \sum_{k=1}^{\infty} \sum_{j=M+1}^{\infty} \frac{|v_j(1-x)| + |v_j(x)|}{\min(x,1-x)} \left\| \int_{\lambda_0}^{+\infty} \frac{e^{-\lambda} \left[(1+\lambda)^{-1} \lambda \right]^j}{(2k\pi)^2 + \lambda} dE(\lambda) e^A f_{c,k} \right\| \le$$

$$\le \frac{2\sqrt{2}}{3} \sum_{k=1}^{\infty} \frac{1}{(2k\pi)^2} \| e^A f_{c,k} \| \sum_{j=M+1}^{\infty} \sup_{\lambda \ge \lambda_0} e^{-\lambda} \left[(1+\lambda)^{-1} \lambda \right]^j =$$

$$= \frac{\sqrt{2}}{6\pi^2} \sum_{k=1}^{\infty} \frac{1}{k^2} \| e^A f_{c,k} \| \sum_{j=M+1}^{\infty} \frac{e^{\frac{1-\sqrt{4j+1}}{2}}}{\sqrt{j}} \sqrt{j} \left(1 - \frac{2}{1+\sqrt{4j+1}} \right)^j \le$$

$$\le \frac{\sqrt{2}}{6\pi^2} \left(\sum_{k=1}^{\infty} \frac{1}{k^4} \right)^{1/2} \left(\sum_{k=1}^{\infty} \| e^A f_{c,k} \|^2 \right)^{1/2} \frac{e^{\frac{1-\sqrt{4(M+1)+1}}{2}}}{\sqrt{M+1}} \sum_{j=1}^{\infty} \sqrt{j} \left(1 - \frac{2}{1+\sqrt{4j+1}} \right)^j \le$$

$$\le \frac{\sqrt{2}}{6\pi^2} \frac{\pi^2}{\sqrt{90}} \left(\sum_{k=1}^{\infty} \| e^A f_{c,k} \|^2 \right)^{1/2} \frac{e^{\frac{1}{2}-\sqrt{M}}}{\sqrt{M}} 3.94 =$$

$$= \frac{3.94\sqrt{e}}{18\sqrt{5}} \frac{e^{-\sqrt{M}}}{\sqrt{M}} \| f_c \|_A \quad (M+1 \ge (\lambda_0 + 1)\lambda_0), \qquad [5.39]$$

where we applied Lemmas 5.2 and 5.5 and used the series

$$\sum_{j=1}^{\infty} \sqrt{j} \left(1 - \frac{2}{1+\sqrt{4j+1}} \right)^j = 3.930506503\ldots < 3.94,$$

which is convergent due to the logarithmic test:

$$\frac{-\ln\left[\sqrt{j}\left(1-\frac{2}{1+\sqrt{4j+1}}\right)^j\right]}{\ln j} = -\frac{1}{2}+\frac{-j\ln\left(1-\frac{2}{1+\sqrt{4j+1}}\right)}{\ln j} \geq l > 1 \quad \forall j \geq j_0.$$

For the fifth summand, we obtain

$$\left\|\frac{D_5}{\min(x,1-x)}\right\| = \left\|-\sum_{k=N+1}^{\infty}\sqrt{2}\left[(2k\pi)^2 I + A\right]^{-1} \times \right.$$

$$\left.\times \sum_{j=1}^{M}\frac{v_j(1-x)+v_j(x)}{\min(x,1-x)}\left[(I+A)^{-1}A\right]^j f_{c,k}\right\| \leq$$

$$\leq \sqrt{2}\sum_{k=N+1}^{\infty}\sum_{j=1}^{M}\frac{|v_j(1-x)|+|v_j(x)|}{\min(x,1-x)}\left\|\int_{\lambda_0}^{+\infty}\frac{\left[(I+\lambda)^{-1}\lambda\right]^j}{(2k\pi)^2+\lambda}dE(\lambda)f_{c,k}\right\| \leq$$

$$\leq \frac{2\sqrt{2}}{3}\sum_{k=N+1}^{\infty}\sum_{j=1}^{M}\frac{1}{(2k\pi)^2}\|f_{c,k}\| = \frac{2\sqrt{2}}{3}\frac{M}{4\pi^2}\sum_{k=N+1}^{\infty}\frac{e^{-\sqrt{k}}}{k^2}\|e^{\sqrt{k}}f_{c,k}\| \leq$$

$$\leq \frac{\sqrt{2}}{6\pi^2}Me^{-\sqrt{N}}\left(\sum_{k=N+1}^{\infty}\frac{1}{k^4}\right)^{1/2}\left(\sum_{k=N+1}^{\infty}\|e^{\sqrt{k}}f_{c,k}\|^2\right)^{1/2} \leq$$

$$\leq \frac{\sqrt{2}}{6\pi^2}Me^{-\sqrt{N}}\left(\int_N^{+\infty}\frac{dx}{x^4}\right)^{1/2}\left(\sum_{k=1}^{\infty}\|e^{\sqrt{k}}f_{c,k}\|^2\right)^{1/2} = \frac{1}{3\sqrt{6}\pi^2}\frac{Me^{-\sqrt{N}}}{N^{3/2}}\|f_c\|_{\infty}. \quad [5.40]$$

Inequalities [5.36]–[5.40] lead to estimate [5.35]. The theorem is proved.

5.1.5. Conclusion

The exact solution $u(x)$ of problem [5.1] is characterized by the weighted estimates in Theorems 5.1 and 5.2. The approximate solutions $u_N(x)$ and $u_{N,M}(x)$ defined by formulas [5.23] and [5.24] respectively are characterized by the weighted estimates in Theorems 5.3–5.6. The weight function $\rho^{-1}(x)$, $\rho(x) = \min(x, 1-x)$, takes into account the distance of the point $x \in (0,1)$ to the boundary points $x = 0$ and $x = 1$.

The weighted estimates proved in Theorems 5.3 and 5.4 demonstrate that if σ increases (i.e. the Fourier coefficients decay more quickly, i.e. $f(x)$ is more regular), the convergence rate of the approximate methods [5.23] and [5.24] is automatically higher. Thus, these methods are methods without saturation of accuracy in the sense of Babenko (2002). The weighted estimates proved in Theorems 5.5 and 5.5 show that under assumption of the exponential smoothness of the coefficients $f_0, f_{c,k}, f_{s,k}, k = 1, 2, \ldots,$ the approximate methods [5.23] and [5.24] have exponential (moreover, super-exponential) rate of convergence.

5.2. Exact and approximate solutions of the BVP in a Banach space

In this section, we continue to study the Caley transform method of solving BVP for the homogeneous and inhomogeneous equations, but this time with a strongly positive operator in a Banach space.

5.2.1. The BVP for the homogeneous equation

In a Banach space E, we consider the problem

$$\frac{d^2 u(x)}{dx^2} - Au(x) = 0, \quad x \in (0,1),$$

$$u(0) = 0, \quad u(1) = u_1,$$

[5.41]

where $u(x): [0,1] \to E$ is an unknown vector-valued function, $u_1 \in E$ is a given vector, $A: E \to E$ is a closed operator with the dense domain $D(A) \subset E$ and the resolvent set $\rho(A)$. We assume (e.g. Pazy (1983, p. 69)) that there exists $\varphi \in (0, \pi/2)$, $\gamma > 0$, and $L > 0$ such that

$$\Sigma \equiv \{z \in \mathbb{C}: \varphi \leq |\arg z| \leq \pi\} \cup \{z \in \mathbb{C}: |z| \leq \gamma\} \subset \rho(A),$$

$$\|(zI - A)^{-1}\| \leq \frac{L}{1 + |z|} \quad \forall z \in \Sigma.$$

[5.42]

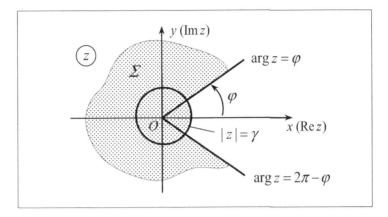

Figure 5.1. *To the definition of a strongly positive operator*

Note that a densely defined closed operator A satisfying conditions [5.42] is called a *strongly positive operator* (e.g. Gavrilyuk and Makarov (1996, 1999)). In Pazy (1983, p. 69), it is shown that the important examples of such operators are produced by the strongly elliptic operators of the order $2m$ in the space $L_p(\Omega)$, $1 \leq p \leq +\infty$, where $\Omega \subset \mathbb{R}^n$ is a bounded domain with the smooth boundary $\partial\Omega$.

In Gavrilyuk and Makarov (1999), it is shown that for $u_1 \in A^\sigma$, $\sigma > 1$, the solution $u(x)$ can be represented by the series

$$u(x) = \operatorname{sh}^{-1}(\sqrt{A})\operatorname{sh}(x\sqrt{A})u_1 = \sum_{k=0}^{\infty} v_k(x) y_k \qquad [5.43]$$

with

$$y_k = (I+A)^{-1} A y_{k-1} = \left[(I+A)^{-1} A\right]^k u_1, \qquad [5.44]$$

where $(I+A)^{-1}A$ is a linear fractional transformation of the operator A, and the functions $v_k(x)$ are defined by the recurrent sequence of the integral equations [5.10]. In section 5.1, it was mentioned that the polynomials $v_k(x)$ are closely related to the Meixner polynomials (Meixner 1934; Makarov 2019) and play the same role as the Laguerre polynomials in the Cayley transform method for solving the Cauchy problem for a first-order abstract differential equation with a strongly positive operator coefficient (Gavrilyuk and Makarov 2004).

Next, we will need some notation and terminology related to certain classes of vectors from E (Gorbachuk and Kniaziuk 1989). Let $C^\infty(A) = \bigcap_{n=0}^{\infty} D(A^n)$ be the set of all infinitely differentiable vectors of the operator A. If a densely defined closed linear operator A has at least one regular point (i.e. $\rho(A) \neq \varnothing$), then $C^\infty(A)$ is dense in E: $\overline{C^\infty(A)} = E$.

Let $(m_n)_{n=0}^{\infty}$ be a non-decreasing sequence of positive numbers and let $\nu > 0$. We denote by $C(A, (m_n), \nu)$ the Banach space of vectors $f \in C^\infty(A)$ with the norm

$$\|f\|_{C(A,(m_n),\nu)} = \sup_n \frac{\|A^n f\|}{\nu^n m_n}. \qquad [5.45]$$

The class $C(A, (m_n)) \equiv \bigcup_{\nu > 0} C(A, (m_n), \nu)$ for various sequences (m_n) is discussed in Gorbachuk and Knyazyuk (1989). For example, vectors from $C(A, (n^n))$ with $m_n = n^n$ are called *analytic* for the operator A (Nelson 1959); vectors from the Gevrey class of Roumieu type $C(A, (n^{n\beta}))$ with $m_n = n^{n\beta}$, $\beta > 1$, are called *ultradifferentiable* (Beals 1972), and vectors from $C(A, (1))$ with $m_n \equiv 1$ are known as *vectors of exponential type* (Radyno 1985).

For convenience, we prove some auxiliary inequalities which we will need further in the text (Makarov and Mayko 2021).

LEMMA 5.6.– *For a strongly positive operator A, the following estimate holds true:*

$$\left\|\left(I + \frac{A}{j}\right)^j A^{-j}\right\| \leq (L+1)^j \quad (j \in \mathbb{N}). \qquad [5.46]$$

PROOF.– From [5.42], we easily have $\|A^{-1}\| \le L$. Then

$$\left\|\left(I+\frac{A}{j}\right)^j A^{-j}\right\| = \left\|\sum_{s=0}^{j}\binom{j}{s}\left(\frac{A}{j}\right)^s A^{-j}\right\| = \left\|\sum_{s=0}^{j}\binom{j}{s}\frac{A^{-(j-s)}}{j^s}\right\| \le$$

$$\le \sum_{s=0}^{j}\binom{j}{s}\frac{\|A^{-(j-s)}\|}{j^s} \le \sum_{s=0}^{j}\binom{j}{s}\frac{\|A^{-1}\|^{j-s}}{j^s} \le \sum_{s=0}^{j}\binom{j}{s}\frac{L^{j-s}}{j^s} = \left(L+\frac{1}{j}\right)^j \le (L+1)^j,$$

which proves the lemma.

LEMMA 5.7.– *For $n > 0$, $\alpha > 0$, the following estimate holds true:*

$$\max_{t \ge 1}\left[\left(1-\frac{1}{t}\right)^n t^{-\alpha}\right] \le \left(\frac{\alpha}{e}\right)^\alpha n^{-\alpha}.$$

PROOF.– For the function

$$\varphi(t) = \left(1-\frac{1}{t}\right)^n t^{-\alpha}, \quad t \ge 1,$$

we find the derivative $\varphi'(t) = (t-1)^{n-1} t^{-\alpha-n-1}(n+\alpha-\alpha t)$. Therefore, $\varphi(t)$ takes its maximum value at the point $t = \dfrac{n+\alpha}{\alpha}$:

$$\max_{t \ge 1}\varphi(t) = \varphi\left(\frac{n+\alpha}{\alpha}\right) = \left(1-\frac{\alpha}{n+\alpha}\right)^n \left(\frac{n+\alpha}{\alpha}\right)^{-\alpha} =$$

$$= \left(1-\frac{\alpha}{n+\alpha}\right)^n \left(1+\frac{\alpha}{n}\right)^{-\alpha} \alpha^\alpha n^{-\alpha} = \left(\frac{n}{n+\alpha}\right)^{n+\alpha} \alpha^\alpha n^{-\alpha} \le e^{-\alpha}\alpha^\alpha n^{-\alpha}.$$

The lemma is proved.

LEMMA 5.8.– *For $n > \alpha > 0$, the following estimate holds true:*

$$\max_{t > 0}\left[\left(\frac{t}{t+1}\right)^n t^{-\alpha}\right] \le \alpha^\alpha n^{-\alpha}.$$

PROOF.– The function

$$\varphi(t) = \left(\frac{t}{t+1}\right)^n t^{-\alpha}, \quad t > 0,$$

has the derivative $\varphi'(t) = t^{n-\alpha-1}(t+1)^{-n-1}(n-\alpha-\alpha t)$, which shows that $\varphi(t)$ takes its maximum value at the point $t = \dfrac{n-\alpha}{\alpha}$:

$$\max_{t>0} \varphi(t) = \varphi\left(\frac{n-\alpha}{\alpha}\right) = \left(\frac{n-\alpha}{n}\right)^n \left(\frac{n-\alpha}{\alpha}\right)^{-\alpha} = \left(\frac{n-\alpha}{n}\right)^{n-\alpha} \alpha^\alpha n^{-\alpha} < \alpha^\alpha n^{-\alpha}.$$

This proves the lemma.

LEMMA 5.9.– *The following estimate holds true:*

$$\max_{t \geq 0} \left(\frac{t}{\left(\frac{t}{k}+1\right)(t+1)}\right)^k = \left(\frac{\sqrt{k}}{\left(\frac{1}{\sqrt{k}}+1\right)(\sqrt{k}+1)}\right)^k \leq e^{1-2\sqrt{k}} \quad (k \in \mathbb{N}).$$

This inequality is unimprovable, namely:

$$\left(\frac{\sqrt{k}}{\left(\frac{1}{\sqrt{k}}+1\right)(\sqrt{k}+1)}\right)^k \sim ee^{-2\sqrt{k}} \text{ as } k \to \infty.$$

PROOF.– We consider the function

$$g(t) = \left(\frac{t}{\left(\frac{t}{k}+1\right)(t+1)}\right)^k, \quad t \geq 0,$$

and find the derivative $\dfrac{d \ln g(t)}{dt} = \dfrac{k(k-t^2)}{t(t+k)(t+1)}$, which means that $g(t)$ takes its maximum value at the point $t = \sqrt{k}$. Therefore, we have

$$\max_{t\geq 0} g(t) = g(\sqrt{k}) = \left(\frac{\sqrt{k}}{\left(\frac{1}{\sqrt{k}}+1\right)(\sqrt{k}+1)}\right)^k = \left(1+\frac{1}{\sqrt{k}}\right)^{-2k}.$$

Using the inequality

$$\left(1+\frac{1}{\sqrt{k}}\right)^{-2k} = e^{-2k\ln\left(1+\frac{1}{\sqrt{k}}\right)} \leq e^{-2k\left(\frac{1}{\sqrt{k}}-\frac{1}{2k}\right)} = ee^{-2\sqrt{k}}$$

and the asymptotic relation

$$\left(1+\frac{1}{\sqrt{k}}\right)^{-2k} = e^{-2k\ln\left(1+\frac{1}{\sqrt{k}}\right)} = e^{-2k\left(\frac{1}{\sqrt{k}}-\frac{1}{2k}+o\left(\frac{1}{k}\right)\right)} = e^{1-2\sqrt{k}+o(1)} \sim e^{1-2\sqrt{k}}$$

as $k \to \infty$, we complete the proof of the lemma.

In the next two lemmas, we obtain the inequalities for the norm $\|y_k\|$.

LEMMA 5.10.– *Under the assumptions*

$$\sigma > 0, \quad 0 < \varepsilon_2 < \min\{1,\sigma\}, \quad k > \sigma - \varepsilon_2, \quad u_1 \in D(A^\sigma),$$

the estimate

$$\|y_k\| \leq \frac{C_2}{k^{(\sigma-\varepsilon_2)/2}} \|A^\sigma u_1\|, \qquad [5.47]$$

holds true with $C_2 = \dfrac{L}{\sin(\pi\varepsilon_2)}(\sigma-\varepsilon_2)^{(\sigma-\varepsilon_2)/2}$.

PROOF.– To represent y_k, we use formula [5.44] and integrate along the path $\Gamma = \Gamma_+ \cup \Gamma_-$ consisting of two rays on the complex plane:

$$\Gamma_\pm = \{z \in \mathbb{C} : z = \rho e^{\pm i\varphi}, \ \rho \in [0,+\infty)\}. \qquad [5.48]$$

Then, we have

$$\|y_k\| = \left\|\left[(I+A)^{-1}A\right]^k u_1\right\| = \left\|\frac{1}{2\pi i}\int_\Gamma \left(\frac{z}{1+z}\right)^k z^{-\sigma}(zI-A)^{-1}A^\sigma u_1 dz\right\| \leq$$

$$\leq \frac{1}{2\pi}\int_\Gamma \left|\frac{z}{1+z}\right|^k |z|^{-\sigma}\frac{L}{1+|z|}|dz| \, \|A^\sigma u_1\|.$$

Making use of the relations

$$|z| = |\rho e^{\pm i\varphi}| = \rho, \quad |dz| = |d(\rho e^{\pm i\varphi})| = |e^{\pm i\varphi}|d\rho = d\rho,$$

$$\left|\frac{z}{1+z}\right|^2 = \left|\frac{\rho e^{\pm i\varphi}}{1+\rho e^{\pm i\varphi}}\right|^2 = \frac{\rho^2}{1+2\rho\cos\varphi+\rho^2} \leq \frac{\rho^2}{1+\rho^2},$$

we have

$$\|y_k\| \leq \frac{L}{\pi}\int_0^{+\infty}\left(\frac{\rho^2}{1+\rho^2}\right)^{k/2}\frac{\rho^{-\sigma}}{1+\rho}d\rho \, \|A^\sigma u_1\| =$$

$$= \frac{L}{\pi}\int_0^{+\infty}\left(\frac{\rho^2}{1+\rho^2}\right)^{k/2}(\rho^2)^{-(\sigma-\varepsilon_2)/2}\frac{d\rho}{(1+\rho)\rho^{\varepsilon_2}} \, \|A^\sigma u_1\| \leq$$

$$\leq \frac{L}{\pi}\sup_{\rho>0}\left[\left(\frac{\rho^2}{1+\rho^2}\right)^{k/2}(\rho^2)^{-(\sigma-\varepsilon_2)/2}\right]\int_0^{+\infty}\frac{d\rho}{(1+\rho)\rho^{\varepsilon_2}} \, \|A^\sigma u_1\| \leq$$

$$\leq \frac{L}{\pi}\sup_{t>0}\left[\left(\frac{t}{1+t}\right)^{k/2}t^{-(\sigma-\varepsilon_2)/2}\right]\frac{\pi}{\sin(\pi\varepsilon_2)} \, \|A^\sigma u_1\|.$$

Applying here Lemma 5.8 for $n = \dfrac{k}{2}$, $\alpha = \dfrac{\sigma-\varepsilon_2}{2}$, we come to the inequality

$$\|y_k\| \leq \frac{L}{\pi}\left(\frac{\sigma-\varepsilon_2}{k}\right)^{(\sigma-\varepsilon_2)/2}\frac{\pi}{\sin(\pi\varepsilon_2)} \, \|A^\sigma u_1\|$$

($k > \sigma - \varepsilon_2$, $0 < \varepsilon_2 < \min\{1, \sigma\}$),

which yields estimate [5.47] and therefore proves the lemma.

LEMMA 5.11.– *Let $u_1 \in E$ satisfy the condition $u_1 \in D(A^k)$ for all $k \in \mathbb{N}$. Then, the following estimate holds true:*

$$\| y_k \| \leq \frac{Le}{\sqrt{2}} e^{-2\sqrt{k}} \frac{(\sqrt{k}+1)^2}{\sqrt{k}} \| u_1 \|_{C(A,(I),v)}, \quad v = \frac{\cos\varphi}{L+1}. \qquad [5.49]$$

PROOF.– Making use of relation [5.44] and integrating along the path described in [5.48], we obtain

$$\| y_k \| = \left\| \left[(I+A)^{-1} A \right]^k u_1 \right\| =$$

$$= \left\| \frac{1}{2\pi i} \int_\Gamma \left(\frac{z}{1+z} \right)^k \left(1+\frac{z}{k} \right)^{-k} (zI-A)^{-1} \left(I+\frac{A}{k} \right)^k u_1 dz \right\| \leq$$

$$\leq \frac{1}{2\pi} \int_\Gamma \left| \frac{z}{1+z} \right|^k \left| 1+\frac{z}{k} \right|^{-k} \frac{L}{1+|z|} |dz| \left\| \left(I+\frac{A}{k} \right)^k u_1 \right\| =$$

$$= \frac{2L}{2\pi} \int_0^{+\infty} \left(\frac{\rho \cos\varphi}{\sqrt{1+2\rho\cos\varphi+\rho^2} \sqrt{1+2\frac{\rho}{k}\cos\varphi+\left(\frac{\rho}{k}\right)^2}} \right)^{k-1} \times$$

$$\times \frac{\rho\cos\varphi}{\sqrt{1+2\rho\cos\varphi+\rho^2} \sqrt{1+2\frac{\rho}{k}\cos\varphi+\left(\frac{\rho}{k}\right)^2}} \frac{d\rho}{1+\rho} \cos^{-k}\varphi \left\| \left(I+\frac{A}{k} \right)^k A^{-k} A^k u_1 \right\|.$$

Applying here Lemma 5.6 and using the relations

$$\sqrt{1+2\rho\cos\varphi+\rho^2} \geq \sqrt{1+2\rho\cos\varphi+\rho^2\cos^2\varphi} = \sqrt{(1+\rho\cos\varphi)^2} = 1+\rho\cos\varphi,$$

$$\sqrt{1+2\frac{\rho}{k}\cos\varphi+\left(\frac{\rho}{k}\right)^2} \geq \sqrt{1+\left(\frac{\rho}{k}\right)^2} = \sqrt{\frac{2\rho}{k}},$$

$$|dz| = |d(\rho e^{\pm i\varphi})| = |e^{\pm i\varphi}|\,d\rho = d\rho,$$

we come to the representation

$$\|y_k\| = \frac{L\sqrt{k}}{\pi\sqrt{2}} \int_0^{+\infty} \left(\frac{\rho\cos\varphi}{(1+2\rho\cos\varphi)\left(1+\frac{\rho}{k}\cos\varphi\right)}\right)^{k-1} \frac{d\rho}{\sqrt{\rho(1+\rho)}} \times$$

$$\times \cos^{-k}\varphi \left\|\left(I+\frac{A}{k}\right)^k A^{-k}\right\| \|A^k u_1\|.$$

Next, applying Lemma 5.9 and norm [5.45], we obtain

$$\|y_k\| \leq \frac{L\sqrt{k}}{\pi\sqrt{2}} \max_{t\geq 0}\left[\frac{t}{(1+t)\left(1+\frac{t}{k}\right)}\right]^{k-1} \underbrace{\int_0^{+\infty} \frac{d\rho}{\sqrt{\rho(1+\rho)}}}_{=\pi} \cos^{-k}\varphi (L+1)^k \|A^k u_1\| \leq$$

$$\leq \frac{L\sqrt{k}}{\sqrt{2}}\left[\frac{\sqrt{k}}{(1+\sqrt{k})\left(1+\frac{1}{\sqrt{k}}\right)}\right]^{k-1} \frac{\|A^k u_1\|}{v^k} \leq$$

$$\leq \frac{L\sqrt{k}}{\sqrt{2}}\left[\frac{\sqrt{k}}{(1+\sqrt{k})\left(1+\frac{1}{\sqrt{k}}\right)}\right]^k \left(\frac{\sqrt{k}+1}{\sqrt{k}}\right)^2 \frac{\|A^k u_1\|}{v^k} \leq$$

$$\leq \frac{L\sqrt{k}}{\sqrt{2}} ee^{-2\sqrt{k}}\left(\frac{\sqrt{k}+1}{\sqrt{k}}\right)^2 \frac{\|A^k u_1\|}{v^k} \leq \frac{Le}{\sqrt{2}} e^{-2\sqrt{k}} \frac{(\sqrt{k}+1)^2}{\sqrt{k}} \|u_1\|_{C(A,(1),v)},$$

which finally gives estimate [5.49] and proves the lemma.

As the approximate solution of problem [5.41], we take the Nth partial sum of series [5.43]:

$$u_N(x) = \sum_{k=0}^{N} v_k(x) y_k .\qquad [5.50]$$

In the next two theorems, we study the error

$$u(x) - u_N(x) = \sum_{k=N+1}^{\infty} v_k(x) y_k$$

under various assumptions about the smoothness of the input vector u_1.

THEOREM 5.7.– Let $u_1 \in D(A^\sigma)$, $\sigma > 1$. Then the accuracy of the approximate solution [5.50] is characterized by the weighted estimate

$$\left\| \frac{u(x) - u_N(x)}{\min(x, 1-x)} \right\| \leq \frac{C}{N^{(\sigma-1-\varepsilon)/2}} \| A^\sigma u_1 \|, \quad x \in (0,1) \quad (N \geq \sigma - 1), \qquad [5.51]$$

where $\varepsilon > 0$ is an arbitrary small number and C is a positive constant independent of N.

PROOF.– For $\sigma > 1$, $0 < \varepsilon_1 < 1$, $0 < \varepsilon_2 < 1$, $1 + \varepsilon_1 + \varepsilon_2 < \sigma$, $N + 1 > \sigma - \varepsilon_2$, the conditions of Lemmas 5.2 and 5.10 are satisfied. Then, we have

$$\left\| \frac{u(x) - u_N(x)}{\min(x, 1-x)} \right\| = \left\| \sum_{k=N+1}^{\infty} \frac{v_k(x)}{\min(x, 1-x)} y_k \right\| \leq \sum_{k=N+1}^{\infty} \left| \frac{v_k(x)}{\min(x, 1-x)} \right| \| y_k \| \leq$$

$$\leq \sum_{k=N+1}^{\infty} \frac{C_1}{k^{(1-\varepsilon_1)/2}} \frac{C_2}{k^{(\sigma-\varepsilon_2)/2}} \| A^\sigma u_1 \| = C_1 C_2 \sum_{k=N+1}^{\infty} \frac{1}{k^{(1+\sigma-\varepsilon_1-\varepsilon_2)/2}} \| A^\sigma u_1 \| \leq$$

$$\leq C_1 C_2 \int_N^{+\infty} \frac{dx}{x^{(1+\sigma-\varepsilon_1-\varepsilon_2)/2}} \| A^\sigma u_1 \| = \frac{2 C_1 C_2}{\sigma - 1 - \varepsilon_1 - \varepsilon_2} \frac{1}{N^{(\sigma-1-\varepsilon_1-\varepsilon_2)/2}} \| A^\sigma u_1 \|,$$

which leads to estimate [5.51], and therefore the assertion of the lemma.

THEOREM 5.8.– Let $u_1 \in C(A, (1), v)$, $v = \dfrac{\cos \varphi}{L+1}$. Then, the accuracy of the approximate solution [5.50] is characterized by the weighted estimate

$$\left\|\frac{u(x)-u_N(x)}{\min(x,1-x)}\right\| \leq \frac{Ce^{-\sqrt{N+1}}}{(N+1)^{1/2-\varepsilon}} \|u_1\|_{C(A,(1),v)}, \quad x \in [0,1] \quad (N \in \mathbb{N}), \qquad [5.52]$$

where $\varepsilon > 0$ is an arbitrary small number and C is a constant independent of N.

PROOF.– Applying Lemmas 5.2 and 5.11, we obtain

$$\left\|\frac{u(x)-u_N(x)}{\min(x,1-x)}\right\| = \left\|\sum_{k=N+1}^{\infty} \frac{v_k(x)}{\min(x,1-x)} y_k\right\| \leq \sum_{k=N+1}^{\infty} \left|\frac{v_k(x)}{\min(x,1-x)}\right| \|y_k\| \leq$$

$$\leq \sum_{k=N+1}^{\infty} \frac{C_1}{k^{(1-\varepsilon_1)/2}} \frac{Le}{\sqrt{2}} e^{-2\sqrt{k}} \frac{\left(\sqrt{k}+1\right)^2}{\sqrt{k}} \|u_1\|_{C(A,(1),v)} \leq$$

$$\leq \frac{C_1 Le}{\sqrt{2}} \frac{e^{-\sqrt{N+1}}}{(N+1)^{(1-\varepsilon_1)/2}} \sum_{k=1}^{\infty} e^{-\sqrt{k}} \frac{\left(\sqrt{k}+1\right)^2}{\sqrt{k}} \|u_1\|_{C(A,(1),v)}.$$

Denoting by S the sum of the convergent series

$$S = \sum_{k=1}^{\infty} e^{-\sqrt{k}} \frac{\left(\sqrt{k}+1\right)^2}{\sqrt{k}} = 8.152349342\ldots$$

and putting $C = \dfrac{eC_1 LS}{\sqrt{2}}$, we finally come to inequality [5.52] and thus prove the theorem.

5.2.2. *The BVP for the inhomogeneous equation*

In a Banach space E, we consider the BVP for the inhomogeneous equation:

$$\frac{d^2 u(x)}{dx^2} - Au(x) = -f(x), \quad x \in (0,1),$$
$$u(0) = 0, \quad u(1) = 0, \qquad [5.53]$$

with the operator A under the same assumption as in problem [5.41].

Omitting the details of obtaining formula [5.18], we again present the solution $u(x)$ in the form

$$u(x) = \sum_{k=1}^{\infty} \sqrt{2} \sin(2k\pi x) \left[(2k\pi)^2 I + A\right]^{-1} f_{s,k} +$$

$$+ A^{-1} \left\{ I - \sum_{j=0}^{\infty} \left[v_j(1-x) + v_j(x) \right] \left[(I+A)^{-1} A\right]^j \right\} f_0 + \qquad [5.54]$$

$$+ \sum_{k=1}^{\infty} \sqrt{2} \left[(2k\pi)^2 I + A\right]^{-1} \left\{ \cos(2k\pi x) I - \sum_{j=0}^{\infty} \left[v_j(1-x) + v_j(x) \right] \left[(I+A)^{-1} A\right]^j \right\} f_{c,k}$$

with the vectors $f_{s,k}, f_0, f_{c,k}$ defined in [5.17].

As the approximate solution of problem [5.53], we take the partial sum

$$u_{N,M}(x) = \sum_{k=1}^{N} \sqrt{2} \sin(2k\pi x) \left[(2k\pi)^2 I + A\right]^{-1} f_{s,k} +$$

$$+ A^{-1} \left\{ I - \sum_{j=0}^{M} \left[v_j(1-x) + v_j(x) \right] \left[(I+A)^{-1} A\right]^j \right\} f_0 + \qquad [5.55]$$

$$+ \sum_{k=1}^{N} \sqrt{2} \left[(2k\pi)^2 I + A\right]^{-1} \times$$

$$\times \left\{ \cos(2k\pi x) I - \sum_{j=0}^{M} \left[v_j(1-x) + v_j(x) \right] \left[(I+A)^{-1} A\right]^j \right\} f_{c,k}.$$

To study the accuracy of the approximate solution $u_{N,M}(x)$, we write the error $u(x) - u_{N,M}(x)$ as the sum

$$u(x) - u_{N,M}(x) = \sum_{k=1}^{5} D_k \qquad [5.56]$$

with the terms D_k defined in [5.26].

In the next two propositions, we consider error [5.56] under various assumptions about the smoothness of the function $f(x)$ in terms of its Fourier coefficients f_0, $f_{c,k}$, $f_{s,k}$ given in [5.17].

THEOREM 5.9.– *Let the following conditions be satisfied:*

$$M = N, \quad \sigma > 0, \quad f_0 \in D(A^\sigma), \quad f_{c,k} \in D(A^\sigma) \quad \forall k \in \mathbb{N},$$

$$\| f_s \|_\sigma \equiv \left(\sum_{k=1}^{\infty} k^{\sigma+1} \| f_{s,k} \|^2 \right)^{1/2} < \infty, \qquad [5.57]$$

$$\| f_c \|_\sigma \equiv \left(\sum_{k=1}^{\infty} k^{\sigma+1} \| f_{c,k} \|^2 \right)^{1/2} < \infty, \quad \| f_c \|_{A^\sigma} \equiv \sum_{k=1}^{\infty} \| A^\sigma f_{c,k} \| < \infty.$$

Then, the accuracy of the approximate solution [5.55] is characterized by the estimate

$$\left\| \frac{u(x) - u_{N,N}(x)}{\min(x, 1-x)} \right\| \leq \frac{C}{N^{(\sigma-\varepsilon)/2}} \left(\| f_s \|_\sigma + \| f_c \|_\sigma + \| A^\sigma f_0 \| + \| f_c \|_{A^\sigma} \right),$$

$$x \in [0,1] \quad (N \geq \sigma),$$

where $\varepsilon > 0$ *is an arbitrary small number and* C *is a positive constant independent of N.*

PROOF.– We will now analyze each summand D_k in [5.56]. Throughout the proof, we will integrate along the path described in [5.48]. We will also need the relations

$$\left| (2k\pi)^2 + z \right| = \left| (2k\pi)^2 + \rho e^{\pm i\varphi} \right| = \sqrt{\left[(2k\pi)^2 + \rho \cos\varphi \right]^2 + (\rho \sin\varphi)^2} =$$

$$= \sqrt{(2k\pi)^4 + 2(2k\pi)^2 \rho \cos\varphi + (\rho \cos\varphi)^2 + (\rho \sin\varphi)^2} =$$

$$= \sqrt{(2k\pi)^4 + 2(2k\pi)^2 \rho \cos\varphi + \rho^2} \geq \sqrt{(2k\pi)^4 + \rho^2} \geq \sqrt{2(2k\pi)^2 \rho} = 2k\pi\sqrt{2\rho},$$

$$\left| (2k\pi)^2 + z \right| = \sqrt{(2k\pi)^4 + 2(2k\pi)^2 \rho \cos\varphi + \rho^2} \geq \rho,$$

$$\left|\frac{z}{1+z}\right|^2 = \left|\frac{\rho e^{\pm i\varphi}}{1+\rho e^{\pm i\varphi}}\right|^2 = \frac{\rho^2}{1+2\rho\cos\varphi+\rho^2} \leq \frac{\rho^2}{1+\rho^2}$$

and the Cauchy–Bunyakovsky inequality for infinite number series.

For the summand D_1, we have

$$\left\|\frac{D_1}{\min(x,1-x)}\right\| = \left\|\sum_{k=N+1}^{\infty}\frac{\sqrt{2}\sin(2k\pi x)}{\min(x,1-x)}\left[(2k\pi)^2 I + A\right]^{-1} f_{s,k}\right\| \leq \quad [5.58]$$

$$\leq \sum_{k=N+1}^{\infty}\frac{\sqrt{2}|\sin(2k\pi x)|}{\min(x,1-x)}\left\|\frac{1}{2\pi i}\int_{\Gamma}\frac{1}{(2k\pi)^2+z}(zI-A)^{-1}dz\, f_{s,k}\right\| \leq$$

$$\leq \sum_{k=N+1}^{\infty}\frac{\sqrt{2}\,2k\pi}{2\pi}\int_{\Gamma}\frac{1}{|(2k\pi)^2+z|}\frac{L}{1+|z|}|dz|\,\|f_{s,k}\| \leq$$

$$\leq \sum_{k=N+1}^{\infty}\frac{\sqrt{2}\,2k\pi}{2\pi}\frac{2L}{2k\pi\sqrt{2}}\int_{0}^{+\infty}\frac{d\rho}{\sqrt{\rho}(1+\rho)}\|f_{s,k}\| = L\sum_{k=N+1}^{\infty}\|f_{s,k}\| =$$

$$= L\sum_{k=N+1}^{\infty}\frac{1}{k^{(\sigma+1)/2}}k^{(\sigma+1)/2}\|f_{s,k}\| \leq L\left(\sum_{k=N+1}^{\infty}\frac{1}{k^{\sigma+1}}\right)^{1/2}\left(\sum_{k=N+1}^{\infty}k^{\sigma+1}\|f_{s,k}\|^2\right)^{1/2} \leq$$

$$\leq L\left(\int_{N}^{\infty}\frac{dx}{x^{\sigma+1}}\right)^{1/2}\left(\sum_{k=N+1}^{\infty}k^{\sigma+1}\|f_{s,k}\|^2\right)^{1/2} \leq \frac{L}{\sigma^{1/2}}\frac{1}{N^{\sigma/2}}\left(\sum_{k=1}^{\infty}k^{\sigma+1}\|f_{s,k}\|^2\right)^{1/2} =$$

$$= \frac{L}{\sigma^{1/2}}\frac{1}{N^{\sigma/2}}\|f_s\|_{\sigma} \quad (N\in\mathbb{N},\,\sigma > 0).$$

To estimate the summand D_2, we take

$$0 < \varepsilon_1 < 1,\quad 0 < \varepsilon_3 < 1,\quad \varepsilon_1 + \varepsilon_3 < \sigma,\quad M > \sigma - \varepsilon_3$$

and then apply Lemmas 5.2 and 5.8 with $n = \dfrac{j}{2}$ and $\alpha = \dfrac{1+\sigma-\varepsilon_3}{2}$. We have

$$\left\| \frac{D_2}{\min(x,1-x)} \right\| = \left\| -\sum_{j=M+1}^{\infty} \frac{v_j(1-x) + v_j(x)}{\min(x,1-x)} A^{-1} \left[(I+A)^{-1} A \right]^j f_0 \right\| \leq \qquad [5.59]$$

$$\leq \sum_{j=M+1}^{\infty} \frac{|v_j(1-x)| + |v_j(x)|}{\min(x,1-x)} \left\| \frac{1}{2\pi i} \int_{\Gamma} \left(\frac{z}{1+z} \right)^j z^{-(1+\sigma)} (zI-A)^{-1} dz\, A^{\sigma} f_0 \right\| \leq$$

$$\leq \sum_{j=M+1}^{\infty} \frac{2C_1}{2\pi\, j^{(1-\varepsilon_1)/2}} \int_{\Gamma} \left| \frac{z}{1+z} \right|^j |z|^{-(1+\sigma)} \frac{L}{1+|z|} |dz| \, \|A^{\sigma} f_0\| \leq$$

$$\leq \frac{2C_1 L}{\pi} \sum_{j=M+1}^{\infty} \frac{1}{j^{(1-\varepsilon_1)/2}} \int_0^{\infty} \left(\frac{\rho^2}{1+\rho^2} \right)^{j/2} (\rho^2)^{-(1+\sigma-\varepsilon_3)/2} \frac{d\rho}{\rho^{\varepsilon_3}(1+\rho)} \|A^{\sigma} f_0\| \leq$$

$$\leq \frac{2C_1 L}{\pi} \sum_{j=M+1}^{\infty} \frac{1}{j^{(1-\varepsilon_1)/2}} \sup_{t>0} \left[\left(\frac{t}{1+t} \right)^{j/2} t^{-(1+\sigma-\varepsilon_3)/2} \right] \int_0^{\infty} \frac{d\rho}{\rho^{\varepsilon_3}(1+\rho)} \|A^{\sigma} f_0\| \leq$$

$$\leq \frac{2C_1 L (1+\sigma-\varepsilon_3)^{(1+\sigma-\varepsilon_3)/2}}{\pi} \frac{\pi}{\sin(\pi\varepsilon_3)} \sum_{j=M+1}^{\infty} \frac{1}{j^{(1-\varepsilon_1)/2}} \frac{1}{j^{(1+\sigma-\varepsilon_3)/2}} \|A^{\sigma} f_0\| =$$

$$= \frac{2C_1 L (1+\sigma-\varepsilon_3)^{(1+\sigma-\varepsilon_3)/2}}{\sin(\pi\varepsilon_3)} \sum_{j=M+1}^{\infty} \frac{1}{j^{(2+\sigma-\varepsilon_1-\varepsilon_3)/2}} \|A^{\sigma} f_0\| \leq$$

$$\leq \frac{2C_1 L (1+\sigma-\varepsilon_3)^{(1+\sigma-\varepsilon_3)/2}}{\sin(\pi\varepsilon_3)} \int_M^{+\infty} \frac{dx}{x^{(2+\sigma-\varepsilon_1-\varepsilon_3)/2}} \|A^{\sigma} f_0\| =$$

$$= \frac{2C_1 L (1+\sigma-\varepsilon_3)^{(1+\sigma-\varepsilon_3)/2}}{\sin(\pi\varepsilon_3)} \frac{2}{\sigma-\varepsilon_1-\varepsilon_3} \frac{1}{M^{(\sigma-\varepsilon_1-\varepsilon_3)/2}} \|A^{\sigma} f_0\| =$$

$$= \frac{C_3}{M^{(\sigma-\varepsilon_1-\varepsilon_3)/2}} \|A^{\sigma} f_0\|,$$

where $C_3 = \dfrac{4 C_1 L (1+\sigma-\varepsilon_3)^{(1+\sigma-\varepsilon_3)/2}}{\sin(\pi\varepsilon_3)(\sigma-\varepsilon_1-\varepsilon_3)}$

$(\sigma > 0,\; 0 < \varepsilon_1 < 1,\; 0 < \varepsilon_3 < 1,\; \varepsilon_1 + \varepsilon_3 < \sigma,\; M > \sigma - \varepsilon_3)$.

The summand D_3 can be estimated in the same way as the summand D_1, namely:

$$\left\| \frac{D_3}{\min(x,1-x)} \right\| = \left\| \sum_{k=N+1}^{\infty} \frac{\sqrt{2}(\cos(2k\pi x)-1)}{\min(x,1-x)} \left[(2k\pi)^2 I + A\right]^{-1} f_{c,k} \right\| \leq \quad [5.60]$$

$$\leq L \left(\int_N^{\infty} \frac{dx}{x^{\sigma+1}} \right)^{1/2} \left(\sum_{k=N+1}^{\infty} k^{\sigma+1} \| f_{c,k} \|^2 \right)^{1/2} \leq \frac{L}{\sigma^{1/2}} \frac{1}{N^{\sigma/2}} \left(\sum_{k=1}^{\infty} k^{\sigma+1} \| f_{c,k} \|^2 \right)^{1/2}$$

$$(N \in \mathbb{N}, \sigma > 0).$$

We estimate D_4 similarly to D_2 taking

$$0 < \varepsilon_1 < 1, \quad 0 < \varepsilon_3 < 1, \quad \varepsilon_1 + \varepsilon_3 < \sigma, \quad M > \sigma - \varepsilon_3$$

and applying Lemmas 5.2 and 5.8 with $n = \dfrac{j}{2}$ i $\alpha = \dfrac{1+\sigma-\varepsilon_3}{2}$, therefore

$$\left\| \frac{D_4}{\min(x,1-x)} \right\| = \quad [5.61]$$

$$= \left\| -\sum_{k=1}^{\infty} \sqrt{2} \left[(2k\pi)^2 I + A\right]^{-1} \sum_{j=M+1}^{\infty} \frac{v_j(1-x)+v_j(x)}{\min(x,1-x)} \left[(I+A)^{-1} A\right]^j f_{c,k} \right\| \leq$$

$$\leq \sum_{k=1}^{\infty} \sum_{j=M+1}^{\infty} \frac{|v_j(1-x)|+|v_j(x)|}{\min(x,1-x)} \left\| \frac{\sqrt{2}}{2\pi i} \int_{\Gamma} \left(\frac{z}{1+z}\right)^j \frac{z^{-\sigma}}{(2k\pi)^2+z} (zI-A)^{-1} dz\, A^\sigma f_{c,k} \right\| \leq$$

$$\leq \sum_{k=1}^{\infty} \sum_{j=M+1}^{\infty} \frac{\sqrt{2}\, 2C_1}{2\pi\, j^{(1-\varepsilon_1)/2}} \int_{\Gamma} \left|\frac{z}{1+z}\right|^j \frac{|z|^{-\sigma}}{|(2k\pi)^2+z|} \frac{L}{1+|z|} |dz| \, \| A^\sigma f_{c,k} \| \leq$$

$$\leq \frac{2\sqrt{2} C_1 L}{\pi} \sum_{k=1}^{\infty} \sum_{j=M+1}^{\infty} \frac{1}{j^{(1-\varepsilon_1)/2}} \int_0^{+\infty} \left(\frac{\rho^2}{1+\rho^2}\right)^{j/2} (\rho^2)^{-(1+\sigma-\varepsilon_3)/2} \frac{d\rho}{\rho^{\varepsilon_3}(1+\rho)} \| A^\sigma f_{c,k} \| \leq$$

$$\leq \frac{2\sqrt{2} C_1 L}{\pi} \sum_{j=M+1}^{\infty} \frac{1}{j^{(1-\varepsilon_1)/2}} \sup_{t>0}\left[\left(\frac{t}{1+t}\right)^{j/2} t^{-(1+\sigma-\varepsilon_3)/2}\right] \int_0^{\infty} \frac{d\rho}{\rho^{\varepsilon_3}(1+\rho)} \sum_{k=1}^{\infty} \| A^\sigma f_{c,k} \| \leq$$

$$\le \frac{2\sqrt{2}C_1 L}{\pi}(1+\sigma-\varepsilon_3)^{(1+\sigma-\varepsilon_3)/2}\frac{\pi}{\sin(\pi\varepsilon_3)}\times$$

$$\times \sum_{j=M+1}^{\infty}\frac{1}{j^{(1-\varepsilon_1)/2}}\frac{1}{j^{(1+\sigma-\varepsilon_3)/2}}\sum_{k=1}^{\infty}\|A^\sigma f_{c,k}\|\le$$

$$\le \frac{2\sqrt{2}C_1 L}{\sin(\pi\varepsilon_3)}(1+\sigma-\varepsilon_3)^{(1+\sigma-\varepsilon_3)/2}\sum_{j=M+1}^{\infty}\frac{1}{j^{(2+\sigma-\varepsilon_1-\varepsilon_3)/2}}\sum_{k=1}^{\infty}\|A^\sigma f_{c,k}\|\le$$

$$\le \frac{2\sqrt{2}C_1 L}{\sin(\pi\varepsilon_3)}(1+\sigma-\varepsilon_3)^{(1+\sigma-\varepsilon_3)/2}\int_{M}^{+\infty}\frac{dx}{x^{(2+\sigma-\varepsilon_1-\varepsilon_3)/2}}\sum_{k=1}^{\infty}\|A^\sigma f_{c,k}\|=$$

$$= \frac{\sqrt{2}C_3}{M^{(\sigma-\varepsilon_1-\varepsilon_3)/2}}\|f_c\|_{A^\sigma} \quad (\sigma>0,\ 0<\varepsilon_1<1,\ 0<\varepsilon_3<1,\ \varepsilon_1+\varepsilon_3<\sigma,\ M>\sigma-\varepsilon_3)$$

with the constant $C_4 = \dfrac{4C_1 L(1+\sigma-\varepsilon_3)^{(1+\sigma-\varepsilon_3)/2}}{\sin(\pi\varepsilon_3)(\sigma-\varepsilon_1-\varepsilon_3)}$ defined in estimate [5.59].

Finally, for the summand D_5, we have

$$\left\|\frac{D_5}{\min(x,1-x)}\right\| = \quad [5.62]$$

$$= \left\|-\sum_{k=N+1}^{\infty}\sqrt{2}\left[(2k\pi)^2 I+A\right]^{-1}\sum_{j=1}^{M}\frac{v_j(1-x)+v_j(x)}{\min(x,1-x)}\left[(I+A)^{-1}A\right]^{j}f_{c,k}\right\|\le$$

$$\le \sum_{k=N+1}^{\infty}\sum_{j=1}^{M}\frac{|v_j(1-x)|+|v_j(x)|}{\min(x,1-x)}\left\|\frac{\sqrt{2}}{2\pi i}\int_{\Gamma}\frac{1}{(2k\pi)^2+z}\left(\frac{z}{1+z}\right)^{j}(zI-A)^{-1}dz f_{c,k}\right\|\le$$

$$\le \sum_{k=N+1}^{\infty}\sum_{j=1}^{M}\frac{2\sqrt{2}}{3\cdot 2\pi}\int_{\Gamma}\left|\frac{z}{1+z}\right|^{j}\frac{1}{|(2k\pi)^2+z|}\frac{L}{1+|z|}|dz|\,\|f_{c,k}\|\le$$

$$\leq \frac{2\sqrt{2}L}{3\pi} \sum_{k=N+1}^{\infty} \sum_{j=1}^{M} \int_{0}^{+\infty} \frac{1}{\sqrt{(2k\pi)^4 + \rho^2}} \left(\frac{\rho^2}{1+\rho^2}\right)^{j/2} \frac{d\rho}{1+\rho} \|f_{c,k}\| \leq$$

$$\leq \frac{2\sqrt{2}L}{3\pi} \sum_{k=N+1}^{\infty} \sum_{j=1}^{M} \int_{0}^{+\infty} \frac{1}{\sqrt{2(2k\pi)^2 \rho}} \frac{d\rho}{1+\rho} \|f_{c,k}\| \leq$$

$$\leq \frac{2\sqrt{2}L}{3\pi} \frac{M}{\sqrt{2}2\pi} \int_{0}^{+\infty} \frac{d\rho}{\sqrt{\rho}(1+\rho)} \sum_{k=N+1}^{\infty} \frac{1}{k} \|f_{c,k}\| =$$

$$= \frac{LM}{3\pi} \sum_{k=N+1}^{\infty} \frac{1}{k} \|f_{c,k}\| = \frac{LM}{3\pi} \sum_{k=N+1}^{\infty} \frac{1}{k^{(3+\sigma)/2}} k^{(1+\sigma)/2} \|f_{c,k}\| \leq$$

$$\leq \frac{LM}{3\pi} \left(\sum_{k=N+1}^{\infty} \frac{1}{k^{3+\sigma}}\right)^{1/2} \left(\sum_{k=N+1}^{\infty} k^{1+\sigma} \|f_{c,k}\|^2\right)^{1/2} \leq \frac{LM}{3\pi} \left(\int_{N}^{+\infty} \frac{dx}{x^{3+\sigma}}\right)^{1/2} \|f_c\|_\sigma \leq$$

$$\leq \frac{LM}{3\pi} \left(\frac{1}{(2+\sigma)N^{2+\sigma}}\right)^{1/2} \|f_c\|_\sigma = \frac{L}{3\pi(2+\sigma)^{1/2}} \frac{M}{N^{1+\sigma/2}} \|f_c\|_\sigma \quad (\sigma > -2).$$

The assertion of the theorem follows now from estimates [5.58]–[5.62] for $M = N$.

In the next theorem, we study error [5.56] under another set of conditions for f_0, $f_{c,k}$, $f_{s,k}$.

THEOREM 5.10.– *Let the following assumptions be fulfilled:*

$$M = N, \quad v = \frac{\cos\varphi}{L+1}, \quad f_0 \in C(A, (1), v), \quad f_{c,k} \in C(A, (1), v) \quad \forall k \in \mathbb{N},$$

$$\|f_c\|_{A^\infty} \equiv \left(\sum_{k=1}^{\infty} \|f_{c,k}\|^2_{C(A,(1),v)}\right)^{1/2} < \infty, \qquad [5.63]$$

$$\|f_s\|_\infty \equiv \sum_{k=1}^{\infty} e^k \|f_{s,k}\| < \infty, \quad \|f_c\|_\infty \equiv \sum_{k=1}^{\infty} e^k \|f_{c,k}\| < \infty.$$

Then, the accuracy of the approximate solution [5.55] is characterized by the weighted estimate

$$\left\|\frac{u(x)-u_{N,N}(x)}{\min(x,1-x)}\right\| \le \frac{Ce^{-\sqrt{N+1}}}{(N+1)^{1/2-\varepsilon}}\left(\|f_s\|_\infty + \|f_c\|_\infty + \|f_0\|_{C(A,(1),\nu)} + \|f_c\|_{A^\infty}\right),$$

$$x \in [0,1] \quad (N \in \mathbb{N}),$$

where $\varepsilon > 0$ is an arbitrary small number and C is a positive constant independent of N.

PROOF.– For all summands in [5.56] except for D_2, we will use the integration path [5.48]. For brevity, the analysis that is similar to that in Theorem 5.10 and Lemma 5.11 will be omitted.

Taking into account [5.58], we have

$$\left\|\frac{D_1}{\min(x,1-x)}\right\| = \left\|\sum_{k=N+1}^{\infty} \frac{\sqrt{2}\sin(2k\pi x)}{\min(x,1-x)}\left[(2k\pi)^2 I + A\right]^{-1} f_{s,k}\right\| \le \quad [5.64]$$

$$\le L \sum_{k=N+1}^{\infty} \|f_{s,k}\| = L \sum_{k=N+1}^{\infty} e^{-k} e^{k} \|f_{s,k}\| \le$$

$$\le Le^{-(N+1)} \sum_{k=1}^{\infty} e^{k} \|f_{s,k}\| \le L \frac{e^{-\sqrt{N+1}}}{\sqrt{N+1}} \|f_s\|_\infty.$$

For estimating the summand D_2, we take the integration path $\tilde{\Gamma}$ consisting of two rays and a circle arc:

$$\tilde{\Gamma} = \tilde{\Gamma}_- \cup \tilde{\Gamma}_+ \cup \Gamma_\gamma,$$

$$\tilde{\Gamma}_\pm = \{z \in \mathbb{C} : z = \rho e^{\pm i\varphi}, \; \rho \in [\gamma,+\infty)\}, \quad \Gamma_\gamma = \{z \in \mathbb{C} : z = \gamma e^{i\theta}, \; \theta \in [-\varphi,\varphi]\}$$

with $dz = d(\rho e^{\pm i\varphi}) = e^{\pm i\varphi} d\rho$ for the rays $\tilde{\Gamma}_\pm$, and $dz = d(\gamma e^{i\theta}) = i\gamma e^{i\theta} d\theta$ for the arc Γ_γ. Then, we have

$$\left\|\frac{D_2}{\min(x,1-x)}\right\| = \left\|\sum_{j=M+1}^{\infty} \frac{v_j(1-x)+v_j(x)}{\min(x,1-x)} A^{-1}\left[(I+A)^{-1}A\right]^j f_0\right\| \le \quad [5.65]$$

$$\leq \sum_{j=M+1}^{\infty} \frac{|v_j(1-x)|+|v_j(x)|}{\min(x,1-x)} \left\| \frac{1}{2\pi i} \int_{\Gamma} \frac{1}{z}\left(\frac{z}{1+z}\right)^j \left(1+\frac{z}{j}\right)^{-j} (zI-A)^{-1} dz \left(I+\frac{A}{j}\right)^j f_0 \right\| \leq$$

$$\leq \sum_{j=M+1}^{\infty} \frac{2C_1}{2\pi j^{(1-\varepsilon_1)/2}} \int_{\Gamma} \frac{1}{|z|} \left|\frac{z}{1+z}\right|^j \left|1+\frac{z}{j}\right|^{-j} \frac{L}{1+|z|} |dz| \left\| \left(I+\frac{A}{j}\right)^j f_0 \right\| =$$

$$= \frac{2C_1 L}{\pi} \sum_{j=M+1}^{\infty} \frac{1}{j^{(1-\varepsilon_1)/2}} \left\{ \int_{\gamma}^{+\infty} \left[\frac{\rho\cos\varphi}{\sqrt{1+2\rho\cos\varphi+\rho^2}\sqrt{1+2\frac{\rho}{j}\cos\varphi+\left(\frac{\rho}{j}\right)^2}} \right]^j \frac{d\rho}{\rho(1+\rho)} + $$

$$+ \int_0^{\varphi} \left[\frac{\gamma\cos\theta}{\sqrt{1+2\gamma\cos\theta+\gamma^2}\sqrt{1+2\frac{\gamma}{j}\cos\theta+\left(\frac{\gamma}{j}\right)^2}} \right]^j \frac{d\theta}{1+\gamma} \right\} \times$$

$$\times \cos^{-j}\varphi \left\| \left(I+\frac{A}{j}\right)^j A^{-j} A^j f_0 \right\| \leq$$

$$= \frac{2C_1 L}{\pi} \sum_{j=M+1}^{\infty} \frac{1}{j^{(1-\varepsilon_1)/2}} \left\{ \int_{\gamma}^{+\infty} \left[\frac{\rho\cos\varphi}{\left(1+\frac{\rho}{j}\cos\varphi\right)(1+\rho\cos\varphi)} \right]^j \frac{d\rho}{\rho(1+\rho)} + $$

$$+ \int_0^{\varphi} \left[\frac{\gamma\cos\theta}{(1+\gamma\cos\theta)\left(1+\frac{\gamma}{j}\cos\theta\right)} \right]^j \frac{d\theta}{1+\gamma} \right\} \cos^{-j}\varphi \left\| \left(I+\frac{A}{j}\right)^j A^{-j} \right\| \|A^j f_0\| \leq$$

$$\leq \frac{2C_1 L}{\pi} \sum_{j=M+1}^{\infty} \frac{1}{j^{(1-\varepsilon_1)/2}} \max_{t\geq 0} \left[\frac{t}{(1+t)\left(1+\frac{t}{j}\right)} \right]^j \left\{ \int_{\gamma}^{+\infty} \frac{d\rho}{\rho(1+\rho)} + \int_0^{\varphi} \frac{d\theta}{1+\gamma} \right\} \times$$

$$\times \cos^{-j}\varphi (L+1)^j \|A^j f_0\|.$$

Applying now Lemma 5.9, we get the estimate

$$\left\| \frac{D_2}{\min(x,1-x)} \right\| \leq \frac{2C_1 L}{\pi} \left[\ln \frac{1+\gamma}{\gamma} + \frac{\varphi}{1+\gamma} \right] \sum_{j=M+1}^{\infty} \frac{e e^{-2\sqrt{j}}}{j^{(1-\varepsilon_1)/2}} \| f_0 \|_{C(A,(1),v)} \leq$$

$$\leq \frac{2C_1 L e}{\pi} \left[\ln \frac{1+\gamma}{\gamma} + \frac{\varphi}{1+\gamma} \right] \sum_{j=1}^{\infty} e^{-\sqrt{j}} \frac{e^{-\sqrt{M+1}}}{(M+1)^{(1-\varepsilon_1)/2}} \| f_0 \|_{C(A,(1),v)} =$$

$$= \frac{C_4 e^{-\sqrt{M+1}}}{(M+1)^{(1-\varepsilon_1)/2}} \| f_0 \|_{C(A,(1),v)}$$

with the constant $C_4 = \frac{2C_1 L \tilde{S} e}{\pi} \left[\ln \frac{1+\gamma}{\gamma} + \frac{\varphi}{1+\gamma} \right]$, where $\tilde{S}$ is the sum of the convergent series $\sum_{j=1}^{\infty} e^{-\sqrt{j}} = 1.670406818\ldots$.

Similarly to D_1, we estimate the summand D_3:

$$\left\| \frac{D_3}{\min(x,1-x)} \right\| = \left\| \sum_{k=N+1}^{\infty} \frac{\sqrt{2}(\cos(2k\pi x)-1)}{\min(x,1-x)} \left[(2k\pi)^2 I + A\right]^{-1} f_{c,k} \right\| \leq \quad [5.66]$$

$$\leq \sum_{k=N+1}^{\infty} \frac{\sqrt{2} \, 2\sin^2 k\pi x}{\min(x,1-x)} \left\| \frac{1}{2\pi i} \int_{\Gamma} \frac{1}{(2k\pi)^2 + z} (zI - A)^{-1} dz \, f_{c,k} \right\| \leq$$

$$\leq \sum_{k=N+1}^{\infty} \frac{2\sqrt{2} k\pi}{2\pi} \int_{\Gamma} \frac{1}{|(2k\pi)^2 + z|} \frac{L}{1+|z|} |dz| \, \| f_{c,k} \| \leq$$

$$\leq L \sum_{k=N+1}^{\infty} e^{-k} e^k \| f_{c,k} \| \leq L e^{-(N+1)} \sum_{k=1}^{\infty} e^k \| f_{c,k} \| \leq L \frac{e^{-\sqrt{N+1}}}{\sqrt{N+1}} \| f_c \|_{\infty}.$$

For the summand D_4, we have

$$\left\| \frac{D_4}{\min(x,1-x)} \right\| = \quad [5.67]$$

$$= \left\| -\sum_{k=1}^{\infty} \sqrt{2}\left[(2k\pi)^2 I + A\right]^{-1} \sum_{j=M+1}^{\infty} \frac{v_j(1-x)+v_j(x)}{\min(x,1-x)} \left[(I+A)^{-1}A\right]^j f_{c,k} \right\| \le$$

$$\le \sum_{k=1}^{\infty} \sum_{j=M+1}^{\infty} \frac{|v_j(1-x)|+|v_j(x)|}{\min(x,1-x)} \times$$

$$\times \left\| \frac{\sqrt{2}}{2\pi i} \int_\Gamma \frac{1}{(2k\pi)^2+z} \left(\frac{z}{1+z}\right)^j \left(1+\frac{z}{j}\right)^{-j} (zI-A)^{-1} dz \left(I+\frac{A}{j}\right)^j f_{c,k} \right\| \le$$

$$\le \sum_{k=1}^{\infty} \sum_{j=M+1}^{\infty} \frac{\sqrt{2}\, 2C_1}{2\pi\, j^{(1-\varepsilon_1)/2}} \int_\Gamma \frac{1}{|(2k\pi)^2+z|} \left|\frac{z}{1+z}\right|^j \left|1+\frac{z}{j}\right|^{-j} \frac{L}{1+|z|}\, |dz| \left\| \left(I+\frac{A}{j}\right)^j f_{c,k} \right\| \le$$

$$\le \frac{\sqrt{2}\, 2C_1 2L}{2\pi} \sum_{k=1}^{\infty} \sum_{j=M+1}^{\infty} \frac{1}{j^{(1-\varepsilon_1)/2}} \times$$

$$\times \int_0^{+\infty} \frac{1}{\sqrt{(2k\pi)^4+\rho^2}} \left[\frac{\rho\cos\varphi}{\sqrt{1+2\rho\cos\varphi+\rho^2}\sqrt{1+2\frac{\rho}{j}\cos\varphi+\left(\frac{\rho}{j}\right)^2}}\right]^j \frac{d\rho}{1+\rho} \times$$

$$\times \cos^{-j}\varphi \times \left\| \left(I+\frac{A}{j}\right)^j A^{-j} A^j f_{c,k} \right\| \le$$

$$\le \frac{\sqrt{2}\, 2C_1 L}{\pi} \sum_{k=1}^{\infty} \sum_{j=M+1}^{\infty} \frac{1}{j^{(1-\varepsilon_1)/2}} \int_0^{+\infty} \frac{1}{\sqrt{2}(2k\pi)^2\rho} \left[\frac{\rho\cos\varphi}{(\rho\cos\varphi+1)\left(\frac{\rho}{j}\cos\varphi+1\right)}\right]^j \frac{d\rho}{1+\rho} \times$$

$$\times \cos^{-j}\varphi \left\| \left(I+\frac{A}{j}\right)^j A^{-j} \right\| \|A^j f_{c,k}\| \le$$

$$\le \frac{\sqrt{2}\, 2C_1 L}{\pi\sqrt{2}\, 2\pi} \sum_{k=1}^{\infty} \sum_{j=M+1}^{\infty} \frac{1}{k\, j^{(1-\varepsilon_1)/2}} \max_{t\ge 0}\left[\frac{t}{(1+t)\left(1+\frac{t}{j}\right)}\right]^j \int_0^{+\infty} \frac{d\rho}{\sqrt{\rho}(1+\rho)} \times$$

$$\times \cos^{-j}\varphi(L+1)^j \|A^j f_{c,k}\| \le \frac{C_1 L}{\pi} \sum_{j=M+1}^{\infty} \frac{ee^{-2\sqrt{j}}}{j^{(1-\varepsilon_1)/2}} \sum_{k=1}^{\infty} \frac{1}{k} \|f_{c,k}\|_{C(A,(1),\nu)} \le$$

$$\le \frac{C_1 L e}{\pi} \frac{e^{-\sqrt{M+1}}}{(M+1)^{(1-\varepsilon_1)/2}} \sum_{j=1}^{\infty} e^{-\sqrt{j}} \left\{ \sum_{k=1}^{\infty} \frac{1}{k^2} \right\}^{1/2} \left\{ \sum_{k=1}^{\infty} \|f_{c,k}\|^2_{C(A,(1),\nu)} \right\}^{1/2} =$$

$$= \frac{C_5 e^{-\sqrt{M+1}}}{(M+1)^{(1-\varepsilon_1)/2}} \|f_c\|_{A^{\infty}}$$

with the constant $C_5 = \dfrac{C_1 L e \tilde{S}}{\sqrt{6}}$ and the series $\tilde{S} = \sum_{j=1}^{\infty} e^{-\sqrt{j}} = 1.670406818\ldots$ defined in [5.65].

Finally, for the last summand by analogy with [5.62], we obtain

$$\left\| \frac{D_5}{\min(x,1-x)} \right\| = \qquad [5.68]$$

$$= \left\| -\sum_{k=N+1}^{\infty} \sqrt{2}\left[(2k\pi)^2 I + A\right]^{-1} \sum_{j=1}^{M} \frac{v_j(1-x)+v_j(x)}{\min(x,1-x)} \left[(I+A)^{-1}A\right]^j f_{c,k} \right\| \le$$

$$\le \sum_{k=N+1}^{\infty} \sum_{j=1}^{M} \frac{|v_j(1-x)|+|v_j(x)|}{\min(x,1-x)} \left\| \frac{\sqrt{2}}{2\pi i} \int_{\Gamma} \frac{1}{(2k\pi)^2 + z} \left(\frac{z}{1+z}\right)^j (zI-A)^{-1} dz f_{c,k} \right\| \le$$

$$\le \frac{\sqrt{2}}{2\pi} \frac{2}{3} \sum_{k=N+1}^{\infty} \sum_{j=1}^{M} \int_{\Gamma} \frac{1}{|(2k\pi)^2 + z|} \left|\frac{z}{1+z}\right|^j \frac{L}{1+|z|} |dz| \| f_{c,k} \| \le$$

$$\le \frac{\sqrt{2}\, 2L}{3\pi} \sum_{k=N+1}^{\infty} \sum_{j=1}^{M} \int_0^{+\infty} \frac{1}{\sqrt{(2k\pi)^4 + \rho^2}} \left(\frac{\rho^2}{1+\rho^2}\right)^{j/2} \frac{d\rho}{1+\rho} \| f_{c,k} \| \le$$

$$\le \frac{\sqrt{2}\, 2L}{3\pi} \sum_{k=N+1}^{\infty} \sum_{j=1}^{M} \int_0^{+\infty} \frac{1}{\sqrt{2(2k\pi)^2 \rho}} \frac{d\rho}{1+\rho} \| f_{c,k} \| \le$$

$$\leq \frac{\sqrt{22}L}{3\pi} \frac{M}{\sqrt{2}2\pi} \int_0^{+\infty} \frac{d\rho}{\sqrt{\rho}(1+\rho)} \sum_{k=N+1}^{\infty} \frac{1}{k} \| f_{c,k} \| =$$

$$= \frac{LM}{3\pi} \sum_{k=N+1}^{\infty} \frac{1}{k} \| f_{c,k} \| \leq \frac{LM}{3\pi} \frac{1}{N+1} \sum_{k=N+1}^{\infty} e^k e^{-k} \| f_{c,k} \| \leq$$

$$\leq \frac{LM}{3\pi(N+1)} e^{-(N+1)} \sum_{k=1}^{\infty} e^k \| f_{s,k} \| \leq \frac{LM}{3\pi(N+1)} \frac{e^{-\sqrt{N+1}}}{\sqrt{N+1}} \| f_s \|_{\infty}.$$

Now, estimates [5.64]–[5.68] with $M = N$ prove the theorem.

5.2.3. Conclusion

We will now summarize the results obtained above.

In Theorems 5.7 and 5.8 for the homogeneous equation and in Theorems 5.9 and 5.10 for the inhomogeneous equation, the approximate solutions [5.23] and [5.24] are characterized by the weighted estimates with the weight function $\rho^{-1}(x)$, $\rho(x) = \min(x, 1-x)$, which takes into account the distance of the point $x \in (0,1)$ to the boundary points of the interval $[0,1]$.

The weighted estimates obtained in Theorems 5.7 and 5.9 indicate that the accuracy of the corresponding approximate solution is dependent on the power of the parameter σ. Namely, the weighted estimate [5.51] in Theorem 5.7 shows that when σ increases (which means that the smoothness of the vector u_1 improves), the convergence rate of the approximate solution $u_N(x)$ automatically increases. Thus, [5.50] is a method *without saturation of accuracy* in the sense of (Babenko 2002). Similarly, the weighted estimate [5.57] in Theorem 5.9 demonstrates that when the parameter σ increases (i.e. the Fourier coefficients f_0, $f_{c,k}$, $f_{s,k}$, $k = 1, 2, \ldots$, of the function $f(x)$ decay faster, i.e. $f(x)$ has better differential properties), the convergence rate of the approximate solution $u_{N,N}(x)$ increases. Therefore, the method [5.55] is also a method without saturation of accuracy.

Next, the weighted estimate [5.52] in Theorem 5.8 shows that method [5.50] has an exponential rate of convergence, provided that the vector u_1 is a vector of the exponential type in the sense of Radyno (1985). Similarly, the weighted estimate

[5.63] in Theorem 5.10 means method [5.55] has an exponential rate of convergence, provided that the vectors f_0, $f_{c,k}$, $f_{s,k}$, $k=1,2,\ldots,$ in the trigonometric Fourier expansion of the function $f(x)$ have the appropriate exponential rate of decay. Note that methods [5.50] and [5.55] are even super-exponentially convergent.

References

Arov, D.Z. and Gavrilyuk, I.P. (1993). A method for solving initial value problems for linear differential equations in Hilbert space based on the Cayley transform. *Numerical Functional Analysis and Optimization. An International Journal*, 14(5–6), 459–473. https://doi.org/10.1080/01630569308816534.

Arov, D.Z., Gavrilyuk, I.P., Makarov, V.L. (1995). Representation and approximation of solutions of initial value problems for differential equations in Hilbert space based on the Cayley transform. In *Elliptic and Parabolic Problems (Pont-à-Mousson, 1994)*, Bandle, C., Chipot, M., Bemelmans, J., Saint Jean Paulin, J., Shafrir, I. (eds). Chapman and Hall, New York.

Babenko, K.Y. (2002). *The Foundations of Numerical Analysis*. Nauka, Moscow.

Bateman, H. and Erdélyi, A. (1953a). *Higher Transcendental Functions. Volume 1*. McGraw-Hill, New York.

Bateman, H. and Erdélyi, A. (1953b). *Higher Transcendental Functions. Volume 2*. McGraw-Hill, New York.

Beals, R. (1972). Semigroups and abstract Gevrey spaces. *Journal of Functional Analysis*, 10(3), 300–308. https://doi.org/10.1016/0022-1236(72)90028-6.

Bechelova, A.R. (1998). On the convergence of difference schemes for the diffusion equation of fractional order. *Ukrains'kyi Matematychnyi Zhurnal*, 50(7), 994–996. https://umj.imath.kiev.ua/index.php/umj/article/view/4879.

Bitsadze, A.V. (1981). *Some Classes of Equations with Partial Derivatives*. Nauka, Moscow.

Bramble, J.H. and Hilbert, S.R. (1970). Estimation of linear functionals on Sobolev spaces with application to Fourier transforms and spline interpolation. *SIAM Journal on Numerical Analysis*, 7(1), 112–124. https://doi.org/10.1137/0707006.

Bramble, J.H. and Hilbert, S.R. (1971). Bounds for a class of linear functionals with applications to Hermite interpolation. *Numerische Mathematik*, 16(4), 362–369. https://doi.org/10.1007/BF02165007.

Cheung, T.-Y. (1977). Three nonlinear initial value problems of the hyperbolic type. *SIAM Journal on Numerical Analysis*, 14(3), 484–491. https://doi.org/10.1137/0714028.

Diethelm, K., Ford, N.J., Freed, A.D., Luchko, Y. (2005). Algorithms for the fractional calculus: A selection of numerical methods. *Computer Methods in Applied Mechanics and Engineering*, 194(6), 743–773. https://doi.org/10.1016/j.cma.2004.06.006.

Dupont, T. and Scott, R. (1980). Polynomial approximation of functions in Sobolev spaces. *Mathematics of Computation*, 34(150), 441–463. https://doi.org/10.1090/S0025-5718-1980-0559195-7.

Evans, L.C. (2010). *Partial Differential Equations*, 2nd edition. American Mathematical Society, Providence.

Galba, E.F. (1985). On the order of accuracy of the difference scheme for the Poisson equation with the mixed boundary condition. *A Collection of Papers: Optimization of Software Algorithms. Proceedings of the V.M. Glushkov Institute. of Cybernetics AS UkrSSR*, Kyiv, 30–34.

Gavrilyuk, I.P. and Makarov, V.L. (1994). The Cayley transform and the solution of an initial value problem for a first order differential equation with an unbounded operator coefficient in Hilbert space. *Numerical Functional Analysis and Optimization*, 15(5–6), 583–598. https://doi.org/10.1080/01630569408816582.

Gavrilyuk, I.P. and Makarov, V.L. (1996). Representation and approximation of the solution of an initial value problem for a first order differential equation in Banach spaces. *Zeitschrift Für Analysis Und Ihre Anwendung*, 15(2), 495–527. https://doi.org/10.4171/zaa/712.

Gavrilyuk, I.P. and Makarov, V.L. (1999). Explicit and approximate solutions of second order elliptic differential equations in Hilbert and Banach spaces. *Numerical Functional Analysis and Optimization*, 20(7), 695–715. https://doi.org/10.1080/01630569908816919.

Gavrilyuk, I.P. and Makarov, V.L. (2004). *Strongly Positive Operators and Numerical Algorithms Without Saturation of Accuracy*. Institute of Mathematics of NASU, Kyiv.

Gavrilyuk, I.P., Makarov, V.L., Mayko, N.V. (2020). Weighted estimates for boundary value problems with fractional derivatives. *Computational Methods in Applied Mathematics*, 20(4), 609–630. https://doi.org/10.1515/cmam-2018-0305.

Gavrilyuk, I.P., Makarov, V.L., Mayko, N.V. (2021). Weighted estimates of the Cayley transform method for abstract differential equations. *Computational Methods in Applied Mathematics*, 21(1), 53–68. https://doi.org/10.1515/cmam-2019-0120.

Gorbachuk, V.Y. and Gorbachuk, M.L. (1984). *Limit Problem for Differential Operator Equations*. Naukova Dumka, Kyiv.

Gorbachuk, V.Y. and Kniaziuk, A.V. (1989). The limit values of solutions of the operator differential equations. *Uspekhi Matem. Nauk*, 44:3(267), 55–91.

Gorodnii, M.F. (1998). On the approximation of a bounded solution of a linear differential equation in a Banach space. *Ukrains'kyi Matematychnyi Zhurnal*, 50(9), 1268–1271. https://umj.imath.kiev.ua/index.php/umj/article/view/4843.

Gorodnii, M.F., Kutsyk, N.M., Chaikovskyi, A.V. (2004). On one generalization of the concept of a sectorial operator. *Visnyk Kyivskoho Universytetu*, 1, 80–86.

Gradshteyn, I.S. and Ryzhik, I.M. (2014). *Table of Integrals, Series, and Products.* Academic Press, Cambridge, MA.

Hartman, P. (1964). *Ordinary Differential Equations.* John Wiley & Sons, New York.

Havu, V. and Malinen, J. (2007). The Cayley transform as a time discretization scheme. *Numerical Functional Analysis and Optimization*, 28(7–8), 825–851.

Jin, B., Lazarov, R., Vabishchevich, P. (2017). Preface: Numerical analysis of fractional differential equations. *Computational Methods in Applied Mathematics*, 17(4), 643–646. https://doi.org/10.1515/cmam-2017-0036.

Jin, B., Lazarov, R., Zhou, Z. (2019). Numerical methods for time-fractional evolution equations with nonsmooth data: A concise overview. *Computer Methods in Applied Mechanics and Engineering*, 346, 332–358. https://doi.org/10.1016/j.cma.2018.12.011.

Jovanović, B.S. and Süli, E. (2014). *Analysis of Finite Difference Schemes: For Linear Partial Differential Equations with Generalized Solutions*, 1st edition. Springer, London.

Jovanović, B.S., Vulkov, L.G., Delić, A. (2013). Boundary value problems for fractional PDE and their numerical approximation. In *Numerical Analysis and Its Applications*, Dimov, I., Faragó, I., Vulkov, L. (eds). Springer, Berlin, Heidelberg.

Kashpirovskii, A.I. and Mytnik, Y.V. (1998). Approximation of solutions of operator-differential equations by operator polynomials. *Ukrains'kyi Matematychnyi Zhurnal*, 50(11), 1506–1516. https://umj.imath.kiev.ua/index.php/umj/article/view/4802.

Kilbas, A.A., Srivastava, H.M., Trujillo, J.J. (2006). *Theory and Applications of Fractional Differential Equations*, 1st edition. Elsevier, New York.

Krein, S.G. (1967). *Linear Differential Equations in a Banach Space.* Nauka, Moscow.

Li, C. and Zeng, F. (2012). Finite difference methods for fractional differential equations. *International Journal of Bifurcation and Chaos*, 22(4), 1230014–1230028. https://doi.org/10.1142/S0218127412300145.

Lions, J.-L. and Magenes, E. (1972). *Non-Homogeneous Boundary Value Problems and Applications.* Springer, Berlin, Heidelberg.

Lubich, C., Sloan, I.H., Thomée, V. (1994). Nonsmooth data error estimates for approximations of an evolution equation with a positive type memory term. *Mathematics of Computations*, 65(213), 1–17. https://www.jstor.org/stable/2153826.

Makarov, V.L. (1989). On a priori estimates of difference schemes giving an account of the boundary effect. *Doklady Bolgarskoĭ Akademii Nauk. Comptes rendus de l'Académie bulgare des sciences*, 42(5), 41–44.

Makarov, V.L. (2019). Meixner's polynomials and their properties. *Dopov. Nats. Akad. Nauk Ukr.*, 7, 3–8. https://doi.org/10.15407/dopovidi2019.07.003.

Makarov, V.L. and Demkiv, L.I. (2003a). The estimates of accuracy of the finite-difference schemes for parabolic equations if the boundary-initial effect is taken into account. *Dopov. Nats. Akad. Nauk Ukr.*, 2, 26–32.

Makarov, V.L. and Demkiv, L.I. (2003b). Accuracy estimates of difference schemes for quasi-linear parabolic equations taking into account the initial-boundary effect. *Computational Methods in Applied Mathematics*, 3(4), 579–595.

Makarov, V.L. and Demkiv, L.I. (2005). Accuracy estimates of difference schemes for quasi-linear elliptic equations with variable coefficients taking into account boundary effect. In *Numerical Analysis and Its Applications*, Li, Z., Vulkov, L., Waśniewski, J. (eds). Springer, Berlin, Heidelberg.

Makarov, V.L. and Demkiv, L.I. (2006). Taking into account the third kind conditions in weight estimates for difference schemes. *Large-Scale Scientific Computing. Lecture Notes in Computer Science*, 3743, 687–694. https://doi.org/10.1007/11666806_79.

Makarov, V.L. and Demkiv, L.I. (2009). Weight uniform accuracy estimates of finite difference method for Poisson equation, taking into account boundary effect. In *Numerical Analysis and Its Applications*, Margenov, S., Vulkov, L.G., Waśniewski, J. (eds). Springer, Berlin, Heidelberg.

Makarov, V.L. and Mayko, N.V. (2019a). The boundary effect in the accuracy estimate for the grid solution of the fractional differential equation. *Computational Methods in Applied Mathematics*, 19(2), 379–394. https://doi.org/10.1515/cmam-2018-0002.

Makarov, V.L. and Mayko, N.V. (2019b). Boundary effect in accuracy estimate of the grid method for solving fractional differential equations. *Cybernetics and Systems Analysis*, 55(1), 65–80. https://doi.org/10.1007/s10559-019-00113-y.

Makarov, V.L. and Mayko, N.V. (2020). Weighted accuracy estimates of the Cayley transform method for abstract boundary-value problems in a Banach space. *Dopov. Nats. Akad. Nauk Ukr.*, 5, 3–9.

Makarov, V.L. and Mayko, N.V. (2021). Weighted estimates of the Cayley transform method for boundary value problems in a Banach space. *Numerical Functional Analysis and Optimization*, 42(2), 211–233. https://doi.org/10.1080/01630563.2020.1871010.

Makarov, V.L. and Ryabichev, V.L. (2002). Unimprovable error estimates of the Cayley transform method for the operator cosine function. *Dopov. Nats. Akad. Nauk Ukr.*, 12, 21–25.

Makarov, V.L., Vasylyk, V.B., Ryabichev, V.L. (2002). Unimprovable order-of-magnitude estimates of the rate of convergence of the Cayley transform method for approximation of an operator exponent. *Cybernetics and Systems Analysis*, 38(4), 632–636. https://doi.org/10.1023/A:1021122622419.

Mayko, N.V. (2013). The boundary effect in the error estimate of the finite-difference scheme for the two-dimensional heat equation. *Journal of Numerical and Applied Mathematics*, 3(113), 91–106.

Mayko, N.V. (2014). Error estimates of the finite-difference scheme for a one-dimensional parabolic equation with allowance for the effect of initial and boundary conditions. *Cybernetics and Systems Analysis*, 50(5), 788–796. https://doi.org/10.1007/s10559-014-9669-6.

Mayko, N.V. (2017). Improved accuracy estimates of the difference scheme for the two-dimensional parabolic equation with regard for the effect of initial and boundary conditions. *Cybernetics and Systems Analysis*, 53(1), 83–91. https://doi.org/10.1007/s10559-017-9909-7.

Mayko, N.V. (2018a). A weighted error estimate for a finite-difference scheme of increased approximation order for a two-dimensional Poisson equation with allowance for the Dirichlet boundary condition. *Cybernetics and Systems Analysis*, 54(1), 130–138. https://doi.org/10.1007/s10559-018-0014-3.

Mayko, N.V. (2018b). The finite-difference scheme of higher order of accuracy for the two-dimensional Poisson equation in a rectangle with regard for the effect of the Dirichlet boundary condition. *Cybernetics and Systems Analysis*, 54(4), 624–635. https://doi.org/10.1007/s10559-018-0063-7.

Mayko, N.V. (2020) Super-exponential rate of convergence of the Cayley transform method for an abstract differential equation. *Cybernetics and Systems Analysis*, 56(3), 492–503. https://doi.org/10.1007/s10559-020-00265-2.

Mayko, N.V. and Ryabichev, V.L. (2005). Accuracy of approximation of a solution to an abstract Cauchy problem. *Cybernetics and Systems Analysis*, 41(3), 437–444. https://doi.org/10.1007/s10559-005-0077-9.

Mayko, N.V. and Ryabichev, V.L. (2009). Approximation theorems for operator exponential and cosine functions. *Cybernetics and Systems Analysis*, 45(5), 800–807. https://doi.org/10.1007/s10559-009-9145-x.

Mayko, N.V. and Ryabichev, V.L. (2016). Boundary effect in the error estimate of the finite-difference scheme for two-dimensional Poisson's equation. *Cybernetics and Systems Analysis*, 52(5), 758–769. https://doi.org/10.1007/s10559-016-9877-3.

Meerschaert, M.M. and Tadjeran, C. (2004). Finite difference approximations for fractional advection–dispersion flow equations. *Journal of Computational and Applied Mathematics*, 172(1), 65–77. https://doi.org/10.1016/j.cam.2004.01.033.

Meixner, J. (1934). Orthogonale polynomsysteme mit einer besondersten gesalt der erzeugenden funktion. *Journal of the London Mathematical Society*, s1-9(1), 6–13. https://doi.org/10.1112/jlms/s1-9.1.6.

Miller, K.S. and Ross, B. (1993). *An Introduction to the Fractional Calculus and Fractional Differential Equations*. John Wiley & Sons, New York.

Molchanov, Y.N. and Galba, E.F. (1990). On the convergence of the difference scheme approximating the Dirichlet problem for an elliptic equation with piecewise constant coefficients. *A Collection of Papers: Numerical Methods and Technology for Developing Application Packages. Proceedings of the V.M. Glushkov Institute of Cybernetics AS UkrSSR*, Kyiv, 161–165.

Morton, K.W. and Mayers, D.F. (2005). *Numerical Solution of Partial Differential Equations*, 2nd edition. Cambridge University Press, Cambridge.

Nakagawa, J., Sakamoto, K., Yamamoto, M. (2010). Overview to mathematical analysis for fractional diffusion equations – New mathematical aspects motivated by industrial collaboration. *Journal of Math-for-Industry (JMI)*, 2A, 99–108.

Nelson, E. (1959). Analytic vectors. *Annals of Mathematics. Second Series*, 70, 572–615.

Oldham, K.B. and Spanier, J. (1974). *The Fractional Calculus*. Academic Press, New York, London.

Pazy, A. (1983). *Semigroups of Linear Operators and Applications to Partial Differential Equations*. Springer-Verlag, New York.

Podlubny, I. (1999). *Fractional Differential Equations*. Academic Press, San Diego.

Pshibikhova, R.A. (2016). Goursat's problem for the fractional telegraph equation with Caputo's derivatives. *Matem. Zametki*, 99(4), 559–563.

Qin, Y. (2016). *Integral and Discrete Inequalities and their Applications. Volume I*. Birkhäuser/Springer, Cham.

Quarteroni, A. and Valli, A. (1994). *Numerical Approximation of Partial Differential Equations*. Springer-Verlag, Berlin.

Radyno, Y.V. (1985). Vectors of the exponential type in the operator calculus and differential equations. *Differents. Uravn.*, 21(9), 1559–1569.

Riesz, F. and Sz.-Nagy, B. (1955). *Functional Analysis*. Fredrick Ungar Publishing Co., New York.

Ryabichev, V.L. and Mayko, N.V. (2004). Unimprovable (in order) estimates of the convergence rate of the Cayley transform method for the operator exponential function. *Visnyk Kyivskoho Universytetu*, 1, 270–278.

Sabatier, J., Agrawal, O.P., Tenreiro Machado, J.A. (2007). *Advances in Fractional Calculus: Theoretical Developments and Applications in Physics and Engineering*. Springer-Verlag, New York.

Samarskii, A.A (2001). *The Theory of Difference Schemes*. CRC Press, Boca Raton.

Samarskii, A.A., Lazarov, R.D., Makarov, V.L. (1987). *Finite-Difference Schemes for Differential Equations with Generalized Solution*. Vysshaia Shkola, Moscow.

Samko, S.G., Kilbas, A.A., Marichev, O.I. (1993). *Fractional Integrals and Derivatives. Theory and Applications*. Gordon and Breach, Yverdon.

Schiesser, W.E. and Griffiths, G.W. (2009). *A Compendium of Partial Differential Equation Models*. Cambridge University Press, Cambridge.

Strang, G. (1960). Difference methods for mixed boundary-value problems. *Duke Mathematical Journal*, 27, 221–232.

Süli, E. and Mayers, D.F. (2003). *An Introduction to Numerical Analysis*. Cambridge University Press, Cambridge.

Thomée, V. (2001). From finite differences to finite elements: A short history of numerical analysis of partial differential equations. *Journal of Computational and Applied Mathematics*, 128(1), 1–54. https://doi.org/10.1016/S0377-0427(00)00507-0.

Torba, S.M. (2007). Direct and inverse theorems of approximate methods for the solution of an abstract Cauchy problem. *Ukrains'kyi Matematychnyi Zhurnal*, 59(6), 838–852.

Volkov, E.A. (1965). On the differential properties of solutions of the boundary value problems for Laplace's and Poisson's equations in a rectangle. *Trudy MYAN SSSR*, 77, 89–112.

Index

A, B

adjoint operator, 6, 7, 85, 228
approximation error, 3, 10, 19, 27, 28, 32, 43, 49, 57, 58, 70, 73, 83, 88, 89, 103, 104, 106, 119, 122, 137, 140, 163, 202, 209
Banach space, 221, 236, 238, 248, 282, 284, 292
boundary value problem (BVP), 1, 18, 19, 31, 46, 47, 66, 71, 115, 116, 123–125, 129–131, 145, 150, 166, 181, 249, 251, 257, 282, 292
Bramble–Hilbert lemma, 17, 18, 29, 67

C, D

Cauchy problem, 213, 247, 248, 284
Cauchy–Bunyakovsky inequality, 51, 149, 153, 295
Cayley transform, 213, 214, 222, 226, 237, 247, 248, 283
closed operator, 282, 283
comparison theorem, 42, 106
convergence rate, 43, 282, 305
Dirichlet boundary condition, 17, 18, 31, 67, 101, 102
discrete analogue, 1, 32, 48, 212

E, F

eigenfunction, 33, 34
eigenvalue, 33, 218, 224, 226
error, 3, 8, 17–19, 31, 32, 46, 49, 70, 71, 77, 83, 86, 101, 103, 104, 118, 121, 123, 137, 139, 140, 143, 163, 166, 173, 177, 181, 215, 226, 270, 277, 291, 293, 294, 299
finite-difference
  derivative, 24, 47, 54
  scheme, 1–3, 16–19, 30–32, 43, 45, 46, 48, 69, 70, 76, 80–83, 100, 102, 111, 162
fixed-point iteration method, 152, 156, 160, 165, 183, 186, 189
fractional derivative, 115, 123, 124, 145, 166, 181
Fredholm integral equation of the second kind, 116, 130, 149, 167

G, H

Gevrey class, 284
Green's function, 250–252, 258
Hilbert space, 213, 232, 247, 249, 250
homogeneous equation, 128, 248, 257, 282, 305

I, K

inhomogeneous equation, 282, 292, 305
initial value problem (IVP), 232, 236
inner product, 3, 6, 7, 20, 23, 52, 70, 83, 103, 213, 249, 250
inverse operator, 5, 22, 51, 84
kernel, 149, 150, 167, 182, 189
Kronecker delta symbol, 8, 24, 54, 85, 216, 218, 226

L, M

Lagrange interpolating polynomial, 121, 139, 177
Laguerre polynomial, 213, 216, 224, 233, 237, 283
Laplace operator, 1, 18, 31, 82, 102, 166
logarithmically sectorial operator, 238, 248
maximum principle, 34, 128
Meixner polynomial, 253, 283
mesh function, 3, 6, 7, 20, 23, 32, 49, 50, 52, 70, 83, 103

N, O

node, 37–40, 70, 71, 75, 89, 92, 95, 97, 109, 110
norm, 3, 5–7, 20, 23, 32, 43, 46, 49, 52, 67, 70, 72, 83, 88, 89, 101, 103, 112, 113, 115, 119, 123, 124, 137, 138, 145, 151, 162, 166, 168, 173, 181, 212, 213, 249, 250, 284, 287, 290
ordinary differential equation (ODE), 123

P, R

positive definite operator, 213, 227, 232, 249
recurrence relation, 106, 233
recurrent sequence, 183, 186, 189, 252, 257, 283
resolvent set, 237, 282
Riemann–Liouville derivative, 124, 126, 146
Roumieu type, 284

S, U

sectorial operator, 238
Sobolev space, 145
spectral set, 249
step, 8, 24, 54, 85
strongly positive operator, 237, 282–284
symmetric operator, 50, 228
uniform norm, 166, 181

V, W

Volterra integral equation of the second kind, 182, 189
weight function, 30, 123, 166, 181, 212, 216, 224, 282, 305
weighted estimate, 16, 18, 30, 46, 66, 67, 76, 80, 100, 115, 118, 119, 122–124, 126, 130, 132, 136, 137, 139, 140, 143, 145, 149, 153, 157, 161, 165, 166, 168, 170, 173, 178, 181, 211, 212, 262–265, 267, 270, 274, 277, 281, 282, 291, 300, 305

Other titles from

in

Mathematics and Statistics

2023

KOROLIOUK Dmitri, SAMOILENKO Igor
Asymptotic and Analytic Methods in Stochastic Evolutionary Systems

PARROCHIA Daniel
Graphs, Orders, Infinites and Philosophy

2022

CHAKRAVARTHY Srinivas R.
Introduction to Matrix-Analytic Methods in Queues 1: Analytical and Simulation Approach – Basics
Introduction to Matrix-Analytic Methods in Queues 2: Analytical and Simulation Approach – Queues and Simulation

DE SAPORTA Benoîte, ZILI Mounir
Martingales and Financial Mathematics in Discrete Time

LESFARI Ahmed
Integrable Systems

RADCHENKI Vadym M.
General Stochastic Measures: Integration, Path Properties and Equations

SIMON Jacques
Distributions
(Analysis for PDEs Set – Volume 3)

2021

KOROLIOUK Dmitri, SAMOILENKO Igor
Random Evolutionary Systems: Asymptotic Properties and Large Deviations

MOKLYACHUK Mikhail
Convex Optimization: Introductory Course

POGORUI Anatoliy, SWISHCHUK Anatoliy, RODRÍGUEZ-DAGNINO Ramón M.
Random Motions in Markov and Semi-Markov Random Environments 1: Homogeneous Random Motions and their Applications
Random Motions in Markov and Semi-Markov Random Environments 2: High-dimensional Random Motions and Financial Applications

PROVENZI Edoardo
From Euclidean to Hilbert Spaces: Introduction to Functional Analysis and its Applications

2020

BARBU Vlad Stefan, VERGNE Nicolas
Statistical Topics and Stochastic Models for Dependent Data with Applications

CHABANYUK Yaroslav, NIKITIN Anatolii, KHIMKA Uliana
Asymptotic Analyses for Complex Evolutionary Systems with Markov and Semi-Markov Switching Using Approximation Schemes

KOROLIOUK Dmitri
Dynamics of Statistical Experiments

MANOU-ABI Solym Mawaki, DABO-NIANG Sophie, SALONE Jean-Jacques
Mathematical Modeling of Random and Deterministic Phenomena

2019

BANNA Oksana, MISHURA Yuliya, RALCHENKO Kostiantyn, SHKLYAR Sergiy
Fractional Brownian Motion: Approximations and Projections

GANA Kamel, BROC Guillaume
Structural Equation Modeling with lavaan

KUKUSH Alexander
Gaussian Measures in Hilbert Space: Construction and Properties

LUZ Maksym, MOKLYACHUK Mikhail
Estimation of Stochastic Processes with Stationary Increments and Cointegrated Sequences

MICHELITSCH Thomas, PÉREZ RIASCOS Alejandro, COLLET Bernard, NOWAKOWSKI Andrzej, NICOLLEAU Franck
Fractional Dynamics on Networks and Lattices

VOTSI Irene, LIMNIOS Nikolaos, PAPADIMITRIOU Eleftheria, TSAKLIDIS George
Earthquake Statistical Analysis through Multi-state Modeling (Statistical Methods for Earthquakes Set – Volume 2)

2018

AZAÏS Romain, BOUGUET Florian
Statistical Inference for Piecewise-deterministic Markov Processes

IBRAHIMI Mohammed
Mergers & Acquisitions: Theory, Strategy, Finance

PARROCHIA Daniel
Mathematics and Philosophy

2017

CARONI Chysseis
First Hitting Time Regression Models: Lifetime Data Analysis Based on Underlying Stochastic Processes
(Mathematical Models and Methods in Reliability Set – Volume 4)

CELANT Giorgio, BRONIATOWSKI Michel
Interpolation and Extrapolation Optimal Designs 2: Finite Dimensional General Models

CONSOLE Rodolfo, MURRU Maura, FALCONE Giuseppe
Earthquake Occurrence: Short- and Long-term Models and their Validation
(Statistical Methods for Earthquakes Set – Volume 1)

D'AMICO Guglielmo, DI BIASE Giuseppe, JANSSEN Jacques, MANCA Raimondo
Semi-Markov Migration Models for Credit Risk
(Stochastic Models for Insurance Set – Volume 1)

GONZÁLEZ VELASCO Miguel, del PUERTO GARCÍA Inés, YANEV George P.
Controlled Branching Processes
(Branching Processes, Branching Random Walks and Branching Particle Fields Set – Volume 2)

HARLAMOV Boris
Stochastic Analysis of Risk and Management
(Stochastic Models in Survival Analysis and Reliability Set – Volume 2)

KERSTING Götz, VATUTIN Vladimir
Discrete Time Branching Processes in Random Environment
(Branching Processes, Branching Random Walks and Branching Particle Fields Set – Volume 1)

MISHURA YULIYA, SHEVCHENKO Georgiy
Theory and Statistical Applications of Stochastic Processes

NIKULIN Mikhail, CHIMITOVA Ekaterina
Chi-squared Goodness-of-fit Tests for Censored Data
(Stochastic Models in Survival Analysis and Reliability Set – Volume 3)

SIMON Jacques
Banach, Fréchet, Hilbert and Neumann Spaces
(Analysis for PDEs Set – Volume 1)

2016

CELANT Giorgio, BRONIATOWSKI Michel
Interpolation and Extrapolation Optimal Designs 1: Polynomial Regression and Approximation Theory

CHIASSERINI Carla Fabiana, GRIBAUDO Marco, MANINI Daniele
Analytical Modeling of Wireless Communication Systems
(Stochastic Models in Computer Science and Telecommunication Networks Set – Volume 1)

GOUDON Thierry
Mathematics for Modeling and Scientific Computing

KAHLE Waltraud, MERCIER Sophie, PAROISSIN Christian
Degradation Processes in Reliability
(Mathematial Models and Methods in Reliability Set – Volume 3)

KERN Michel
Numerical Methods for Inverse Problems

RYKOV Vladimir
Reliability of Engineering Systems and Technological Risks
(Stochastic Models in Survival Analysis and Reliability Set – Volume 1)

2015

DE SAPORTA Benoîte, DUFOUR François, ZHANG Huilong
Numerical Methods for Simulation and Optimization of Piecewise Deterministic Markov Processes

DEVOLDER Pierre, JANSSEN Jacques, MANCA Raimondo
Basic Stochastic Processes

LE GAT Yves
Recurrent Event Modeling Based on the Yule Process
(Mathematical Models and Methods in Reliability Set – Volume 2)

2014

COOKE Roger M., NIEBOER Daan, MISIEWICZ Jolanta
*Fat-tailed Distributions: Data, Diagnostics and Dependence
(Mathematical Models and Methods in Reliability Set – Volume 1)*

MACKEVIČIUS Vigirdas
Integral and Measure: From Rather Simple to Rather Complex

PASCHOS Vangelis Th
*Combinatorial Optimization – 3-volume series – 2nd edition
Concepts of Combinatorial Optimization / Concepts and
Fundamentals – volume 1
Paradigms of Combinatorial Optimization – volume 2
Applications of Combinatorial Optimization – volume 3*

2013

COUALLIER Vincent, GERVILLE-RÉACHE Léo, HUBER Catherine, LIMNIOS Nikolaos, MESBAH Mounir
Statistical Models and Methods for Reliability and Survival Analysis

JANSSEN Jacques, MANCA Oronzio, MANCA Raimondo
Applied Diffusion Processes from Engineering to Finance

SERICOLA Bruno
Markov Chains: Theory, Algorithms and Applications

2012

BOSQ Denis
Mathematical Statistics and Stochastic Processes

CHRISTENSEN Karl Bang, KREINER Svend, MESBAH Mounir
Rasch Models in Health

DEVOLDER Pierre, JANSSEN Jacques, MANCA Raimondo
Stochastic Methods for Pension Funds

2011

MACKEVIČIUS Vigirdas
Introduction to Stochastic Analysis: Integrals and Differential Equations

MAHJOUB Ridha
Recent Progress in Combinatorial Optimization – ISCO2010

RAYNAUD Hervé, ARROW Kenneth
Managerial Logic

2010

BAGDONAVIČIUS Vilijandas, KRUOPIS Julius, NIKULIN Mikhail
Nonparametric Tests for Censored Data

BAGDONAVIČIUS Vilijandas, KRUOPIS Julius, NIKULIN Mikhail
Nonparametric Tests for Complete Data

IOSIFESCU Marius *et al.*
Introduction to Stochastic Models

VASSILIOU PCG
Discrete-time Asset Pricing Models in Applied Stochastic Finance

2008

ANISIMOV Vladimir
Switching Processes in Queuing Models

FICHE Georges, HÉBUTERNE Gérard
Mathematics for Engineers

HUBER Catherine, LIMNIOS Nikolaos *et al.*
Mathematical Methods in Survival Analysis, Reliability and Quality of Life

JANSSEN Jacques, MANCA Raimondo, VOLPE Ernesto
Mathematical Finance

2007
HARLAMOV Boris
Continuous Semi-Markov Processes

2006
CLERC Maurice
Particle Swarm Optimization

Printed and bound by CPI Group (UK) Ltd, Croydon, CR0 4YY
17/03/2024

14471545-0004